Quadratic Forms

Algebra and Applications

Volume 25

Algebra and Applications aims to publish well-written and carefully refereed monographs with up-to-date expositions of research in all fields of algebra, including its classical impact on commutative and noncommutative algebraic and differential geometry, K-theory and algebraic topology, and further applications in related domains, such as number theory, homotopy and (co)homology theory through to discrete mathematics and mathematical physics.

Particular emphasis will be put on state-of-the-art topics such as rings of differential operators, Lie algebras and super-algebras, group rings and algebras, Kac-Moody theory, arithmetic algebraic geometry, Hopf algebras and quantum groups, as well as their applications within mathematics and beyond. Books dedicated to computational aspects of these topics will also be welcome.

More information about this series at http://www.springer.com/series/6253

Michael Barot • Jesús Arturo Jiménez González •
José-Antonio de la Peña

Quadratic Forms

Combinatorics and Numerical Results

Springer

Michael Barot
Kantonsschule Schaffhausen
Switzerland

Jesús Arturo Jiménez González
Instituto de Matemáticas, UNAM
Mexico

José-Antonio de la Peña
Instituto de Matemáticas, UNAM
Miembro de El Colegio Nacional
Mexico

ISSN 1572-5553 ISSN 2192-2950 (electronic)
Algebra and Applications
ISBN 978-3-030-05626-1 ISBN 978-3-030-05627-8 (eBook)
https://doi.org/10.1007/978-3-030-05627-8

Library of Congress Control Number: 2018964709

Mathematics Subject Classification (2010): Primary: 11D09, 11D45, 05C50; Secondary: 15A63, 11E16, 11E20, 11E25, 05C22, 11D85, 11H55, 11R11, 15B36, 16G10, 16G20, 16G60, 17B22, 51F15

This Springer imprint is published by the registered company Springer Nature Switzerland AG
The registered company address is: Gewerbestrasse 11, 6330 Cham, Switzerland

Para Nelia, José Antonio y para México, que me han dado tanto

(M.B.)

A mis padres, Teresa y Arturo

(J.J.)

A Nelia ya mi madre

(J.A.P.)

Preface

This work intends to collect and present combinatorial and numerical results on integral quadratic forms as originally obtained in the context of the representation theory of algebras and derived categories. Apart from their shared source and structure, these results have some features in common:

(a) They are elementary, only requiring some linear algebra for their understanding.
(b) Some of the beautiful results remain practically unknown to students and scholars. As a taste, we mention the following theorem due to Drozd: If $q : \mathbb{Z}^n \to \mathbb{Z}$ is an integral quadratic unit form which is positive definite on the positive cone, then q accepts only finitely many roots, that is, vectors $u \in \mathbb{Z}^n$ with $q(u) = 1$.
(c) The results are scattered in papers written between 1970 and the present day. Some in journals which are scarcely available.
(d) Although some books on representation theory and derived categories have appeared in recent years, filling an important theoretical gap, none of those works touch any of the topics considered in this book (with the possible exception of the first chapter in Ringel's book *Tame algebras and integral quadratic forms*).
(e) Most of the results that we present share an important device, namely the consideration of diagrams (Dynkin diagrams, Euclidean diagrams) and their groups of reflections. These objects have their source in Lie theory, mainly in the classification of root systems in the sense of Bourbaki. For a general account of the history of these diagrams, we refer the reader to the article *The ubiquity of Coxeter–Dynkin diagrams (an introduction to the ADE problem)* by Hazewinkel, Siersma and Veldkamp.

As said before, the main body of the text has almost no prerequisites, besides some basic linear algebra. We have made no attempt to give a self-contained exposition of the subject, although most of the material lends itself to a quite complete presentation. Some of the topics beyond the scope of the book are drawn up in the exercises, and so the reader is encouraged to go through them to gain a wider insight into the matter. We have not committed to give exhaustive references of definitions and results, referring instead to external sources for historical notes.

Readers interested in combinatorial aspects of integral quadratic forms, potentially related to the representations of algebras, Weyl groups and Lie theory, will find these notes helpful and hopefully stimulating. Due to the accessible nature of the topic, we also believe the book is a nice starting point for undergraduate students interested in representation theory.

The chapter titles are mainly drawn from representation theory, referring to arithmetic properties of quadratic forms (positivity, non-negativity and their *weak versions*). In Chap. 1 we review the basic concepts and results used throughout the following chapters, particularly some classical binary integral forms and quadratic forms over the real numbers. In Chaps. 2 and 3 the concepts of positivity and non-negativity of integral quadratic forms are studied, while Chaps. 5 and 6 analyze *weak versions* of these attributes (where one restricts attention to vectors in the positive cone). Here we consider (pre- or semi-)unit forms. In contrast to the unitary case, Chap. 4 is dedicated to presenting analogous results for concealed integral quadratic forms, as well as surveying some group theoretical and spectral properties of such forms. We point out that Chaps. 2 and 3 are not prerequisites for Chaps. 4–6.

There are many new proofs of old results all over the text, as well as generalizations and remarkable new results. Concepts, results and algorithms in the text are illustrated with plenty of examples, exercises, figures and tables. In order to maintain a level of readability in the proofs, some technical steps are left as exercises for the reader.

This work has grown over the course of many years as the authors moved between cities and jobs. It is a pleasure to acknowledge the comments and interest of our students (two groups of them divided by time and distance). We also thank the Instituto de Matemáticas UNAM, Centro de Investigación en Matemáticas A.C. and Instituto Tecnológico Autónomo de México for their support. The second author acknowledges the support of the program FORDECYT CONACYT during his postdoctoral stays in CIMAT A.C. and IMATE Mexico City.

As the first author has expressed in his *Introduction to the representation theory of algebras*, we apologize to the reader for any errors that still remain in the text.

Motivation and Problems

Is it possible to solve $x^2 - xy + y^2 = 2$ within the integers? Since $x^2 - xy + y^2 = (x - \frac{1}{2}y)^2 + \frac{3}{4}y^2$ is a sum of squares we should have $\frac{3}{4}y^2 \leq 2$ or $|y| \leq \sqrt{\frac{8}{3}} < 2$, that is, $y = -1, 0, 1$. By symmetry, $x = -1, 0, 1$. We depict the values of $x^2 - xy + y^2$ for different x and y in the following table:

$$x \left\{ \overbrace{\begin{matrix} 3 & 1 & 1 \\ 1 & 0 & 1 \\ 1 & 1 & 3 \end{matrix}}^{y} \right.$$

and conclude that the question has a negative answer.

How many integer solutions does $x^2 - 3xy + y^2 = 1$ have? It is clear that $(x, y) = \pm(1, 0), \pm(0, 1)$ are solutions. But, for instance, $(x, y) = (3, 8)$ is also a solution. We now reformulate the left-hand side:

$$x^2 - 3xy + y^2 = (x, y) \begin{pmatrix} 1 & -\frac{3}{2} \\ -\frac{3}{2} & 1 \end{pmatrix} \begin{pmatrix} x \\ y \end{pmatrix},$$

and to get rid of the halves, we look for integer solutions of

$$(x, y) \begin{pmatrix} 2 & -3 \\ -3 & 2 \end{pmatrix} \begin{pmatrix} x \\ y \end{pmatrix} = 2.$$

Let $T = \begin{pmatrix} 3 & -1 \\ 8 & -3 \end{pmatrix}$ and observe two things: $\det(T) = -1$ and the first column of T is a solution to our problem. In other words, T defines a $\mathbb{Z}$-invertible linear transformation $T : \mathbb{Z}^2 \to \mathbb{Z}^2$, that is, a change of basis. Curiously we have $T^t \begin{pmatrix} 2 & -3 \\ -3 & 2 \end{pmatrix} T = \begin{pmatrix} 2 & -3 \\ -3 & 2 \end{pmatrix}$. Thus, if $v = \binom{x}{y}$ is a solution, then so are $Tv, T^2v, \ldots$ But unfortunately $T^2 = \mathrm{Id}$, and it seems that our trick failed. Since the equation is symmetric in x and y we can also switch the coordinates, that is, we consider $S = \begin{pmatrix} 0 & 1 \\ 1 & 0 \end{pmatrix} T$ and then $v, Sv, S^2v, \ldots$ are solutions to our problem. For instance:

$$\begin{pmatrix} 1 \\ 0 \end{pmatrix}, \begin{pmatrix} 8 \\ 3 \end{pmatrix}, \begin{pmatrix} 55 \\ 21 \end{pmatrix}, \begin{pmatrix} 377 \\ 144 \end{pmatrix}, \ldots$$

are all solutions of our original problem. If $v = \binom{v_1}{v_2}$ is a solution with $v_1 > v_2 > 0$ then $(Sv)_1 = (Tv)_2 = 8v_1 - 3v_2 > 5v_1$ and $(Sv)_2 = (Tv)_1 = 3v_1 - v_2 > 2v_1$, thus we see that $S^i v \neq S^j v$ for all $i \neq j$. In conclusion, there are infinitely many solutions to our problem.

We will be interested in studying integral quadratic forms q in n variables, by which we mean a homogeneous polynomial of degree two with coefficients q_{ij} in $\mathbb{Z}$.

$$q(x_1, \ldots, x_n) = \sum_{1 \leq i < j \leq n} q_{ij} x_i x_j.$$

Specifically, we are interested in which integers $m \in \mathbb{Z}$ are represented by q, meaning that we can find some vector $x \in \mathbb{Z}^n$ so that $q(x) = m$.

Some famous problems in mathematics can be traced back to a representability problem. For instance, Pythagorean triples (a, b, c) are integral solutions of the equation

$$x^2 + y^2 - z^2 = 0.$$

Four thousand years ago the Babylonians had already filled clay tablets with lists of triplets solving the above dependence. Not only $(3, 4, 5)$ and $(5, 12, 13)$ appear, but larger numbers such as $(3367, 3456, 4825)$. The chances that such answers were found by trial and error are slim. The Babylonians must have used some kind of elementary number theory to generate their triples.

Among other famous representability problems we mention the following celebrated theorems:

Theorem 1 (Lagrange, 1772). *Any non-negative integer m can be written as a sum of four integer squares.*

The quadratic form $q(x_1, x_2, x_3, x_4) = x_1^2+x_2^2+x_3^2+x_4^2$ referred to in Lagrange's Theorem is an example of a so-called *universal* integral quadratic form.

Theorem 2 (Legendre, 1798). *A non-negative integer m can be written as a sum of three integer squares if and only if m is not of the form $4^a(8k + 7)$ for some $a, k \in \mathbb{Z}$.*

Theorem 3 (Fermat, 1640). *A prime number $p > 2$ can be written as a sum of two integer squares if and only if $p \equiv 1 (mod\ 4)$.*

While we can ask about which integers are represented by a given quadratic form q, it is also interesting to ask in how many ways it is represented by q. For particular cases, there are also some classical answers to this problem. Let $r_q(m)$ count the number of ways of representing m by q. We recall:

Theorem 4 (Jacobi, 1828). *If $q = x^2 + y^2 + z^2 + w^2$ and m is a positive integer, then*

$$r_q(m) = 8 \sum_{\substack{0<d|m \\ d\neq 4k}} d.$$

As it turns out, it is a very difficult problem to determine $r_q(m)$ for general quadratic forms. More modest goals are still quite useful. We formulate two main questions:

Question 1. Can we describe which integers m are represented by q? (When is $r_q(m) > 0$?)

Question 2. In about how many ways is m represented by q? (How big is $r_q(m)$? Is it finite?)

An important thing to notice is that two quadratic forms q_1 and q_2 look the same (so we will call them *equivalent*) if they only differ by some invertible (linear) change of variables with integer coefficients. For convenience, we write this equivalence as $q_1 \sim q_2$. Notice that if q_1 and q_2 are equivalent, then $r_{q_1}(m) = r_{q_2}(m)$ for all integers m.

It will also be important to keep track of how many ways a form q is equivalent to itself. We call these self-equivalences the automorphisms of q, and collectively refer to them as **Aut**(q). Since for answering our two main questions we cannot distinguish between equivalent quadratic forms, it makes sense to consider the representation problem for classes of equivalent forms.

In fact, associated to a quadratic form q there is a symmetric n by n integer matrix M_q defined as

$$q(v) = \frac{1}{2} v^t M_q v.$$

Two quadratic forms q_1 and q_2 are equivalent when $M_{q_1} = T^t M_{q_2} T$ for some $\mathbb{Z}$-invertible matrix T. Observe that, in particular, in this situation we have $\det(M_{q_1}) = \det(M_{q_2})$. Sometimes we will denote $x^t M_q y$ by $q(x|y)$, which yields the n by n matrix

$$M_q = (q(e_i|e_j))_{i,j=1}^n ,$$

in the canonical basis $e_1, \ldots, e_n$ of the group $\mathbb{Z}^n$.

Representation problems will frequently occur in this book, sometimes in unexpected ways. For instance, we will consider the following quadratic form:

$$q_0(x_1, x_2, x_3, x_4) = \sum_{i=1}^{4} x_i^2 - \sum_{1 \le i < j \le 4} x_i x_j,$$

and observe that the numbers -2 and -3 are represented by q_0:

$$q_0(1, 1, 1, 1) = -2 \quad \text{and} \quad q_0(1, 1, 2, 2) = -3.$$

This simple fact has the following important consequences. Consider a quadratic form $q(x_1, \ldots, x_n)$ whose restriction to the first coordinates is q_0. We claim that for any sincere integral solution $u = (u_1, \ldots, u_n)$ of $q(x) = 1$, there is another integral vector $v \neq u$ majorizing u with $q(v) = 1$. We indicate the steps of the proof.

Indeed, assume that u is a sincere integral vector satisfying $q(u) = 1$ and assume that u is not majorized by any other integral solution of $q(x) = 1$. First observe that $q(v|e_i) \ge 0$ for every $i = 1, \ldots, n$ since otherwise the vector $v = u - q(u|e_i)e_i$ majorizes u and $q(v) = 1$. Since $q(u|u) = 2$, there is either one exceptional index a with $q(u|e_a) \neq 0$ and $u_a q(u|e_a) = 2$ or there are two exceptional indices a, b

with $q(u|e_a) \neq 0 \neq q(u|e_b)$ and $u_a q(u|e_a) = 1 = u_b q(u|e_b)$. In the first case, the situation $q(u|e_a) = 2$ is discarded since then $q(2u-e_a) = 1$ and $2u-e_a$ majorizes u.

Consider $w = (1, 1, 1, 1, 0, \ldots, 0)$ and $w' = (1, 1, 2, 2, 0, \ldots, 0)$ in $\mathbb{Z}^n$ satisfying $q(w) = -2$ and $q(w') = -3$. By the above analysis $0 \leq q(u|w) \leq 2$. We distinguish cases depending on the value of $q(u|w)$. If $q(u|w) = 0$ then also $q(u|w') = 0$ and $q(2u + w') = 1$. If $q(u|w) = 2$ then $q(u + w) = 1$. If $q(u|w) = 1$ then taking $w'' = w - e_a$ we have $q(u|w'') = 0$, $q(w'') = q_0(w'') = 0$ and $q(u + w'') = 1$. All constructed vectors majorize u, a contradiction that completes the proof of our claim.

The quadratic forms we consider in this book are *unit forms*. A unit form q is a quadratic form of the shape

$$q(x_1, \ldots, x_n) = x_1^2 + \ldots + x_n^2 + \sum_{i<j} q_{ij} x_i x_j.$$

For the sake of visualization, we depict integral unit forms as diagrams. Indeed, we consider unit forms q defined by graphs with n vertices, where we set s solid edges (resp. s dotted edges) between vertices i and j if $q_{ij} = -s$ (resp. $q_{ij} = s$). A typical example is the following:

which corresponds to the quadratic form q_0 given above.

Our interest in quadratic forms comes from the study of finite-dimensional algebras. Although we cleaned the presented material completely from that background, one can still see its traces, namely the emphasis on roots and properties like weak positivity, which play a central role throughout the book. In the following paragraphs we give a very rough sketch of the suppressed representation theoretical material to provide some understanding for the discussed notions.

If A is the ring of upper triangular $n \times n$-matrices with entries in $\mathbb{C}$, then obviously A is a finite-dimensional $\mathbb{C}$ algebra. Any A-module is automatically a $\mathbb{C}$-vector space, and so are the sets of A-module homomorphisms $\mathbf{Hom}_A(M, N)$ between two fixed (left) modules. The property of finitely generated modules translates into finite dimension of vector spaces over $\mathbb{C}$.

The ring A decomposes as a left module into a sum of projective modules, and in our example we have $A = \bigoplus_{i=1}^n P_i$ where $P_i = AE_{ii}$ and E_{ii} is the matrix with a unique non-zero entry, located in the i-th row and i-th column. For any left A-module M we define $\underline{\mathbf{vec}}(M)$ to be the vector whose i-th entry is $\mathbf{dim}_{\mathbb{C}}\mathbf{Hom}_A(P_i, M)$. Take $C_{ij} = \mathbf{dim}_{\mathbb{C}}\mathbf{Hom}_A(P_j, P_i)$ and define the quadratic form $q(v) = v^t C^{-1} v$ for $C = (C_{ij})_{i,j=1}^n$.

Invoking **Gabriel's Theorem** it turns out that we have $q(\underline{\mathbf{vec}}(M)) = 1$ if and only if M is indecomposable, that is, M cannot be properly decomposed into a direct sum of A-modules. In our example

$$\begin{aligned} q(x) &= x_1^2 - x_1x_2 + x_2^2 - x_2x_3 + x_3^2 + \cdots + x_{n-1}x_n + x_n^2 \\ &= \frac{1}{2}[x_1^2 + (x_1 - x_2)^2 + (x_2 - x_3)^2 + \cdots + (x_{n-1} - x_n)^2 + x_n^2], \end{aligned}$$

and thus we see that if $x \neq 0$ then $q(x) > 0$. As we will see later, this implies that there are only finitely many vectors x for which $q(x) = 1$. In conclusion: there are only finitely many possibilities for $(\mathbf{dim}_{\mathbb{C}}\mathbf{Hom}_A(P_i, M))_{i=1}^n$ if M is indecomposable.

We are only interested in solutions of $q(x) = 1$ for which all entries x_i are non-negative, since we should have $x_i = \mathbf{dim}_{\mathbb{C}}\mathbf{Hom}_A(P_i, M)$. This explains why we study, for instance, if a quadratic form q has finitely or infinitely many vectors x with positive entries and $q(x) = 1$.

Schaffhausen, Switzerland — Michael Barot
Mexico City, Mexico — Jesús Arturo Jiménez González
Mexico City, Mexico — José-Antonio de la Peña

Contents

List of Figures

List of Tables

Chapter 1
Fundamental Concepts

In this chapter we recall the basic concepts and results that will be used throughout the book. Most of the concepts may be found in standard books on matrix theory and linear algebra. We will also introduce the correspondence between integral quadratic forms and bigraphs, which will be an essential tool.

There are many interesting historical surveys on the development of contemporary algebra, in particular in what could be considered as the origin of modern representation theory. In the following paragraphs we briefly discuss some of the highlights that led to the topics considered in this book, based on Katz and Parshall [36], Brechenmacher [14, 15], Gustafson [29] and on comments in Ringel [46], Gabriel and Roiter [26] and notes by the authors.

Particular instances of integral quadratic equations have been considered for a long time: from Pythagorean triples (second millennium B.C.), Brahmagupta who developed solutions for what is now known as Pell's equation (628 A.D.), to Fermat's Theorem on the sum of pairs of square integers (1640). It is now generally accepted that Lagrange's work in 1775 established the current general framework for the study of quadratic forms, considering for the first time equivalences, reduction methods and discriminants, concepts that were further developed by Euler and Legendre in the late eighteenth century. In 1801 Gauss published his *Disquisitiones Arithmeticae*, dedicating extensive analysis to binary integral quadratic forms. Gauss' influential work arguably inspired Dirichlet, Dedekind and Hilbert in the nineteenth century to develop a transition towards algebraic number theory, with concepts such as quadratic (and more general) number fields.

Also in the nineteenth century a monumental revolution in algebra was beginning, triggered by ideas from Hamilton, Cayley, Grassmann and Clifford, among many others. For instance, a huge effort was invested in formulations of structural theorems for different types of semi-simple algebras (commutative algebras by Hilbert and the non-commutative case by Molien and Cartan, and later by

M. Barot et al., *Quadratic Forms*, Algebra and Applications 25,
https://doi.org/10.1007/978-3-030-05627-8_1

Wedderburn using purely algebraic methods). At this stage, a 'representation theoretical' flavor was already prominent: both the canonical form of matrices (Jordan–Weierstrass) and of pencils of matrices (Weierstrass–Kronecker) may now be formulated as problems of representations of algebras.

Noether's work on 'module theoretical' aspects of algebras changed perspectives in the early twentieth century. For instance, it led to a sudden revived interest in representation type, which led Brauer and Thrall to formulate their famous conjectures (mid 40s), considered to be precursors of modern representation theory. Parallel and independent work on the representation type of group algebras over fields with positive characteristic guided Baev, Heller, Reiner and Krugljak in the 1960s to the idea of tame and wild behavior.

In the early 70s, rings of finite representation type attracted the attention of Gabriel, who used diagrammatic methods to classify hereditary algebras with that property when studying earlier results by Yoshii. The emergence of Dynkin diagrams in Gabriel's work, and recently developed ideas on root systems coming from the classification of semi-simple finite-dimensional Lie algebras in the 1950s, allowed Tits, Bernstein, Gelfand and Ponomarev to establish direct connections between representations of algebras and roots system of certain quadratic forms and their associated reflections. Further investigation of hereditary algebras translated the tame-wild dichotomy to associative finite-dimensional algebras (Donovan–Freislich), proven ultimately by Drozd in 1979. These connections, together with such powerful tools as the flourishing homological algebra, sparked a considerable amount of research on properties of quadratic forms in relation to (significantly complex) algebraic structures and their representations. In these notes we intend to collect some remarkable results in this direction, achieved from the 70s to the present day.

1.1 Quadratic and Bilinear Forms

Consider an *integral quadratic form* $q : \mathbb{Z}^n \to \mathbb{Z}$, that is, q is a homogeneous polynomial of second degree in n variables $x_1, \dots, x_n$ with coefficients in $\mathbb{Z}$. Hence a quadratic form has the shape

$$q(x) = \sum_{1 \leq i \leq j \leq n} q_{ij} x_i x_j \in \mathbb{Z}[x_1, \dots, x_n].$$

For convenience, throughout the text we will take $q_{ij} := q_{ji}$ for $i > j$. If we denote the column vector x by $(x_1, \dots, x_n)$, for any matrix $A = (a_{ij})$ with coefficients in $\mathbb{Z}$ we get a quadratic form $x^t A x = \sum_{i=1}^n a_{ii} x_i^2 + \sum_{i<j} (a_{ij} + a_{ji}) x_i x_j$. We say that the *symmetric $n \times n$-matrix A is associated to the quadratic form q* if $x^t A x = q(x)$. Notice that the matrix A is unique (usually denoted by A_q), and its coefficients belong to $\frac{1}{2}\mathbb{Z}$. Define then the *determinant* of q as

$\det(q) = \det(A_q)$. There is always a unique lower triangular matrix T_q with coefficients in $\mathbb{Z}$ such that $q(x) = x^t T_q x$, referred to as the *Gram matrix of* q . Given $r_1, \ldots, r_n \in \mathbb{Z}$ we denote by $\mathbf{diag}(r_1, \ldots, r_n)$ the quadratic form $q(x) = r_1 x_1^2 + \ldots + r_n x_n^2$.

Given a (symmetric) matrix A we have a (symmetric) bilinear form

$$\begin{aligned} \mathbb{Z}^n \times \mathbb{Z}^n &\longrightarrow \mathbb{Z} \\ (x, y) &\longmapsto x^t Ay. \end{aligned}$$

The case $x = y$ produces the quadratic form $q(x) = x^t Ax$. Conversely, given a (symmetric) bilinear form $(-|-) : \mathbb{Z}^n \times \mathbb{Z}^n \to \mathbb{Z}$ define the (symmetric) matrix $\mathbf{SM}_{(-|-)}$ with coefficients $\mathbf{SM}_{(-|-)_{i,j}} = (e_i|e_j)$, where $e_1 = (1, 0, \ldots, 0), \ldots, e_n = (0, \ldots, 0, 1)$ is the canonical basis for $\mathbb{Z}^n$.

For a quadratic form q denote by $q(x|y)$ the symmetric bilinear form given by

$$q(x + y) = q(x) + q(y) + q(x|y).$$

That $q(x|y)$ is a symmetric bilinear form follows directly from the observation $q(x|y) = x^t(T_q + T_q^t)y$, where T_q is the Gram matrix of q. Notice that $q(x|x) = q(2x) - 2q(x) = 2q(x)$ and that $q(x|y) = 2(x^t A_q y)$ for all $x, y \in \mathbb{Z}^n$, that is,

$$\mathbf{SM}_{q(-|-)} = T_q + T_q^t = 2A_q.$$

Lemma 1.1. *For an integral quadratic form* $q : \mathbb{Z}^n \to \mathbb{Z}$*, an arbitrary vector* $v = (v_1, \ldots, v_n)$ *in* $\mathbb{Z}^n$ *and an index* $\ell \in \{1, \ldots, n\}$*, we have*

$$q(v|e_\ell) = 2q_{\ell\ell}v_\ell + \sum_{i \neq \ell} q_{\ell i} v_i = \frac{\partial}{\partial v_\ell} q(v),$$

where $\frac{\partial}{\partial v_\ell}$ *denotes the partial derivative of* q *with respect to the variable* v_ℓ.

Proof. The last equality is clear. To show the first one denote by $(a_{ij})_{i,j=1}^n$ the symmetric matrix A_q associated to q. Writing $q(x) = \sum_{1 \leq i \leq j \leq n} q_{ij}x_i x_j$ and $q_{ij} = q_{ij}$ for $i \neq j$ we have $q_{ij} = 2a_{ij}$ for $i \neq j$ and $q_{ii} = a_{ii}$ for $i = 1, \ldots, n$. Using the expression $q(x|y) = 2(x^t A_q y)$ we therefore have

$$\begin{aligned} q(v|e_\ell) &= q(e_\ell|v) = e_\ell^t(2A_q)v \\ &= (2a_{\ell 1}, \ldots, 2a_{\ell\ell}, \ldots, 2a_{\ell n})v = 2q_{\ell\ell}v_\ell + \sum_{i \neq \ell} q_{\ell i} v_i, \end{aligned}$$

which completes the proof. □

We recall how the coefficients of a bilinear form $\mathbb{Z}^n \times \mathbb{Z}^n \to \mathbb{Z}$, given by $(x|y) \mapsto x^t Ay$ for a matrix A, change under a transformation of variables,

$$x_i = \sum_{j=1}^{n} t_{ij}\xi_j, \qquad \text{for } i \text{ in } \{1, \dots, n\} \text{ and } t_{ij} \text{ in } \mathbb{Z}.$$

In matrix notation we write $x = T\xi$, where x and ξ are the column matrices $(x_1, \dots, x_n)$ and $(\xi_1, \dots, \xi_n)$ respectively, and T is the *change of basis matrix* (t_{ij}). We obtain a new bilinear form in the variables ξ and ν, given by

$$\langle \xi|\nu\rangle = \xi^t \widetilde{A}\nu = \xi^t T^t AT\nu = (T\xi|T\nu),$$

where $x = T\xi$, $y = T\nu$ and $\widetilde{A} = T^t AT$. If T is a $\mathbb{Z}$-invertible matrix we say that the bilinear forms $(-|-)$ and $\langle -|-\rangle$ are equivalent (that is, the matrices A and $\widetilde{A}$ are *congruent*, written $A \sim \widetilde{A}$). We also say that two quadratic forms q and q' are *equivalent* if their associated symmetric matrices A_q and $A_{q'}$ are congruent. Since $q(x|y) = 2(x^t A_q y)$, observe that two integral quadratic forms q and q' are equivalent if and only if so are their corresponding bilinear forms $q(-|-)$ and $q'(-|-)$. Recall that T is $\mathbb{Z}$-invertible if and only if $\det(T) = \pm 1$, in which case (assuming $\widetilde{A} = T^t AT$) we have $\det(\widetilde{A}) = \det(A)\det(T)^2 = \det(A)$.

The *radical* of q is the subset of $\mathbb{Z}^n$ given by

$$\begin{aligned}\mathbf{rad}(q) &= \{v \in \mathbb{Z}^n \mid q(u+v) = q(u) + q(v) \text{ for all } u \text{ in } \mathbb{Z}^n\} \\ &= \{v \in \mathbb{Z}^n \mid q(u|v) = 0 \text{ for all } u \text{ in } \mathbb{Z}^n\} \\ &= \{v \in \mathbb{Z}^n \mid q(v|e_i) = 0 \text{ for } i = 1, \dots, n\}\end{aligned}$$

Clearly $v \in \mathbf{rad}(q)$ if and only if $A_q v = 0$, thus $\mathbf{rad}(q)$ is a subgroup of $\mathbb{Z}^n$ and its rank is called the *corank* of q, written $\mathbf{cork}(q)$. Observe that

$$\mathbf{rad}(q) = \{v \in \mathbb{Z}^n \mid q(u) = q(u+v) \text{ for all } u \in \mathbb{Z}^n\}.$$

We say that q is *regular* if $\mathbf{rad}(q) = 0$, and notice that q is regular if and only if $\det(A) \neq 0$.

If the symmetric matrix A_q of $q : \mathbb{Z}^n \to \mathbb{Z}$ has diagonal block form,

$$A_q = A_1 \oplus A_2 = \begin{bmatrix} A_1 & 0 \\ 0 & A_2 \end{bmatrix},$$

then we say that q *decomposes* as $q = q_1 \oplus q_2$ where $q_1 : \mathbb{Z}^{n_1} \to \mathbb{Z}$ is the form given by $q_1(x_1) = x_1^t A_1 x_1$ (for $x_1 \in \mathbb{Z}^{n_1}$ and $0 \leq n_1 \leq n$) and $q_2 : \mathbb{Z}^{n-n_1} \to \mathbb{Z}$ is given by $q_2(x_2) = x_2^t A_2 x_2$ for $x_2 \in \mathbb{Z}^{n-n_1}$. For an index set I, a quadratic form $q : \mathbb{Z}^I \to \mathbb{Z}$ is *disconnected* if there exists a (nontrivial) partition $I = I_1 \cup I_2$

such that $q_{ij} = 0$ for $i \in I_1$ and $j \in I_2$ (observe that there is an enumeration of indices such that $q = q^1 \oplus q^2$ for suitable quadratic forms $q^1 : \mathbb{Z}^{I_1} \to \mathbb{Z}$ and $q^2 : \mathbb{Z}^{I_2} \to \mathbb{Z}$). Otherwise q is said to be *connected*.

Since integral quadratic forms are the main object of study in subsequent chapters, we establish once and for all a visualization tool through (regularized) signed graphs. Our motivation comes from an algebraic setting, where the alternative terminology *bigraph* is commonly used (see for instance [18] for general concepts in graph theory).

A *bigraph* B is a graph (B_0, B_1) (that is, a set of *vertices* B_0 together with a set of edges B_1, admitting loops and multi-edges) such that B_1 may contain both solid and dotted edges. Throughout, the elements of B_0 will have labels taken from a finite subset of natural numbers (usually $\{1, 2, \dots, n\}$).

Any pair of edges between vertices i, j are called *parallel edges* (possibly $i = j$). We will assume that no solid edge is parallel to a dotted edge. Such bigraphs are called *regular*, and the process of deleting all such pairs (a pair at a time) is called *regularization* of bigraphs. For $1 \leq i \leq j \leq n$, let a_{ij} be the number of solid edges (resp. b_{ij} the number of dotted edges) joining vertices i and j in a bigraph B. We define the *(upper triangular) adjacency matrix T_B of the bigraph B* as follows:

$$(T_B)_{ij} = \begin{cases} a_{ij} - b_{ij}, & \text{if } i \leq j; \\ 0, & \text{if } i > j. \end{cases}$$

Define the *integral quadratic form $q_B : \mathbb{Z}^n \to \mathbb{Z}$ associated to a bigraph B* with n vertices as

$$q_B(x) = x^t(\mathbf{Id}_n - T_B)x,$$

and notice that the symmetric matrix A_{q_B} associated to the quadratic form q_B is given by $A_{q_B} = \mathbf{Id}_n - \frac{1}{2}(T_B + T_B^t)$. Alternatively, take the *(symmetric) adjacency matrix A_B* of B having as coefficients $(A_B)_{ji} = (A_B)_{ij} = a_{ij} - b_{ij}$ if $i \leq j$, and observe that $q_B(x|y) = x^t A_B y$.

Now, given an integral quadratic form $q = \sum_{1 \leq i \leq j \leq n} q_{ij} x_i x_j$ we define the *bigraph B_q associated to q* as follows. Let $(B_q)_0$ be the set $\{1, \dots, n\}$, and for each pair of different vertices i, j the set $(B_q)_1$ contains $|q_{ij}|$ edges connecting i and j. These edges are solid if $q_{ij} < 0$ or dotted if $q_{ij} > 0$. In addition, B_q has $|1 - q_{ii}|$ solid loops attached to the vertex i if $1 - q_{ii} \geq 0$, otherwise there are $|1 - q_{ii}|$ dotted loops on it. By construction B_q is always a regular bigraph, and we have $q_{B_q} = q$. Notice moreover that B_{q_B} is the regularization of the bigraph B.

Example 1.2. Let us consider the quadratic form $q(x_1, x_2, x_3, x_4) = 4x_1^2 + x_2^2 + x_4^2 - 3x_1x_2 - 4x_1x_3 + 4x_2x_4 - 3x_3x_4$. Its associated bigraph and corresponding (upper) adjacency matrix are depicted below:

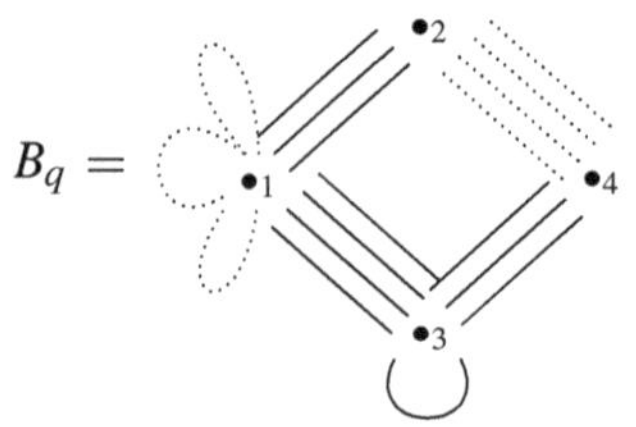

$$T_{B_q} = \begin{pmatrix} -3 & 3 & 4 & 0 \\ 0 & 0 & 0 & -4 \\ 0 & 0 & 1 & 3 \\ 0 & 0 & 0 & 0 \end{pmatrix}$$

We say that q is a *quadratic unit form* or simply *a unitary form* if $a_{ii} = 1$ for all $1 \leq i \leq n$. Notice that q is a unit form if and only if its associated bigraph B_q has no loop. Observe also that if B_q has no loops or multiple edges, then by Lemma 1.1

$$q(v|e_i) = 2v_i + \sum_{j \neq i} q_{ij} v_j = 2v_i - \sum_{\substack{\text{there is} \\ \text{an edge} \\ i-j}} v_j.$$

Consider a quadratic form $q : \mathbb{Z}^I \to \mathbb{Z}$ for an index set I, and for a subset $J \subset I$ take the inclusion $\rho : \mathbb{Z}^J \to \mathbb{Z}^I$ given by $\rho : e_i \mapsto e_i$. The *restriction* q^J of q to the index set J is given by

$$q^J(x) := q(\rho(x)).$$

Throughout the text, for a vertex $i \in I$ we write $q^{(i)}$ instead of $q^{I-\{i\}}$. If $(B_q)_0 = I$ let B' be the *full subbigraph of B_q determined by J* (that is, $(B')_0 = J$ and $(B')_1$ consists of those edges in B_q joining vertices in J), then $q^J = q_{B'}$. Therefore, for a vertex i in a bigraph B, the expression $B^{(i)}$ will denote the bigraph obtained from B by deleting vertex i and all edges containing it.

Exercises 1.3. In the following exercises we consider quadratic forms with coefficients over a general ring R, not necessarily over $\mathbb{Z}$.

1. Take the field with two elements $R = \mathbb{F}_2$. Show that there are exactly eight quadratic forms in the variables x_1, x_2, four of them have two associated symmetric matrices, while the rest have no associated symmetric matrix at all.
2. Let $q(x_1, \ldots, x_n)$ be a quadratic form over the ring R. There is an induced map $Q : R^n \to R$ given by $Q(v) = q(v)$. Show that the mapping Q is quadratic, that is, that it satisfies:

   i) $Q(av) = a^2 Q(v)$ for any $a \in R$ and $v \in R^n$.
   ii) The map $Q(-|-) : R^n \times R^n \to R$ given by

   $$Q(v|w) = Q(v+w) - Q(v) - Q(w)$$

   is R-bilinear and symmetric.

3. Two quadratic maps $Q : R^n \to R$ and $Q' : R^n \to R$ are called isometric if there exists an invertible map $T : R^n \to R^n$ such that $Q(v) = Q'(Tv)$ for all $v \in R^n$. Prove:

   i) Let $q(x_1, \dots, x_n)$ and $q'(x_1, \dots, x_n)$ be two quadratic forms with corresponding quadratic maps $Q, Q' : R^n \to R$. Assume that the matrices A_q and $A_{q'}$ are congruent, then Q and Q' are isometric.
   ii) Let R be a field with $\mathbf{Char}(R) \neq 2$. Assume that Q and Q' are isometric, then the matrices A_q and $A_{q'}$ are congruent.

4. Let R be a unique factorization domain (UFD) and $\beta : R^n \to R$ a linear map. Then $\mathbf{Ker}(\beta)$ is a pure submodule of R^n, that is, if $v \in R^n$ and $0 \neq a \in R$ are such that $av \in \mathbf{Ker}(\beta)$, then $v \in \mathbf{Ker}(\beta)$. Show the following,

   i) A pure submodule of R^n is direct summand of R^n.
   ii) If $R^n = V \oplus R^s$, then V is isomorphic to R^{n-s}.

5. Let $T : R^n \to R^n$ be an isometry for the quadratic forms $q(x_1, \dots, x_n)$ and $q'(x_1, \dots, x_n)$, that is, $q(x) = q'(Tx)$ for all x in R^n. Show that the map T induces an isomorphism between $\mathbf{rad}(q)$ and $\mathbf{rad}(q')$.
6. Let $q = ax_1^2 + bx_1x_2 + cx_2^2$ and $q' = a'x_1^2 + b'x_1x_2 + c'x_2^2$ be two quadratic forms over $\mathbb{Z}$. Show that q and q' are congruent if and only if there exist numbers $\alpha, \beta, \gamma, \delta \in \mathbb{Z}$ with $\alpha\delta - \beta\gamma = \pm 1$ such that

$$\begin{aligned} a' &= a\alpha^2 + b\alpha\gamma + c\gamma^2, \\ b' &= 2a\alpha\beta + b(\alpha\delta + \beta\gamma) + 2c\gamma\delta, \\ c' &= a\beta^2 + b\beta\delta + c\delta^2. \end{aligned}$$

1.2 Reflections

Let $q(x_1, \dots, x_n) : \mathbb{Z}^n \to \mathbb{Z}$ be a quadratic form over the integers. Consider the canonical basis $e_1, \dots, e_n$ for $\mathbb{Z}^n$. We say that $0 \neq x \in \mathbb{Z}^n$ is a *reflection vector* of q if $\frac{q(x|e_i)}{q(x)} \in \mathbb{Z}$ for all $i = 1, \dots, n$. For a reflection vector x define the *reflection morphism at* x , $\sigma_x : \mathbb{Z}^n \to \mathbb{Z}^n$, by

$$\sigma_x(y) = y - \frac{q(x|y)}{q(x)} x.$$

Examples 1.4.

1. If q is a unit form, then all canonical vectors $e_1, \dots, e_n$ are reflection vectors. The corresponding reflection σ_{e_i} is called *simple reflection at* i and is denoted by σ_i.
2. If $q(v) = 1$ then v is a reflection vector.

3. Considering the integral quadratic form $q(x_1, x_2) = x_1^2 + x_2^2 - x_1x_2$, then $v = e_1 - e_2 \in \mathbb{Z}^2$ is a reflection vector with $q(v) = 3$. Observe that $\sigma_v : \mathbb{Z}^2 \to \mathbb{Z}^2$ has associated matrix $\left[\begin{smallmatrix} 0 & 1 \\ 1 & 0 \end{smallmatrix}\right]$, which cannot be obtained as a product of simple reflections σ_i $(1 \leq i \leq n)$ (see Exercise 1.7.1).

In the following lemma we summarize some basic facts about reflections. We stress that the same result holds for more general rings, in particular for $\mathbb{Z}$, $\mathbb{R}$ and $\mathbb{C}$.

Lemma 1.5. *Let $v \in \mathbb{Z}^n$ be a reflection vector for an integral quadratic form q.*

a) The reflection $\sigma_v : \mathbb{Z}^n \to \mathbb{Z}^n$ maps v into $-v$, and acts as the identity in the orthogonal complement of v given by $v^\perp = \{w \in \mathbb{Z}^n \mid q(v|w) = 0\}$.
b) Reflections are involutions, that is, $\sigma_v^2 = \mathbf{Id}_{\mathbb{Z}^n}$.
c) Reflections are q-invariant (or isometries, that is, $q(\sigma_v(x)) = q(x)$ for all $x \in \mathbb{Z}^n$).
d) If w is a reflection vector, then $\sigma_v(w)$ is a reflection vector.
e) If $v \neq w$ are reflection vectors then

$$\sigma_v\sigma_w = \sigma_w\sigma_v \quad \textit{if and only if} \quad q(w|v) = 0.$$

f) For $i \neq j$ we have $\sigma_i\sigma_j = \sigma_j\sigma_i$ if and only if $q_{ij} = 0$.
g) For $i \neq j$ we have $\sigma_i\sigma_j\sigma_i = \sigma_j\sigma_i\sigma_j$ if and only if $q_{ij} \in \{1, -1\}$.

Proof. Points (a) and (b) are clear. For (c) observe that

$$q(\sigma_v(x)) = q(x) + \left(\frac{q(v|x)}{q(v)}\right)^2 q(v) - \frac{q(v|x)}{q(v)} q(v|x) = q(x).$$

For (d) and (e) let w be a reflection vector and $x \in \mathbb{R}^n$. Then

$$q(x|\sigma_v(w)) = q(x|w) - \frac{q(v|w)}{q(v)} q(x|w),$$

where $\frac{q(v|x)}{q(v)} \in \mathbb{Z}$ since v is a reflection vector, and $q(x|w)$ is divisible by $q(w) = q(\sigma_v(w))$ since w is a reflection vector. This shows (d), and (e) follows from the observation

$$\sigma_v(\sigma_w(x)) - \sigma_w(\sigma_v(x)) = \frac{q(v|w)[q(w|x)v - q(v|x)w]}{q(v)q(w)}.$$

Thus if $v \neq w$, then σ_v and σ_w commute if and only if $q(v|w) = 0$. Point (f) follows from (e), and (g) is left as an easy exercise. □

Example 1.6. Let $q = 4x_1^2 + x_2^2 + x_4^2 - 3x_1x_2 - 4x_1x_3 + 4x_2x_4 - 3x_3x_4$ be the integral quadratic form of Example 1.2. Canonical vectors e_2 and e_4 are reflection vectors of q, and the simple reflections σ_2 and σ_4 are respectively given by matrices

$$\begin{pmatrix} 1 & 0 & 0 & 0 \\ 3 & -1 & 0 & -4 \\ 0 & 0 & 1 & 0 \\ 0 & 0 & 0 & 1 \end{pmatrix} \quad \text{and} \quad \begin{pmatrix} 1 & 0 & 0 & 0 \\ 0 & 1 & 0 & 0 \\ 0 & 0 & 1 & 0 \\ 0 & -4 & 3 & -1 \end{pmatrix}.$$

An *isometry* of q is a q-invariant $\mathbb{Z}$-invertible transformation $\sigma : \mathbb{Z}^n \to \mathbb{Z}^n$ (that is, σ satisfies $q(\sigma(x)) = q(x)$ for any vector x in $\mathbb{Z}^n$). In terms of matrices, this is equivalent to having $B^t A_q B = A_q$ for the associated $n \times n$ matrices B of σ and A_q of q. An isometry σ of q satisfies $\det(\sigma)^2 = 1$.

For any reflection vector v, the reflection σ_v is an isometry of q with determinant $\det(\sigma_v) = -1$. We will denote by $\mathcal{O}(q) = \mathcal{O}_{\mathbb{Z}}(q)$ the *group of isometries* of q, and by $\mathbf{W}(q) = \mathbf{W}_{\mathbb{Z}}(q)$ the *Weyl group* of a unit form q, that is, the subgroup of $\mathcal{O}(q)$ generated by the simple reflections $\sigma_1, \dots, \sigma_n$ of q (see Sect. 4.4 for more on Weyl groups).

One of our main interests is to analyze the set $R(q)$ of *roots of a quadratic form* q (or simply q-roots), and the set $\Sigma(q)$ of *isotropic vectors of* q, given by

$$R(q) = \{v \in \mathbb{Z}^n \mid q(v) = 1\} \qquad \text{and} \qquad \Sigma(q) = \{v \in \mathbb{Z}^n \mid q(v) = 0\}.$$

For a unit form q it follows from Lemma 1.5(c) that both sets $R(q)$ and $\Sigma(q)$ are stable under the action of the Weyl group $\mathbf{W}(q)$ associated to q. A root of the shape $y = w(e_i)$ for w in $\mathbf{W}(q)$ and e_i a canonical basis vector is called a *real root* of q. The canonical vectors e_i are sometimes called *simple roots*.

As noted in Example 1.4(3), the Weyl group $\mathbf{W}(q)$ associated to the unit form $q(x_1, x_2) = x_1^2 + x_2^2 - x_1x_2$ is a proper subgroup of $\mathcal{O}(q)$. In fact, $\mathbf{W}(q)$ has order 6 and $\mathcal{O}(q)$ is formed by those integral 2×2 matrices $\begin{bmatrix} a & b \\ c & d \end{bmatrix}$ with either $ad - bc = 1$ and $a - d = b = -c$, or $ad - bc = -1$ and $a = b - c = -d$. In particular, the order of $\mathcal{O}(q)$ is 12.

Exercises 1.7.

1. Let q be the quadratic form of Example 1.4(3) and consider its Weyl group $\mathbf{W}(q)$. Write down those 2×2 matrices corresponding to elements of the Weyl group $\mathbf{W}(q)$. Show that the transformation $\sigma_v = \begin{bmatrix} 0 & 1 \\ 1 & 0 \end{bmatrix}$ does not belong to $\mathbf{W}(q)$. [Hint: $\mathbf{W}(q)$ has $(2 + 1)!$ elements.]
2. Describe all the elements of $\mathcal{O}(q)$ for q as in Exercise 1.
3. Show that $\sigma_i\sigma_j\sigma_i = \sigma_j\sigma_i\sigma_j$ if and only if $q_{ij} \in \{1, -1\}$.
4. Let $\alpha : \mathbb{Z}^n \to \mathbb{Z}^n$ be an isometry of q, and v be a reflection vector of q. Show that $\alpha(v)$ is a reflection vector and that $\alpha\sigma_v\alpha^{-1} = \sigma_{\alpha(v)}$.
5. Let $v \in \mathbb{Z}^n$ be a vector with $q(v) \neq 0$, and consider the lattice $v^{\perp} = \{w \in \mathbb{Z}^n \mid q(v|w) = 0\}$. Show that $\mathbb{Z}^n = \mathbb{Z}v \oplus v^{\perp}$ if and only if v is a reflection vector.

6. Let $q : \mathbb{Z}^n \to \mathbb{Z}$ be a unit form and $1 \leq i \leq n$ an index. Consider the restricted form $q^{(i)} : \mathbb{Z}^{n-1} \to \mathbb{Z}$. Show that $\mathbf{W}(q^{(i)})$ may be seen as a proper subgroup of $\mathbf{W}(q)$. [Hint: Suppose $i = 1$ and let σ'_j be a reflection of $q^{(1)}$ for $2 \leq j \leq n$. Notice that the matrix of the reflection σ_j has the shape

$$\sigma_j = \begin{bmatrix} 1 & 0 \\ * & \sigma'_j \end{bmatrix}.$$

If a product of reflections $\sigma_{j_1} \ldots \sigma_{j_s}$ with $2 \leq j_1, \ldots, j_s \leq n$ has a matrix of shape

$$\begin{bmatrix} 1 & 0 \\ * & \mathbf{Id}_{n-1} \end{bmatrix},$$

then $\sigma_{j_1} \ldots \sigma_{j_s} = \mathbf{Id}_{\mathbb{Z}^n}$.]
7. Let B_1 and B_2 be bigraphs and $B = B_1 \sqcup B_2$ their disjoint union. Let $q_i = q_{B_i}$ for $i = 1, 2$ and $q = q_B$. Show that $\mathbf{W}(q) = \mathbf{W}(q_1) \times \mathbf{W}(q_2)$ is the direct product of groups $\mathbf{W}(q_1)$ and $\mathbf{W}(q_2)$.

1.3 Representability

We say that a quadratic form q *represents an integer* a if there is a nonzero vector $v \in \mathbb{Z}^n$ such that $q(v) = a$. We write

$$\mathbf{D}(q) = \mathbf{D}_{\mathbb{Z}}(q) = \{q(v) \text{ such that } v \in \mathbb{Z}^n \text{ and } v \neq 0\}$$

for the set of elements which are represented by q. A quadratic form $q : \mathbb{Z}^n \to \mathbb{Z}$ is called *isotropic* (over the integers $\mathbb{Z}$) if $0 \in \mathbf{D}_{\mathbb{Z}}(q)$, otherwise q is called *anisotropic*. If v is a nonzero vector with $q(v) = 0$, we call v an *isotropic vector* for q, and denote the set of isotropic vectors of q by $\Sigma(q)$. Clearly a quadratic form $q : \mathbb{Z} \to \mathbb{Z}$ given by $x \mapsto ax^2$ is anisotropic if and only if $a \neq 0$.

An integral quadratic form $q : \mathbb{Z}^n \to \mathbb{Z}$ is said to be *universal* (over the positive integers) if $D_{\mathbb{Z}}(q) = \mathbb{Z}$ (resp. $D_{\mathbb{Z}}(q) = \mathbb{N}$). A quadratic form q is said to be *positive* if $q(x) > 0$ for any nonzero vector $x \in \mathbb{Z}^n$. By a *universal positive form* we mean a positive form q which is universal over the positive integers. The *Conway–Schneeberger Fifteen Theorem* states that a positive integral form q whose associated symmetric matrix A_q has integer coefficients (referred to in this section as *matrix-integral form*) is universal if and only if it represents all positive numbers up to 15. A positive integral quadratic form accepting half integers in its symmetric matrix is universal whenever it represents all positive numbers up to 290. The original proof of Conway and Schneeberger's Fifteen Theorem, based on many specific computations, was never published. A shorter proof given by Bhargava [11] is based in a *escalation* process, which we describe in what follows, where we assume that all quadratic forms are positive and matrix-integral.

The *truant* of a nonuniversal positive quadratic form q is the smallest positive integer not represented by q. An *escalation* of a nonuniversal quadratic form $q(x_1, \ldots, x_n)$ is a positive quadratic form $\widetilde{q}(x_1, \ldots, x_n, x_{n+1})$ such that the restriction to the first n variables is q, and $\widetilde{q}(0, \ldots, 0, 1)$ is the truant of q. For instance, the direct sum

$$q(x_1, \ldots, x_n) + ax_{n+1}^2$$

is an escalation of q, where a is the truant of q. An *escalator form* is either $q_1(x_1) = x_1^2$ or a quadratic form obtained as an escalation of a nonuniversal escalator form. The fundamental step in the proof of the Fifteen Theorem is to show that there are only finitely many escalator forms (all of them in at most five variables).

Observe first that the truant of q_1 is 2, thus any escalation of the quadratic form $q_1(x_1) = x_1^2$ is determined by

$$q_2(x_1, x_2) = x_1^2 + 2bx_1x_2 + 2x_2^2,$$

and by the Cauchy–Schwarz inequality,

$$|b|^2 = |q_2((1, 0)|(0, 1))|^2 \leq q_2(1, 0)q_2(0, 1) = 2.$$

All escalators in two variables are listed below,

$$\begin{aligned} q_{2,1}(x_1, x_2) &= x_1^2 + 2x_2^2, \\ q_{2,2}(x_1, x_2) &= x_1^2 + 2x_1x_2 + 2x_2^2 = (x_1 + x_2)^2 + x_2^2, \\ q_{2,3}(x_1, x_2) &= x_1^2 - 2x_1x_2 + 2x_2^2 = (x_1 - x_2)^2 + x_2^2. \end{aligned}$$

Notice that there are only two isomorphism classes of escalators in two variables, namely those with associated symmetric matrices

$$A_1 = \begin{pmatrix} 1 & 0 \\ 0 & 2 \end{pmatrix} \quad \text{and} \quad A_2 = \begin{pmatrix} 1 & 0 \\ 0 & 1 \end{pmatrix},$$

both nonuniversal with truants 3 and 5 respectively. Any escalation of the form q_{A_1} is isomorphic to a form with associated symmetric matrix

$$B_{1,1} = \begin{pmatrix} 1 & 0 & 0 \\ 0 & 1 & 0 \\ 0 & 0 & 1 \end{pmatrix}, \quad B_{1,2} = \begin{pmatrix} 1 & 0 & 0 \\ 0 & 1 & 0 \\ 0 & 0 & 2 \end{pmatrix}, \quad B_{1,3} = \begin{pmatrix} 1 & 0 & 0 \\ 0 & 1 & 0 \\ 0 & 0 & 3 \end{pmatrix}.$$

Similarly, any escalation of the form q_{A_2} is isomorphic to a form with symmetric matrix

$$B_{2,1} = \begin{pmatrix} 1 & 0 & 0 \\ 0 & 2 & 0 \\ 0 & 0 & 2 \end{pmatrix}, \quad B_{2,2} = \begin{pmatrix} 1 & 0 & 0 \\ 0 & 2 & 0 \\ 0 & 0 & 3 \end{pmatrix}, \quad B_{2,3} = \begin{pmatrix} 1 & 0 & 0 \\ 0 & 2 & 1 \\ 0 & 1 & 4 \end{pmatrix},$$

$$B_{2,4} = \begin{pmatrix} 1 & 0 & 0 \\ 0 & 2 & 0 \\ 0 & 0 & 4 \end{pmatrix}, \quad B_{2,5} = \begin{pmatrix} 1 & 0 & 0 \\ 0 & 2 & 1 \\ 0 & 1 & 5 \end{pmatrix}, \quad B_{2,6} = \begin{pmatrix} 1 & 0 & 0 \\ 0 & 2 & 0 \\ 0 & 0 & 5 \end{pmatrix}.$$

These nine quadratic forms are all nonuniversal, and can be escalated to give rise to 207 nonisomorphic quadratic forms in four variables. Only six out of these 207 quadratic forms are nonuniversal, and their escalations produce all remaining escalator forms (1630 isomorphism classes in five variables, all of them universal). Thus arguing as above, the following theorem was shown in [11].

Theorem 1.8 (Fifteen Theorem). *Let q be an integral positive form with integral associated symmetric matrix A_q. Then q is a universal if and only if q represents the numbers* $1, \dots, 15$.

1.4 Binary Integral Quadratic Forms

This section deals with *binary integral quadratic forms* $q = ax_1^2 + bx_1x_2 + cx_2^2$ (with a, b, c in $\mathbb{Z}$), which will be denoted by $q = (a, b, c)$, and their classification through the determinant,

$$\det(q) = ac - \frac{b^2}{4}.$$

The most prominent examples of binary forms are *Pell forms* $q_{[r]} = (1, 0, -r)$ and *Kronecker forms* $q_m = (1, -m, 1)$ for $r, m \in \mathbb{Z}$. Here we briefly describe the general theory of binary forms (due originally to Gauss [28]), collecting some results from Conway [19] and Buell [17].

First notice that $d = -4\det(q)$ is an integer satisfying $d \equiv 0, 1 \bmod (4)$, usually known as the *discriminant* of q. Indeed, if $b = 2k+\delta$ for some $k \in \mathbb{Z}$ and $\delta \in \{0, 1\}$, then $b^2 = 4(k^2 + \delta k) + \delta$, and therefore $d \equiv \delta \bmod (4)$. Moreover, for any integer $d \equiv 0 \bmod (4)$ (resp. $d \equiv 1 \bmod (4)$) there exists a binary quadratic form q with discriminant $d = -4\det(q)$, namely $q = (1, 0, -d/4)$ (resp. $q = (1, 1, -(d - 1)/4)$). In any case this is called the *principal form* of discriminant d. We say that a binary form $q = (a, b, c)$ is respectively *definite*, *semi-definite* or *indefinite* if its discriminant $d = b^2 - 4ac$ satisfies respectively $d < 0$, $d = 0$ or $d > 0$. Notice that if $d \geq 0$ then either $a, c \geq 0$ or $a, c \leq 0$. We say that q is *positive (semi-) definite* in the first case, and *negative (semi-)definite* in the second.

Two binary forms q and q' are called *properly equivalent* if there exists an integral transformation $T : \mathbb{Z}^2 \to \mathbb{Z}^2$ with $\det(T) = 1$ such that $q' = qT$. If u and v are the columns of T (which will be written as $T = [u|v]$), then we have

$$qT(x_1, x_2) = q(ux_1 + vx_2) = q(u)x_1^2 + q(u|v)x_1x_2 + q(v)x_2^2,$$

that is, $qT = (q(u), q(u|v), q(v))$. For instance, a Kronecker form q_m is definite if and only if $|m| \leq 1$, while a Pell form $q_{[r]}$ is definite precisely when $r < 0$. Moreover, q_{2t} and $q_{[t^2-1]}$ are properly equivalent, and so are q_m and q_{-m} for all $m \in \mathbb{Z}$.

The following algorithm describes how to construct a sequence of binary forms, starting with a given form $q = (a, b, c)$ having nonsquare discriminant. This algorithm is used in [19] when b is an even number to define so-called *reduced forms*. Here we focus on binary forms with nonsquare discriminant, commenting on the quadratic case at the end of this section.

Algorithm 1.9 (Binary Reduction). *Let $q = (a, b, c)$ be a binary form with $c \neq 0$. Define a new binary form $\overrightarrow{q} = (\overrightarrow{a}, \overrightarrow{b}, \overrightarrow{c})$ as follows. Consider the equation*

$$b + \overrightarrow{b} \equiv 0 \bmod 2c. \qquad (*)$$

i) Take $\overrightarrow{a} = c$.

ii) If equation $(*)$ has a solution $\overrightarrow{b}$ satisfying $(\overrightarrow{b})^2 - d \leq 0$, take $\overrightarrow{b}$ as such a solution with maximal value.

iii) If equation $(*)$ has no solution $\overrightarrow{b}$ satisfying $(\overrightarrow{b})^2 - d \leq 0$, take $\overrightarrow{b}$ as the solution of $(*)$ with minimal absolute value, choosing the sign of $\overrightarrow{b}$ opposite to that of b in case of a tie.

iv) Take $\overrightarrow{c} = \frac{(\overrightarrow{b})^2 - d}{4c}$.

Then the binary form $\overrightarrow{q}$, referred to as the (right) shift *of the binary form q, is properly equivalent to q.*

Proof. Consider the matrix $T = [u|v] = \left(\begin{smallmatrix} 0 & -1 \\ 1 & \alpha \end{smallmatrix}\right)$ where $\alpha = \frac{b + \overrightarrow{b}}{2c}$. We know that $qT = (q(u), q(u|v), q(v))$, and clearly $q(u) = q((0, 1)) = c = \overrightarrow{a}$. We compute,

$$\begin{aligned} q(v) &= q((-1, \alpha)) = a - b\alpha + c\alpha^2 \\ &= a - \frac{b^2 + b\overrightarrow{b}}{2c} + \frac{b^2 + 2b\overrightarrow{b} + (\overrightarrow{b})^2}{4c} \\ &= a + \frac{(\overrightarrow{b})^2 - b^2}{4c} = \frac{(\overrightarrow{b})^2 - d}{4c} = \overrightarrow{c}. \end{aligned}$$

We also have

$$\begin{aligned}
q(u|v) &= q(u+v) - q(u) - q(v) \\
&= q((-1, 1+\alpha)) - \overrightarrow{a} - \overrightarrow{c} \\
&= a - b(1+\alpha) + c(1+\alpha)^2 - c - \overrightarrow{c} \\
&= a - b\alpha + \overrightarrow{b} + c\alpha^2 - \overrightarrow{c} \\
&= \overrightarrow{b} + a - \frac{b^2 + b\overrightarrow{b}}{2c} + \frac{b^2 + 2b\overrightarrow{b} + (\overrightarrow{b})^2}{4c} - \frac{(\overrightarrow{b})^2 - b^2 + 4ac}{4c} = \overrightarrow{b}.
\end{aligned}$$

Hence $qT = (\overrightarrow{a}, \overrightarrow{b}, \overrightarrow{c}) = \overrightarrow{q}$, and the result follows since $\det(T) = 1$. □

As a mild generalization of Buell's definition in [17], we say that a binary form q is *(binary) reduced* if it satisfies the following conditions,

$$|b| \leq \min\{|a|, |c|\}, \text{ if } q \text{ is definite,}$$
$$|\sqrt{d} - 2|a|| < b < \sqrt{d}, \text{ if } q \text{ is indefinite.}$$

We say that a binary form q' is an *iterated (right) shift* of q if there is a sequence of binary forms $q_0, \ldots, q_r$ with $r \geq 0$ such that $q = q_0$, $q' = q_r$ and $q_{i+1} = \overrightarrow{q_i}$ for $0 \leq i < r$. If $r > 0$ we say that q' is a *nontrivial iterated (right) shift* of q. Observe that if q has nonsquare discriminant, then we may iterate Algorithm 1.9 indefinitely, since none of the binary forms q_i appearing in this sequence has zero third term. We start by analyzing Algorithm 1.9 when applied to positive definite forms (the negative definite case can be treated analogously).

Theorem 1.10. *If q is a positive definite binary form, then there is an iterated shift q' of q such that q' is binary reduced. Moreover, if q itself is binary reduced, then so is $\overrightarrow{q}$.*

Proof. Take $q = (a, b, c)$. By assumption we have $d = b^2 - 4ac < 0$ and $0 \leq a, c$. Therefore in Algorithm 1.9 we always take $\overrightarrow{b}$ with minimal absolute value (point (iii)).

Let us first show that if $|b| < c$ then $\overrightarrow{q} = (c, -b, a)$. By minimality it is clear that $\overrightarrow{b} = -b$ (since any other solution of equation $(*)$ satisfies $|\overrightarrow{b}| > c > |b|$). Then $\overrightarrow{c} = \frac{b^2-d}{4c} = a$. In particular, if q is binary reduced then so is $\overrightarrow{q} = (c, -b, a)$, and $q = \overrightarrow{\overrightarrow{q}}$.

Let us now assume that q is not binary reduced. By the above we may assume that $c < |b|$. We proceed by induction on $|b|$. Notice that $c < |b|$ implies that $|\overrightarrow{b}| < |b|$. Indeed, if $|\overrightarrow{b}| \geq |b| < c$ then we may choose a sign $\pm$ such that $\overrightarrow{b} \pm 2c$ is a solution of $(*)$ with smaller absolute value than $\overrightarrow{b}$, which is impossible by minimality of $\overrightarrow{b}$. Hence, if $|\overrightarrow{b}| \leq \overrightarrow{c}$ then $\overrightarrow{q}$ is a binary reduced form. Use induction otherwise. □

Corollary 1.11. *For every negative integer d there is a finite number of proper equivalence classes of positive definite binary forms with discriminant d.*

Proof. Assume $q = (a, b, c)$ is a binary reduced positive definite form with discriminant $d = b^2 - 4ac \leq 0$. Observe that, as shown in the proof of Theorem 1.10, if $q = (a, b, c)$ is binary reduced then $\overrightarrow{q} = (c, -b, a)$. Hence we may assume that $a \leq c$.

First notice that $3b^2 \leq -d$. This follows from the equation

$$4b^2 \leq 4a^2 \leq 4ac = b^2 - d,$$

which holds since q is binary reduced. This property also implies that $ac - b^2 > 0$, therefore we have $3ac \leq -d$. Since both a and c are positive, there is a finite number of possible values for a, b and c. This shows that there is a finite number of binary reduced positive definite forms with discriminant d, and the result follows from Theorem 1.10. □

As shown in the proof of Theorem 1.10, for any positive definite binary form $q = (a, b, c)$ the sequence of iterated right shifts $q, \overrightarrow{q}, \overrightarrow{\overrightarrow{q}}, \ldots$ is ultimately periodic, for $q = \overrightarrow{\overrightarrow{q}}$ if q is binary reduced. We point out that $q = \overrightarrow{q}$ if and only if $b = 0$ and $a = c$. For instance, the Kronecker forms q_1 and q_{-1} are periodic positive definite binary forms, and the same holds for the Pell equations q_r and $\overrightarrow{q_r} = (-r, 0, 1)$ for $r < 0$. Similarly it can be shown that

$$(13, 10, 2)(2, 2, 1)(1, 0, 1)$$

is a sequence of iterated shifts of the positive definite form $(13, 10, 2)$, where $q = (1, 0, 1)$ is binary reduced with $q = \overrightarrow{q}$. In Table 1.1 we describe all positive definite binary reduced forms with discriminant $-50 \leq d < 0$.

The same periodicity phenomenon on iterated sequences of shifts is presented in the indefinite case with nonsquare discriminant, which we consider next. In the following preliminary result we show some alternative definitions for binary reduced indefinite forms (compare with [17] and [19]).

Lemma 1.12. *For an indefinite binary form $q = (a, b, c)$ with discriminant d the following conditions are equivalent.*

i) $|\sqrt{d} - 2|a|| < b < \sqrt{d}$.
ii) $|\sqrt{d} - 2|c|| < b < \sqrt{d}$.
iii) $0 < b < \sqrt{d} < \min\{b + 2|a|, b + 2|c|\}$.

Proof. Assume (i) holds. Then $|\sqrt{d} - b| = \sqrt{d} - b < 2|a|$, and since

$$|\sqrt{d} - b| \cdot |\sqrt{d} + b| = 4|a| \cdot |c|,$$

Table 1.1 Positive definite reduced binary forms (a, b, c) with small discriminant and $a \leq c$

Discriminant	h	Primitive forms	Nonprimitive forms
-3	1	$(1, 1, 1)$	–
-4	1	$(1, 0, 1)$	–
-7	1	$(1, 1, 2)$	–
-8	1	$(1, 0, 2)$	–
-11	1	$(1, 1, 3)$	–
-12	2	$(1, 0, 3)$	$(2, 2, 2)$
-15	2	$(1, 1, 4)$, $(2, 1, 1)$	–
-16	2	$(1, 0, 4)$	$(2, 0, 2)$
-19	1	$(1, 1, 5)$	–
-20	2	$(1, 0, 5)$, $(2, 2, 3)$	–
-23	3	$(1, 1, 6)$, $(2, \pm 1, 3)$	–
-24	2	$(1, 0, 6)$, $(2, 0, 3)$	–
-27	2	$(1, 1, 7)$	$(3, 3, 3)$
-28	2	$(1, 0, 7)$	$(2, 2, 4)$
-31	3	$(1, 1, 8)$, $(2, \pm 1, 4)$	–
-32	3	$(1, 0, 8)$, $(3, 2, 3)$	$(2, 0, 4)$
-35	2	$(1, 1, 9)$, $(3, 1, 3)$	–
-36	3	$(1, 0, 9)$, $(2, 2, 5)$	$(3, 0, 3)$
-39	4	$(1, 1, 10)$, $(2, \pm 1, 5)$, $(3, 3, 4)$	–
-40	2	$(1, 0, 10)$, $(2, 0, 5)$	–
-43	1	$(1, 1, 11)$	–
-44	4	$(1, 0, 11)$, $(3, \pm 2, 4)$	$(2, 2, 6)$
-47	5	$(1, 1, 12)$, $(2, \pm 1, 6)$, $(3, \pm 1, 4)$	–
-48	4	$(1, 0, 12)$, $(3, 0, 4)$	$(2, 0, 6)$, $(4, 4, 4)$

The corresponding cycle of binary reduced forms is given by $(a, b, c)(c, -b, a)$. The number of proper equivalence classes of forms with discriminant d is denoted by $h = h(d)$

we have $2|c| < |\sqrt{d} + b| = \sqrt{d} + b$. Then $-b < \sqrt{d} - 2|c|$. Similarly, since $-b < \sqrt{d} - 2|a|$ (that is, $2|a| < \sqrt{d} + b = |\sqrt{d} + b|$), we must have $\sqrt{d} - b = |\sqrt{d} - b| < 2|c|$, that is, $\sqrt{d} - 2|c| < b$. This shows that (i) implies (ii), and the converse follows applying the above to the binary form (c, b, a).

Now, if (ii) holds then $\sqrt{d} < b + 2|c|$, and since (i) also holds we have $\sqrt{d} < b + 2|a|$. Therefore (ii) implies (iii). Assume finally that (iii) holds. Since $\sqrt{d} < \min\{b + 2|a|, b + 2|c|\}$ we have

$$\sqrt{d} - 2|a| < b \qquad \text{and} \qquad \sqrt{d} - b < 2|c|.$$

Considering that $0 < b < \sqrt{d}$ we have $|\sqrt{d} - b| < 2|c|$, which again implies that $\sqrt{d} + b = |\sqrt{d} + b| > 2|a|$ since $|\sqrt{d} - b| \cdot |\sqrt{d} + b| = 4|a| \cdot |c|$. Then $-b < \sqrt{d} - 2|a|$ and (i) holds. □

Lemma 1.13. *Let $q_0, q_1, q_2, \dots$ be a sequence of iterated right shifts of indefinite binary forms with nonsquare discriminant d. Taking $q_i = (a_i, b_i, c_i)$, there is an $r \geq 0$ such that $|b_r| < \sqrt{d}$. Moreover, for any $r \geq 0$ with $|b_r| < \sqrt{d}$ we have*

$$|b_i| < \sqrt{d} < b_i + 2|a_i|, \qquad \textit{for all } i > r.$$

Proof. Let $q = (a, b, c)$ be an indefinite form with nonsquare discriminant d. Notice that $|b| < \sqrt{d}$ if and only if $ac < 0$. If $|b| < \sqrt{d}$ then there is a solution $\overrightarrow{b}$ to equation $b + \overrightarrow{b} \equiv 0 \bmod 2c$ with $(\overrightarrow{b})^2 - d < 0$, namely $\overrightarrow{b} = -b$. Therefore $\overrightarrow{c} = \frac{(\overrightarrow{b})^2 - d}{4c}$ has the opposite sign to $c = \overrightarrow{a}$, that is, $|\overrightarrow{b}| < \sqrt{b}$. Let us suppose that $\sqrt{d} > \overrightarrow{b} + 2|c|$. Since $|\overrightarrow{b}| < \sqrt{d}$ we actually have $\sqrt{d} > |\overrightarrow{b} + 2|c||$, that is, $(\overrightarrow{b} + 2|c|)^2 - d < 0$, in contradiction with the maximality of $\overrightarrow{b}$. Hence $\sqrt{d} < \overrightarrow{b} + 2|c|$, which shows the second claim.

Assuming now that $\sqrt{d} < |b|$ we have $|b| > \min\{|a|, |c|\} =: m$, since

$$(|\overrightarrow{b}| + m)(|\overrightarrow{b}| - m) = b^2 - m^2 \geq b^2 - 4ac > 0.$$

Take $q = q_0$ and its sequence of iterated right shifts $q_0, q_1, \dots$ and assume there is an $s > 0$ such that $|b_i| > \sqrt{d}$ for $0 \leq i \leq s$,

$$(a_0, b_0, c_0)(a_1, b_1, c_1)(a_2, b_2, c_2) \dots (a_{s-1}, b_{s-1}, c_{s-1})(a_s, b_s, c_s).$$

Let us suppose that there is a solution b_{i+1} to Eq. $(*)$ $b_i + b_{i+1} \equiv \bmod 2c_i$ with $b_{i+1}^2 - d < 0$, as in Algorithm 1.9(ii). Then $c_{i+1} = \frac{b^2 - d}{4c}$ has the opposite sign to $c_i = a_{i+1}$. In particular, $|b_{i+1}| < \sqrt{d}$ and $i > s$. Therefore there is no solution of $(*)$ satisfying $b_{i+1}^2 - d < 0$ for $0 \leq i < s$. Then we must choose b_{i+1} with minimal absolute value, which implies that $|b_{i+1}| \leq |c_i| = |a_{i+1}|$. Since $\sqrt{d} < |b_{i+1}|$, then we have $|b_{i+1}| > \min\{|a_{i+1}|, |c_{i+1}|\} = |c_{i+1}|$. Then the sequence above satisfies

$$|c_0| = |a_1| \geq |b_1| > |c_1| = \dots \geq |c_{s-1}| = |a_s| \geq |b_s| > |c_s|,$$

which means that $s < |c_0|$. This bound for the value of s shows the first claim. □

Theorem 1.14. *If q is an indefinite binary form with nonsquare discriminant, then there is an iterated shift q' of q such that q' is binary reduced. Moreover, if q itself is binary reduced, then so is $\overrightarrow{q}$.*

Proof. Take $q = (a, b, c)$. For the first claim we proceed by induction on $|a|$. Using Lemma 1.13 we may assume that $|b| < \sqrt{d} < b + 2|a|$. If $|a| \leq |c|$, then we have $|b| < \sqrt{d} < \min\{b + 2|a|, b + 2||c\}$. Observe that if $b \leq 0$ then necessarily $\sqrt{d} < 2|a|$ and $\sqrt{d} < 2|c|$, which implies that $d < (2|a|)(2|c|) = -4ac$, that is, $b^2 < 0$, a contradiction. Then $0 < b < \sqrt{d} < \min\{b + 2|a|, b + 2||c\}$, and by Lemma 1.12 the binary form q is binary reduced. If $|a| > |c|$, take $\overrightarrow{q} = (\overrightarrow{a}, \overrightarrow{b}, \overrightarrow{c})$ the right shift of q, and since $|a| > |c| = |\vec{a}|$ we may use induction to complete the result. This shows the first statement.

Assume now that q is binary reduced. By Lemmas 1.12 and 1.13 we have

$$|\vec{b}| < \sqrt{d} < \vec{b} + 2|\vec{a}|.$$

We show that $b + \vec{b} > 0$. That $b + \vec{b} \geq 0$ follows from maximality of $\vec{b}$, since $-b$ is a solution of (∗) with $(-b)^2 - d < 0$ (cf. Algorithm 1.9). If $\vec{b} = -b$ then $|\sqrt{d} + b| = \sqrt{d} + b < 2|c|$. As has been argued before, this implies that $2|a| < |\sqrt{d} - b| = \sqrt{d} - b$, that is, $\sqrt{d} > b + 2|a|$, in contradiction with q being a binary reduced form (see Lemma 1.12 above).

Hence there exists $k > 0$ such that $b + \vec{b} = 2|c|k$, and we have the following inequalities,

$$2|c| - \vec{b} = b - 2|c|(k-1) \leq b < \sqrt{d},$$

that is, $-\vec{b} < \sqrt{d} - 2|c| = \sqrt{d} - 2|\vec{a}|$. Since $\sqrt{d} - 2|\vec{a}| < \vec{b}$, we conclude that $\vec{b} > 0$ and that $\vec{q}$ is binary reduced. □

Observe that if $q = (a, b, c)$ is a binary reduced indefinite form with nonsquare discriminant d, then $0 < b < \sqrt{d}$ and $0 \leq (2|a|)(2|c|) = d - b^2 < d$. Therefore there are only finitely many such forms. This remark, together with Theorem 1.14, implies the following result.

Corollary 1.15. *For every positive nonsquare integer d there is a finite number of proper equivalence classes of indefinite binary forms with discriminant d.*

It also follows from Theorem 1.14 and the remark above that any sequence of iterated right shifts of indefinite binary forms with nonsquare discriminant is eventually periodic. In general it is possible that different binary forms q and q' with nonsquare discriminant satisfy $\vec{q} = \vec{q'}$. Take for instance $q = (1, 0, 2)$ and $q' = (3, -2, 1)$ for the positive definite case, or $q = (1, 3, 1)$ and $q' = (-1, 1, 1)$ for the indefinite case. However, this cannot happen if q and q' are binary reduced.

Lemma 1.16. *Let q and q' be binary reduced forms with nonsquare discriminant d such that $\vec{q} = \vec{q'}$. Then $q = q'$.*

Proof. Since $q = \vec{\vec{q}}$ if q is a binary reduced definite form, the claim is clear for the definite case.

Assume now that q is indefinite. Take $q = (a, b, c)$, $q' = (a', b', c')$ and $\vec{q} = \vec{q'} = (\vec{a}, \vec{b}, \vec{c})$. Then we have $c = c' = \vec{a}$. We also have

$$b + \vec{b} \equiv 0 \bmod 2|c| \qquad \text{and} \qquad b' + \vec{b} \equiv 0 \bmod 2|c|.$$

Hence $b - b' = (\sqrt{d} - b') - (\sqrt{d} - b) \equiv 0 \bmod 2|c|$. Since both q and q' are binary reduced, we have

$$0 < \sqrt{d} - b < 2|c| \qquad \text{and} \qquad 0 < \sqrt{d} - b' < 2|c|,$$

which implies that $|(\sqrt{d} - b') - (\sqrt{d} - b)| < 2|c|$. Then $b = b'$, and necessarily $a = a'$ since q and q' are properly equivalent. □

By a *cycle of reduced forms* we mean a sequence of iterated shifts $q_0, q_1, \ldots, q_r$ with $r \geq 0$ such that $\overrightarrow{q_r} = q_0$. We say that a binary form q with nonsquare discriminant is *cyclic* if it is a nontrivial iterated right shift of itself.

Corollary 1.17. *Let q be a binary form with nonsquare discriminant. Then q is binary reduced if and only if q is a cyclic form.*

Proof. If q is cyclic, then q is binary reduced by Theorems 1.10 and 1.14, and an obvious version of Theorem 1.10 for the negative definite case.

Let q be a binary reduced form. If q is definite, then $q = \overrightarrow{\overrightarrow{q}}$ (as shown in the proof of Theorem 1.10 for the positive case, and similarly for the negative case). Hence q is a cyclic form.

Assume now that q is a binary reduced indefinite form. Since the sequence of iterated shifts starting with q is ultimately periodic, by Theorem 1.14 we may assume that $\overrightarrow{q}$ is cyclic. Then there is a binary reduced form q' with $\overrightarrow{q'} = \overrightarrow{q}$. By Lemma 1.16 we have $q = q'$, that is, q is a cyclic binary form. □

It is clear that if the iterations of indefinite binary forms q and q' with nonsquare discriminant lead to the same cycle of reduced forms, then q and q' are properly equivalent. Buell has shown that the converse is also true [17], that is, there is a bijective correspondence between proper classes of binary forms with nonsquare discriminant and cycles of binary reduced forms. Let us see some examples before considering forms with square discriminant.

Applying Algorithm 1.9 to the Kronecker form $q_m = (1, -m, 1)$ with $|m| \geq 3$ (which has positive nonsquare discriminant $(|m| + 2)(|m| - 2)$), we get the binary reduced form $p = \overrightarrow{q_m} = (1, |m| - 2, 2 - |m|)$. The corresponding cycle of reduced forms has period two (that is, $\overrightarrow{\overrightarrow{p}} = p$), and consists of the forms

$$p = (1, |m| - 2, 2 - |m|) \quad \text{and} \quad \overrightarrow{p} = (2 - |m|, |m| - 2, 1).$$

Consider now the binary form $q = (3, 10, 6)$ with discriminant $d = 28$. Construct the sequence of iterated shifts $q^0 = q, q^1, q^2, \ldots$ and observe that q^0 and $q^1 = (6, 2, -1)$ are not binary reduced forms, and that $q^6 = q^2$, that is the forms $q^2, \ldots, q^5$ constitute the cycle of binary reduced forms associated to q^0,

$$(-1, 4, 3)(3, 2, -2)(-2, 2, 3)(3, 4, -1).$$

Another example of sequence of indefinite binary reduced forms is given by

$$(3, 10, -4)(-4, 6, 7)(7, 8, -3)(-3, 10, 4)(4, 6, -7)(-7, 8, 3),$$

with discriminant $d = 148$. We follow Conway's notation in [19], where $(1\,{}^{4}\,3\,{}^{2}\,2\,{}^{2}\,3\,{}^{4})$ and $(3\,{}^{10}\,4\,{}^{6}\,7\,{}^{8})$ denote the cycles above. We stress that this notation expresses one, two or four cycles, those obtained by setting alternating signs in the lower row. In Table 1.2 we exhibit all cycles of indefinite binary reduced forms with discriminant $0 < d < 50$.

Table 1.2 Cycles of indefinite binary reduced forms of small discriminant $0 < d < 50$

Discriminant	h	Primitive forms	Nonprimitive forms
1	1	$(0\,{}^{1})$	–
4	2	$(0\,{}^{2}\,1\,{}^{2})$	$(0\,{}^{2})$
5	1	$(1\,{}^{1})$	–
8	1	$(1\,{}^{2})$	–
9	3	$(0\,{}^{3}\,1\,{}^{3})$	$(0\,{}^{3})$
12	2	$(1\,{}^{2}\,2\,{}^{2})$	–
13	1	$(1\,{}^{3})$	–
16	4	$(0\,{}^{4}\,1\,{}^{4})$	$(0\,{}^{4})$, $(0\,{}^{4}\,2\,{}^{4})$
17	1	$(1\,{}^{3}\,2\,{}^{1}\,2\,{}^{3})$	–
20	2	$(1\,{}^{4})$	$(2\,{}^{2})$
21	2	$(1\,{}^{3}\,3\,{}^{3})$	–
24	2	$(1\,{}^{4}\,2\,{}^{4})$	–
25	5	$(0\,{}^{5}\,1\,{}^{5})$; $(0\,{}^{5}\,2\,{}^{3}\,2\,{}^{5})$	$(0\,{}^{5})$
28	2	$(1\,{}^{4}\,3\,{}^{2}\,2\,{}^{2}\,3\,{}^{4})$	–
29	1	$(1\,{}^{5})$	–
32	3	$(1\,{}^{4}\,4\,{}^{4})$	$(2\,{}^{4})$
33	2	$(1\,{}^{5}\,2\,{}^{3}\,3\,{}^{3}\,2\,{}^{5})$	–
36	6	$(0\,{}^{6}\,1\,{}^{6})$	$(0\,{}^{6})$; $(0\,{}^{6}\,2\,{}^{6})$; $(0\,{}^{6}\,3\,{}^{6})$
37	1	$(1\,{}^{5}\,3\,{}^{1}\,3\,{}^{5})$	–
40	2	$(1\,{}^{6})$; $(2\,{}^{4}\,3\,{}^{2}\,3\,{}^{4})$	–
41	1	$(1\,{}^{5}\,4\,{}^{3}\,2\,{}^{5}\,2\,{}^{3}\,4\,{}^{5})$	–
44	2	$(1\,{}^{6}\,2\,{}^{6})$	–
45	3	$(1\,{}^{5}\,5\,{}^{5})$	$(3\,{}^{3})$
48	4	$(1\,{}^{6}\,3\,{}^{6})$	$(2\,{}^{4}\,4\,{}^{4})$
49	7	$(0\,{}^{7}\,1\,{}^{7})$; $(0\,{}^{7}\,2\,{}^{5}\,3\,{}^{7})$	$(0\,{}^{7})$

The number of (proper) equivalence classes of such forms with discriminant d is denoted by $h = h(d)$

A binary form (a, b, c) is called *primitive* if $\gcd(a, b, c) = 1$. Notice that if q is nonprimitive and properly equivalent to q', then q' is also nonprimitive, for if an integer s divides a, b and c, then s divides $q(u)$, $q(u|v)$ and $q(v)$. Taking $s = \gcd(a, b, c)$, the form $p = \frac{1}{s}q$ is a primitive binary form with $s^2 \det(p) = \det(q)$. Clearly the form q is binary reduced if and only if so is p, and $\frac{1}{s}\overrightarrow{q} = \overrightarrow{p}$. Hence we may restrict our attention to primitive forms, as we do next when considering binary forms with square discriminant. Let us first analyze the case of zero discriminant.

Lemma 1.18. *Let q be a primitive binary form with zero discriminant. Then there exists $a \geq 0$ and $c \in \mathbb{Z}$ with $\gcd(a, c) = 1$ such that*

$$q = (a^2, 2ac, c^2) \qquad \text{or} \qquad q = (-a^2, 2ac, -c^2).$$

In particular, any such form is properly equivalent to $(\pm 1, 0, 0)$.

Proof. Assume first that $q = (a', b', c')$ is a primitive form with zero discriminant, that is, $(b')^2 = 4ac$. We may also assume that $a' \geq 0$, and therefore $c' \geq 0$. Then $\gcd(a', c') = 1$, and considering prime decompositions of a' and c', this implies that there are relatively prime integers a and c with $a' = a^2$, $c' = c^2$ and $b' = 2ac$. Of course we may choose $a \geq 0$. Moreover, since $\gcd(a, c) = 1$, there are integers u_1 and u_2 such that $au_1 + b_u 2 = 1$. Taking the matrix $T = \begin{pmatrix} u_1 & -b \\ u_2 & a \end{pmatrix}$ a direct calculation verifies that $\det(T) = 1$ and $qT = (1, 0, 0)$.

If $a' \leq 0$ then $c' \leq 0$, and the result follows similarly. □

Next we complete Algorithm 1.9 to include binary forms (a, b, c) with $c = 0$.

Algorithm 1.19 (Binary Reduction). *For a primitive binary form $q = (a, b, 0)$ with $b \neq 0$ define a new binary form $\overrightarrow{q} = (0, b, c)$ taking c with minimal absolute value (positive in case of a tie) such that*

$$ac \equiv 1 \bmod b.$$

Then q and $\overrightarrow{q}$ are properly equivalent forms. In particular, $\overrightarrow{q}$ is primitive.

Proof. Let $T = [u|v]$ be a linear transformation with $u = (u_1, u_2)$, $v = (v_1, v_2)$ and $\det(T) = 1$. A direct calculation shows that

$$qT = (q(u), q(u|v), q(v)) = (u_1(au_1 + bu_2), b + 2v_1(au_1 + bu_2), v_1(av_1 + bv_2)).$$

By hypothesis there is a v_2 with $ac + bv_2 = 1$, therefore taking $u = (-b, a)$ and $v = (-c, -v_2)$, we have $\det(T) = bv_2 + ac = 1$ and

$$qT = (-b(-ab + ab), b + 2v_1(-ab + ab), -c(-ac - bv_2)) = (0, b, c).$$

Hence $qT = \overrightarrow{q}$, that is, q and $\overrightarrow{q}$ are properly equivalent. That $\overrightarrow{q}$ is primitive is clear. □

It is easy to verify that applying Algorithm 1.9 to an indefinite form q with square discriminant d we eventually obtain a binary form $(a, \sqrt{d}, 0)$ for some integer a (see Exercise 7 below). In addition to binary forms satisfying any of the equivalent conditions in Lemma 1.12, those forms $(a, b, 0)$ and $(0, b, a)$ satisfying $-b < 2a \le b$ are also called *binary reduced*. Extending the shifting process with Algorithm 1.19, it can easily be shown that sequences of iterated shifts are also ultimately periodic for indefinite forms with square discriminant, and the same correspondence between cyclic forms and binary reduced forms holds in this case. It is important to notice at this point that the usual definition of *cycle* in the case of binary forms with square discriminant does not include the construction given in Algorithm 1.19. Instead, in this case the cycle associated to a binary form is just a finite sequence of reduced forms, justified by the relation between *continued fractions* and reduction of binary forms (see Buell [17]). For instance, the form $q = (-7, 4, 3)$ yields the sequence

$$(-7, 4, 3)(3, 8, -3)(-3, 10, 0)(0, 10, 3),$$

where the right shift of $(0, 10, 3)$ is $(3, 8, -3)$, that is, the last three forms are binary reduced. See more examples in Table 1.2, where we use a similar notation as in the indefinite case with nonsquare discriminant, that is, the sequence of reduced forms above is denoted by $\left(0^{10}3^{\ 8}\,3^{10}\right)$. For this case we have to be careful with the signs in the lower row, considering the rule $ac \equiv 1 \bmod b$ when completing a period in the cycle.

The following criterion is an easy consequence of our discussion above.

Proposition 1.20. *A binary form is isotropic if and only if it has square discriminant.*

Proof. Let q be a binary form with discriminant b^2. By the discussion above, q is equivalent to a form $(a, b, 0)$ for some $a \in \mathbb{Z}$. In particular, q is isotropic.

Suppose that $u = (u_1, u_2)$ is an isotropic vector for q. We may assume that u_1 and u_2 are relatively prime. Choose integers v_1 and v_2 such that $u_1v_2 - v_1u_2 = 1$, and consider the matrix $T = [u|v]$, where v is the vector (v_1, v_2). Since $qT = (q(u), q(u|v), q(v)) = (0, q(u|v), q(v))$, the forms qT and q have square discriminant $d = q(u|v)^2$. This completes the proof. □

We end this section with a characterization of definite and indefinite binary forms, whose generalization to multiple variables will play a fundamental role in subsequent chapters.

Proposition 1.21. *Let q be a binary form.*

a) The form q is positive definite if and only if $q(x) > 0$ for any $0 \ne x$ in $\mathbb{Z}^2$.
b) The form q is negative definite if and only if $q(x) < 0$ for any $0 \ne x$ in $\mathbb{Z}^2$.
c) The form q is indefinite if and only if there are vectors x and y in $\mathbb{Z}^2$ such that $q(x) > 0$ and $q(y) < 0$.

Proof. Since all mentioned properties are invariant under proper equivalence, we may assume that $q = (a, b, c)$ is a binary reduced form. If q is indefinite with nonsquare discriminant, then a and c have opposite sign (for $0 < b < \sqrt{d}$). Then point (c) follows by taking x and y to be the canonical vectors e_1 and e_2 in $\mathbb{Z}^2$. If q has square discriminant then $ac = 0$ and $0 \leq a, c < b$. Taking $x = e_1 + e_2$ and $y = e_1 - e_2$ in this case, we complete the proof of (c).

Assume now that q is positive definite. Then there are integers $0 \leq r_a \leq r_c$ such that $a = |b| + r_a$ and $c = |b| + r_c$, and we have

$$q(x_1, x_2) = (|b|+r_a)x_1^2+bx_1x_2+(|b|+r_c)x_2^2 = \frac{|b|}{2}[x_1^2+(x_1 \pm x_2)^2+x_2^2]+r_a x_1^2+r_c x_2^2,$$

where $\pm$ is the sign of b. This shows (a), since we have expressed q as sum of squares with nonnegative coefficients. Point (b) can be shown similarly. □

Exercises 1.22.

1. Give an explicit proper equivalence between (a, a, c) and $(a, -a, c)$.
2. Find the cycle of reduced forms associated to the following quadratic forms.

   i) $(1, 3, -2)$.
   ii) $(1, 35, -22)$.
   iii) $(22, 9, -14)$.

3. Give an example of a positive definite binary form q such that $\overrightarrow{q}$ is not binary reduced.
4. How many proper equivalence classes of indefinite binary forms with discriminant $d = 52$ are there?
5. How many proper equivalence classes of positive definite binary forms with discriminant $d = -51$ are there?
6. Find the cycle of reduced forms corresponding to the Pell form q_r with $r > 0$.
7. For primitive binary forms $q = (a, b, 0)$ and $q' = (a', b, 0)$, show that q is properly equivalent to q' if and only if $a + a' \equiv 0 \bmod b$. Conclude that for every $b > 0$ there is a finite number of proper equivalence classes of binary forms with discriminant b^2.
8. Let $(f_n)_{n\in\mathbb{N}}$ be the sequence of Fibonacci numbers defined recursively as $f_0 = 0$, $f_1 = 1$ and $f_{i+1} = f_i + f_{i-1}$ for $i \geq 1$. For $n > 0$, define the quadratic form

   $$q_{2n} = f_{2n-1}x_1^2 - 2f_{2n}x_1x_2 + f_{2n+1}x_2^2.$$

   Prove that $q_{2n} \sim x_1^2 + x_2^2$.
9. Let $(f_n)_{n\in\mathbb{N}}$ be the sequence of Fibonacci numbers. For $n > 0$ consider the quadratic form q_{2n} defined in Exercise 8, and

   $$q_{2n-1} = f_{2n-1}x_1^2 - (f_{2n} + f_{2n-2})x_1x_2 + f_{2n-1}x_2^2.$$

   Prove that $1 \in \mathbf{D}_{\mathbb{Z}}(q_{2n})$ for any $n > 0$, also $1 \in \mathbf{D}_{\mathbb{Z}}(q_1)$ but $1 \notin \mathbf{D}_{\mathbb{Z}}(q_{2n-1})$ for $n \geq 2$.

1.5 Kronecker Forms and the Pell Equation

In this section we give explicit solutions to equation $q(x) = 1$ with vectors x in $\mathbb{Z}^2$ for Kronecker forms q_m (with associated bigraphs $\mathbb{K}_m$ as below) and for Pell forms $q_{[r]}$ (the so-called *positive Pell equation*).

$\mathbb{K}_m =$ 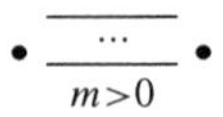 and $\mathbb{K}_m =$ 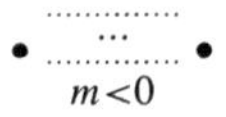

As direct consequence of Proposition 1.20 we have that a Kronecker form q_m is isotropic if and only if $|m| = 2$, and similarly a Pell form $q_{[r]}$ is isotropic if and only if r is a square integer.

Consider matrices $T = \begin{pmatrix} 0 & 1 \\ -1 & m \end{pmatrix}$ and $S = \begin{pmatrix} 0 & 1 \\ 1 & 0 \end{pmatrix}$, and their corresponding linear transformations

$$T: \quad \mathbb{Z}^2 \longrightarrow \mathbb{Z}^2, \qquad (x_1, x_2) \longmapsto (x_2, mx_2 - x_1), \qquad\qquad S: \quad \mathbb{Z}^2 \longrightarrow \mathbb{Z}^2, \qquad (x_1, x_2) \longmapsto (x_2, x_1).$$

Observe that if A_{q_m} is the symmetric matrix associated to q_m then

$$T^t A_{q_m} T = \begin{pmatrix} 0 & -1 \\ 1 & m \end{pmatrix} \begin{pmatrix} 1 & -\frac{m}{2} \\ -\frac{m}{2} & 1 \end{pmatrix} \begin{pmatrix} 0 & 1 \\ -1 & m \end{pmatrix} = A_{q_m},$$

and similarly $S^t A_{q_m} S = A_{q_m}$ and $(T^{-1})^t A_{q_m} T^{-1} = A_{q_m}$, where $T^{-1} = \begin{pmatrix} m & -1 \\ 1 & 0 \end{pmatrix}$.

Take $\varepsilon \in \{+1, -1\}$ such that $m = \varepsilon|m|$ and define recursively $a_0 = 0$, $a_1 = 1$ and $a_{s+1} = |m|a_s - a_{s-1}$ for $s \geq 1$. In the following proposition we show that the set of roots $R(q_m)$ of the Kronecker form q_m is

$$R(q_m) = \{\pm(a_s, \varepsilon^s a_{s+1}), \pm(\varepsilon^s a_{s+1}, a_s)\}_{s \geq 0}.$$

With a direct calculation one can show that

$$\{T^s e_i\}_{s \in \mathbb{Z},\, i=1,2} = \{\pm(a_s, \varepsilon^s a_{s+1}), \pm(\varepsilon^s a_{s+1}, a_s)\}_{s \geq 0},$$

where e_1 and e_2 are the canonical vectors in $\mathbb{Z}^2$. Observe also that if $m = 0$ then $\{T^s e_i\}_{s \in \mathbb{Z},\, i=1,2} = \{\pm e_i\}_{i=1,2}$, and that for $m = 1$ we have

$$\{T^s e_i\}_{s \in \mathbb{Z},\, i=1,2} = \{\pm e_i, \pm(e_1 + e_2)\}_{i=1,2}.$$

Proposition 1.23. *For $m \in \mathbb{Z}$ the set of roots $R(q_m)$ of the Kronecker form q_m is given by*

$$R(q_m) = \{T^s e_i\}_{s \in \mathbb{Z},\, i=1,2}.$$

In particular, q_m has infinitely many roots if and only if $|m| > 1$.

Proof. The claim is clear for $m = 0, 1$ since $R(q_0) = \{\pm e_i\}_{i=1,2}$ and

$$R(q_1) = \{\pm e_i, \pm(e_1 + e_2)\}_{i=1,2}.$$

Assume that $m \geq 2$. For arbitrary $s \in \mathbb{Z}$ and $i = 1, 2$ by the above we have

$$q_m(T^s e_i) = e_i^t (T^s)^t A_{q_m} T^s e_i = e_i^t A_{q_m} e_i = q(e_i) = 1.$$

Therefore $\{T^s e_i\}_{s\in\mathbb{Z},\, i=1,2} \subseteq R(q_m)$.

Assume now that $x = (x_1, x_2)$ is a root of q_m with $x_1 x_2 \neq 0$, and observe that we may suppose that $0 < x_1 < x_2$ (for $m \geq 2$). We claim that if $T^{-1}x = (y_1, y_2)$ then $0 \leq y_1 < y_2 = x_1$. Indeed, if $y_1 \geq x_1$ then $y_1 = mx_1 - x_2$ implies that $(m-1)x_1 \geq x_2$ and

$$1 = x_1^2 + x_2^2 - (m-1)x_1x_2 - x_1x_2 \leq x_1(x_1 - x_2) < 0,$$

a contradiction. We conclude that there is an $s < 0$ with $T^s x = (0, 1)$, hence $x \in \{T^s e_i\}_{s\in\mathbb{Z},\, i=1,2}$. For negative m we proceed similarly. □

It is interesting to observe that the simple reflections associated to a Kronecker form q_m with $m > 0$ can be obtained as $\sigma_1 = ST$ and $\sigma_2 = TS$. In particular, the transformation $\sigma_2\sigma_1$ given by

$$\sigma_2\sigma_1 = T^2 = \begin{pmatrix} -1 & m \\ -m & m^2 - 1 \end{pmatrix}$$

is known as the Coxeter transformation of the Kronecker form (cf. Sect. 4.6).

The binary integral quadratic form $q_{[d]} = x_1^2 - dx_2^2$ has a long history, and is still a source of active research (see for instance [4]). Here we are interested in finding its roots, that is, solutions to the *Pell equation*,

$$x_1^2 - dx_2^2 = 1,$$

for integer values of x_1 and x_2. A solution with $x_1x_2 = 0$ is called *trivial*. Notice that for $d < 0$ or d a square integer, the Pell equation has only trivial solutions (since in the latter case, $1 = (x_1+\sqrt{d}x_2)(x_1-\sqrt{d}x_2)$ implies $x_1+\sqrt{d}x_2 = x_1-\sqrt{d}x_2 = \pm 1$). Hence we will assume that $d > 0$ is not a square integer.

It will be convenient to consider the (real) quadratic ring $\mathbb{Z}[\sqrt{d}]$ (see for instance [17]). Its elements have the form $\alpha = x_1 + \sqrt{d}x_2$ with integers x_1 and x_2. Sums and products are given by

$$(x_1 + \sqrt{d}x_2) + (x_1' + \sqrt{d}x_2') = (x_1 + x_1') + \sqrt{d}(x_2 + x_2'),$$

$$(x_1 + \sqrt{d}x_2)(x_1' + \sqrt{d}x_2') = (x_1x_1' + dx_2x_2') + \sqrt{d}(x_1x_2' + x_1'x_2).$$

The conjugate of α is $\overline{\alpha} = x_1 - \sqrt{d}x_2$, and the function given by

$$\mathbf{N}(\alpha) = \alpha\overline{\alpha} = x_1^2 - dx_2^2$$

is usually known as the *norm* of α. We will identify the ring $\mathbb{Z}[\sqrt{d}]$ with $\mathbb{Z}^2$ via $x_1 + \sqrt{d}x_2 \mapsto (x_1, x_2)$ (which is well defined since $\sqrt{d}$ is an irrational number). Under this identification a norm one quadratic number is a solution to the Pell equation. Observe that $\mathbf{N} : \mathbb{Z}[\sqrt{d}] \to \mathbb{Z}$ is a multiplicative function (Exercise 2). Therefore, given a solution α to the Pell equation $N(\alpha) = 1$, all of the quadratic numbers $\pm\alpha^n$ for $n \in \mathbb{Z}$ are also solutions to the Pell equation. A solution $\alpha > 1$ such that any solution to the Pell equation has the form $\pm\alpha^n$ with $n \in \mathbb{Z}$ is called a *fundamental solution*.

Proposition 1.24. *If the Pell equation $x_1^2 - dx_2^2 = 1$ has a nontrivial solution, then it has a fundamental solution.*

Proof. Multiplying by (-1) and taking the conjugate if necessary, we may assume there is a solution $y_1 + \sqrt{d}y_2$ to the Pell equation with $y_1 > 0$ and $y_2 > 0$. Since $\{x_1 + \sqrt{d}x_2\}_{x_1, x_2 \in \mathbb{N}}$ is a discrete subset of $\mathbb{R}$, we may also assume that $\alpha = y_1 + \sqrt{d}y_2$ is the minimal solution greater than 1 (see Exercise 3). If β is a solution greater than 1, choose $n \geq 1$ such that $\alpha^n < \beta \leq \alpha^{n+1}$. Then

$$1 < \alpha^{-n}\beta \leq \alpha,$$

and by minimality, $\beta = \alpha^{n+1}$. Therefore any solution has the form $\pm\alpha^n$ with $n \in \mathbb{Z}$, that is, α is a fundamental solution to the Pell equation $x_1^2 - dx_2^2 = 1$. □

We show now the existence of nontrivial solutions to the Pell equation $x_1^2 - dx_2^2 = 1$. The following technical lemma due to Dirichlet, in terms of elementary modular arithmetic, establishes the fundamental step in the proof of Theorem 1.26 (attributed to Lagrange, cf. [4]).

Lemma 1.25. *There exists a nonzero integer m with solutions (x_1, x_2) and (y_1, y_2) to the equation $q_{[d]}(x) = m$, with $(x_1, x_2) \neq \pm(y_1, y_2)$ and*

$$x_1 \equiv x_2 \bmod |m| \quad \textit{and} \quad y_1 \equiv y_2 \bmod |m|.$$

Proof. Observe first that there exists an $M > 0$ such that $|x_1^2 - dx_2^2| < M$ has infinitely many solutions. Indeed, by Exercise 6 there are infinitely many $\frac{p}{q}$ with $q > 0$ such that $|\sqrt{d} - \frac{p}{q}| < \frac{1}{q^2}$, thus $|p - \sqrt{d}q| < \frac{1}{q}$ and

$$\begin{aligned} |p^2 - dq^2| &= |p - \sqrt{d}q||p + \sqrt{d}q| < \frac{1}{q}(|p - \sqrt{d}q| + 2\sqrt{d}q) \\ &< 1 + 2\sqrt{d} = M. \end{aligned}$$

Therefore there exists an m with $|m| < M$ and infinitely many $p + \sqrt{d}q$ with $\mathbf{N}(p + \sqrt{d}q) = m$ (observe that $m \neq 0$, see Exercise 4).

Now, there are only m^2 options for p modulo $|m|$ and for q modulo $|m|$, therefore we may find (p, q) and (p', q') satisfying the claim. □

Theorem 1.26. *For d a positive nonsquare integer there exists a nontrivial solution to the Pell equation $x_1^2 - dx_2^2 = 1$.*

Proof. Let m, (x_1, x_2) and (y_1, y_2) be as in Lemma 1.25 and take $\alpha = x_1 + \sqrt{d}x_2$ and $\beta = y_1 + \sqrt{d}y_2$. Notice that $\alpha\overline{\beta}$ has the form

$$\alpha\overline{\beta} = (x_1 + \sqrt{d}x_2)(y_1 - \sqrt{d}y_2) = (x_1y_1 - dx_2y_2) + \sqrt{d}(y_1x_2 - x_1y_2),$$

where

$$x_1y_1 - dx_2y_2 \equiv x_1^2 - dx_2^2 \equiv 0 \bmod |m|,$$

$$y_1x_2 - x_1y_2 \equiv x_1x_2 - x_1x_2 \equiv 0 \bmod |m|.$$

Hence there exists a $w_1 + \sqrt{d}w_2$ such that $\alpha\overline{\beta} = m(w_1 + \sqrt{d}w_2)$, therefore

$$m^2\mathbf{N}(w_1 + \sqrt{d}w_2) = \mathbf{N}(m(w_1 + \sqrt{d}w_2)) = \mathbf{N}(\alpha\overline{\beta}) = m^2,$$

and since $m \neq 0$ we have $\mathbf{N}(w_1 + \sqrt{d}w_2) = 1$. Finally $w_2 \neq 0$, for otherwise $y_1x_2 = x_1y_2$, which implies that α is a multiple of β, contradicting $\mathbf{N}(\alpha) = \mathbf{N}(\beta)$ and $\alpha \neq \pm\beta$. □

Combining Proposition 1.24 and Theorem 1.26 we now describe all solutions to $x_1^2 - dx_2^2 = 1$. By the above, starting with a fundamental solution $\alpha = x_1 + \sqrt{d}x_2$, all solutions have the shape $\pm\alpha^n$ $(n \in \mathbb{Z})$. For $n > 0$ we have

$$\alpha^n = \left[\sum_{j=0}^{\lfloor \frac{n}{2} \rfloor} \binom{n}{2j} d^j x_1^{n-2j} x_2^{2j}\right] + \sqrt{d}\left[\sum_{j=0}^{\lfloor \frac{n-1}{2} \rfloor} \binom{n}{2j+1} d^j x_1^{n-2j-1} x_2^{2j+1}\right],$$

and $\alpha^{-n} = \overline{\alpha^n}$.

There is a nice relation between the roots of certain Kronecker forms and Pell forms, and the *Chebyshev polynomials* of the first kind $T_n(t)$ and second kind $U_n(t)$. Define the polynomials $U_{-1}(\ell) = 0$ and $U_0(\ell) = 1$ in the variable ℓ, and take recursively for $s \geq 0$,

$$U_{s+1}(\ell) = 2\ell U_s(\ell) - U_{s-1}.$$

Notice that by construction and Proposition 1.23, for $\ell > 0$ the roots of $q_{2\ell}$ with nonnegative entries are given by the following vectors in $\mathbb{Z}^2$,

$$(U_s(\ell), U_{s+1}(\ell)), \qquad \text{and} \qquad (U_{s+1}(\ell), U_s(\ell)).$$

Consider now the linear transformation $W : \mathbb{Z}^2 \to \mathbb{Z}^2$ with matrix $W = \begin{pmatrix} 1 & -\ell \\ 0 & 1 \end{pmatrix}$. A direct calculation shows that $q_{[\ell^2-1]}W = q_{2\ell}$. Therefore, if we take $T_{-1}(\ell) = 0$ and $T_s(\ell) = U_s(\ell) - \ell U_{s-1}(\ell)$ for $s \geq 0$, then the following are solutions to the Pell equation $q_{[\ell^2-1]} = 1$,

$$(T_{s+1}(\ell), U_s(\ell)).$$

We stress that the Chebyshev polynomial of the first kind T_s may be given recursively by setting $T_0(\ell) = \ell$ and for $s \geq 0$,

$$T_{s+1}(\ell) = 2\ell T_s(\ell) - T_{s-1}(\ell).$$

See the exercises below and Chap. 4 for alternative definitions and properties of Chebyshev polynomials.

Exercises 1.27.

1. Show that for $s \geq 0$ the Chebyshev polynomials are given by

$$T_s(\ell) = \cos(s \arccos(\ell)) \quad \text{and} \quad U_s(t) = \frac{\sin((s+1)\arccos(\ell))}{\sqrt{1-\ell^2}}.$$

2. Show that if $\alpha, \beta \in \mathbb{Z}[\sqrt{d}]$, then $\mathbf{N}(\alpha\beta) = \mathbf{N}(\alpha)\mathbf{N}(\beta)$.
3. Consider the order in $\mathbb{Z}[\sqrt{d}]$ induced from the order in $\mathbb{R}$, and let (x_1, x_2) be a nontrivial solution to the Pell equation $x_1^2 - dx_2^2 = 1$. Show that

   i) $x_1, x_2 > 0$ if and only if $x_1 + \sqrt{d}x_2 > 1$.
   ii) $x_1 > 0$ and $x_2 < 0$ if and only if $0 < x_1 + \sqrt{d}x_2 < 1$.

4. Show that if $\mathbf{N}(\alpha) = 0$ for $\alpha \in \mathbb{Z}[\sqrt{d}]$ then $\alpha = 0$.
5. Find the fundamental solution to the equation $x^2 - 5y^2 = 1$.
6. **Dirichlet's approximation theorem.** For $z \in \mathbb{R}$ a *D-approximation* to z is a rational number $\frac{p}{q}$ with $q > 0$ (and $\gcd(p, q) = 1$) such that $|z - \frac{p}{q}| < \frac{1}{q^2}$.

   i) Show that any positive real number has a D-approximation.
   ii) For $z > 0$, show that z is irrational if and only if z has infinitely many D-approximations.

   [Hint: for $n > 0$ consider the numbers $iz - \lfloor iz \rfloor$ for $i = 0, \ldots, n$, where $\lfloor w \rfloor$ denotes the integral part of w. Use the Pigeonhole Principle for these $n + 1$ numbers within the set of n semi-closed intervals $[\frac{j}{n}, \frac{j+1}{n})$ for $j = 0, \ldots, n-1$.]
7. Find the roots of the Chebyshev polynomials T_s and U_s.
8. Prove the following explicit formula for U_s,

$$U_s(\ell) = \sum_{i=1}^{\lfloor s/2 \rfloor} (-1)^i \binom{s-i}{i} (2\ell)^{s-2i}.$$

1.6 Quadratic Forms with Real Coefficients

Here we consider the case of quadratic forms $q(x_1, \ldots, x_n) = \sum_{i,j=1}^n q_{ij}x_i x_j$ with q_{ij} in the field of real numbers $\mathbb{R}$ for $1 \le i \le j \le n$. We use throughout general results in linear algebra, as can be found for instance in [27] or [33].

Proposition 1.28. *Let A be a symmetric $n \times n$ real matrix and $q(x) = x^t Ax$.*

a) All the eigenvalues of A are real numbers.
b) If $\lambda_1 \le \lambda_2 \le \cdots \le \lambda_n$ are the eigenvalues of A, then

$$\lambda_1 = \min\{q(x) \mid x \in \mathbb{R}^n \text{ with } \|x\| = 1\},$$

$$\lambda_n = \max\{q(x) \mid x \in \mathbb{R}^n \text{ with } \|x\| = 1\}.$$

Proof.

(a) Let $\lambda \in \mathbb{C}$ and $0 \ne v \in \mathbb{C}^n$ with $Av = \lambda v$. Then $\overline{\lambda}\|v\|^2 = \overline{\lambda}\overline{v}^t v = (\overline{Av})^t v = \overline{v}^t \overline{A}^t v = \overline{v}^t Av = \lambda\|v\|^2$, where $\overline{\lambda}$ denotes the complex conjugate of λ. Hence $\lambda = \overline{\lambda} \in \mathbb{R}$.

(b) The well-known *Gram–Schmidt orthonormalisation process* yields an $n \times n$ matrix U with orthonormal columns such that

$$A = U^t DU,$$

where $D = \operatorname{diag}(\lambda_1, \ldots, \lambda_n)$ is the diagonal matrix with entries $\lambda_1, \ldots, \lambda_n$. Hence for $x \in \mathbb{R}^n$ with $\|x\| = 1$ and $y = Ux$, we get $\|y\| = 1$ and $q(x) = x^t Ax = y^t Dy = \sum_{i=1}^n \lambda_i y_i$. Therefore we have

$$\lambda_1 = \lambda_1\|y\|^2 = \lambda_1 \sum_{i=1}^n y_i^2 \le \sum_{i=1}^n \lambda_i y_i^2 = q(x) \le \lambda_n \sum_{i=1}^n y_i^2 = \lambda_n\|y\|^2 = \lambda_n.$$

If u^1 and u^n are vectors in $\mathbb{R}^n$ with $||u^1|| = 1 = ||u^n||$ and $Au^i = \lambda_i u^i$ for $i = 1, n$, then

$$q(u^i) = (u^i)^t Au^i = \lambda_i ||u^i||^2 = \lambda_i,$$

for $i = 1, n$. This completes the proof. □

It can be shown that a real quadratic form $q(x_1, \ldots, x_n)$ can be represented in an infinite number of ways as

$$q(x) = \sum_{i=1}^r a_i X_i^2,$$

with nonzero coefficients a_i, and where $X_i = \sum_{j=1}^{n} \alpha_{ij} x_j$ for $i = 1, \dots, r$ are linearly independent linear forms in the variables $x_1, \dots, x_n$. Here r is the *rank of the quadratic form* q, that is, the rank of the symmetric matrix A_q associated to q.

We show that the number of coefficients $a_i > 0$ (resp. $a_i < 0$) is an *invariant* of the quadratic form q (see for instance [27, X§2]).

Theorem 1.29 (Sylvester's Law of Inertia). *Given two representations of the quadratic form $q(x)$ as*

$$q(x) = \sum_{i=1}^{r} a_i X_i^2 \quad \text{and} \quad q(x) = \sum_{i=1}^{r} b_i Y_i^2,$$

where X_i (resp. Y_i) are linearly independent real linear forms in the variables $x_1, \dots, x_n$ with $a_1 \geq a_2 \geq \cdots \geq a_s > 0 > a_{s+1} \geq \cdots \geq a_r$ and $b_1 \geq b_2 \geq \cdots \geq b_t > 0 > b_{t+1} \geq \cdots \geq b_r$, then we have $s = t$.

Proof. With the above notation assume that $s < t$. Since X_i (resp. Y_i) are real linear forms in the variables $x_1, \dots, x_n$, we may consider the following equations

$$X_1 = 0,\ X_2 = 0,\ \dots\ X_s = 0,\ Y_{t+1} = 0,\ \dots\ Y_r = 0,$$

as a system of $r - (t - s)$ equations in the r variables $Y_1, \dots, Y_r$. Because $r - (t - s) < r$, there exists a nonzero solution $Y_0 = (Y_1, Y_2, \dots, Y_r)$ for this system. Consequently there is a nonzero $v \in \mathbb{R}^n$ such that for $q(x) = \sum_{i=1}^{r} a_i X_i^2$ we get $q(v) \leq 0$, while for $q(x) = \sum_{i=1}^{r} b_i Y_i^2$ we get $q(v) > 0$. A contradiction showing that $s = t$. □

The difference $\boldsymbol{\sigma}(q)$ between the number of positive squares and the number of negative squares in a representation $q(x) = \sum_{i=1}^{r} a_i X_i^2$ is called the *signature* of q. The number $n_0 := n - r$ is known as the *corank of* q. The common number $n_+ := s = t$ in Theorem 1.29 is referred to as the *positive index of inertia* of q (and similarly for $n_- := r - n_+$), and the triple (n_+, n_0, n_-) is also called the *signature* of q.

There are simple methods to reduce a real quadratic form to a sum of independent squares. We will describe here an algorithm due to Lagrange (cf. [27, X§3]).

Algorithm 1.30 (Lagrange's Method). *Let $q(x) = \sum_{i,j=1}^{n} a_{ij} x_i x_j$ be a given real quadratic form with $a_{ij} = a_{ji}$ for $i < j$. We consider two cases:*

(1) Assume $a_{11} \neq 0$. Then

$$q(x) = \frac{1}{a_{11}} \left(\sum_{j=1}^{n} a_{1j} x_j \right)^2 + q_1(x_2, \dots, x_n),$$

where q_1 is a quadratic form in the variables $x_2, \dots, x_n$.

(2) Assume $a_{11} = 0 = a_{22}$ and $a_{12} \neq 0$. Then

$$q(x) = \frac{1}{2a_{12}}\left[\sum_{j=1}^{n}(a_{1j}+a_{2j})x_j\right]^2 - \frac{1}{2a_{12}}\left[\sum_{j=1}^{n}(a_{1j}-a_{2j})x_j\right]^2 + q_2(x_3,\ldots,x_n),$$

where q_2 is a quadratic form in the variables $x_3, \ldots, x_n$. Observe that the linear forms $\sum_{j=1}^{n} a_{1j}x_j$ and $\sum_{j=1}^{n} a_{2j}x_j$ are linearly independent.

By successive application of Steps (1) and (2), the form $q(x)$ can always be reduced to a sum of squares. Moreover, the linear forms obtained are linearly independent since at each step the constructed linear form contains a variable which does not appear in the remaining quadratic form.

Let us consider some examples.

i) Consider the matrix $B = \begin{pmatrix} a & r \\ r & b \end{pmatrix}$ and the binary quadratic form $q(x) = x^t Bx$. Using Algorithm 1.30 we take $q(x)$ to the form

$$q(x) = \begin{cases} \frac{1}{a}(ax_1 + rx_2)^2 + \left(b - \frac{r^2}{a}\right)x_2^2, & \text{if } a \neq 0; \\ \frac{1}{2r}(rx_1 + rx_2)^2 - \frac{1}{2r}(rx_1 - rx_2)^2, & \text{if } a = 0 = b. \end{cases}$$

ii) The quadratic form $q(x_1, x_2, x_3, x_4) = 4x_1^2 + x_2^2 + x_4^2 - 3x_1x_2 - 4x_1x_3 + 4x_2x_4 - 3x_3x_4$ of Example 1.2 is taken to the following sum of squares (see Exercise 6(iv) below)

$$\frac{1}{4}(4x_1 - \frac{3}{2}x_2 - 2x_3)^2 + \frac{16}{7}(\frac{7}{16}x_2^2 - \frac{3}{4}x_3 + 2x_4)^2 - \frac{7}{16}(-\frac{16}{7}x_3 + \frac{27}{14}x_4)^2 - \frac{417}{64}x_4.$$

Let $A = (a_{ij})$ be an $n \times n$ symmetric real matrix and take $1 \leq i_1 < i_2 < \cdots < i_s \leq n$. The *principal minor* $A(i_1, \ldots, i_s)$ of A is defined as the determinant of the matrix

$$\begin{pmatrix} a_{i_1i_1} & a_{i_1i_2} & \cdots & a_{i_1i_s} \\ a_{i_2i_1} & a_{i_2i_2} & \cdots & a_{i_2i_s} \\ \vdots & \vdots & \ddots & \vdots \\ a_{i_si_1} & a_{i_si_2} & \cdots & a_{i_si_s} \end{pmatrix}.$$

For $i \leq j$, we denote by $A^{(i,j)}$ the minor $A(1, \ldots, \hat{i}, \ldots, \hat{j}, \ldots, n)$ obtained by omitting the i-th row and the j-th column of the matrix A. The i-th *consecutive principal minor* is the determinant $D_i = A(1, \ldots, i)$. The *adjugate matrix* $\mathbf{ad}(A)$ of A is defined as the $n \times n$-matrix with entries

$$\mathbf{ad}(A)_{ij} = (-1)^{i+j}A^{(i,j)}.$$

It is well known that $A\,\mathbf{ad}(A) = \det(A)\mathbf{Id}_n = \mathbf{ad}(A)\,A$.

Let r be the rank of A and assume $D_i = A(1,\ldots,i) \neq 0$ for $1 \le i \le r$. It can be shown (see Exercise 6 below) that the matrix A is congruent to

$$D = \operatorname{diag}\left(\frac{1}{D_1}, \frac{D_1}{D_2}, \ldots, \frac{D_{r-1}}{D_r}, 0, \ldots, 0\right).$$

In particular, the quadratic form q admits the following expression, known as *Jacobi's Formula* (cf. [27, X§3]),

$$q(x) = \sum_{i=1}^{r} \frac{D_{i-1}}{D_i} X_i^2, \quad (D_0 = 1),$$

for linearly independent functions $X_1, \ldots, X_r$. In fact, this formula can be rewritten as

$$q(x) = \sum_{i=1}^{r} \frac{Y_i^2}{D_{i-1} D_i},$$

where $Y_i = c_{ii}x_i + c_{i,i+1}x_{i+1} + \cdots + c_{in}x_n$ (for $i = 1, \ldots, r$), and

$$c_{ij} = \det\begin{pmatrix} a_{11} & \ldots & a_{1\,i-1} & a_{1\,j} \\ \vdots & \ddots & \vdots & \vdots \\ a_{i-1\,1} & \ldots & a_{i-1\,i-1} & a_{i-1\,j} \\ a_{i\,1} & \ldots & a_{i\,i-1} & a_{ij} \end{pmatrix} \quad \left(=: A\begin{pmatrix} 1 \ldots i-1\; i \\ 1 \ldots i-1\; j \end{pmatrix}\right).$$

Exercises 1.31.

1. Let U be a $n \times n$ real matrix. Show that the following are equivalent:

   (i) $U^t U = I_n$ is the identity matrix.
   (ii) U is nonsingular and $U^t = U^{-1}$.
   (iii) The rows of U form an orthonormal basis of $\mathbb{R}^n$.
   (iv) For all $x \in \mathbb{R}^n$, the norm of $U(x)$ is the same as that of x, that is, $(U(x))^t U(x) = x^t x$.

   A matrix with these properties is called *real orthogonal.*

2. For $\theta \in \mathbb{R}$, take $T(\theta) = \begin{pmatrix} \cos\theta & \sin\theta \\ -\sin\theta & \cos\theta \end{pmatrix}$. Show that a 2×2-matrix U is real orthogonal if and only if $U = T(\theta)$ or $U = \begin{pmatrix} 1 & 0 \\ 0 & -1 \end{pmatrix} T(\theta)$, for some $\theta \in \mathbb{R}$.

3. Let U be a 3×3 real orthogonal matrix. Show that if $\det U > 0$, then U is congruent to $\begin{pmatrix} 1 & 0 \\ 0 & T(\theta) \end{pmatrix}$ for some $\theta \in \mathbb{R}$.

4. Show that any real symmetric matrix A is congruent to a diagonal matrix

$$\mathrm{diag}\,(1, 1, \ldots, 1, 0, 0, \ldots, 0, -1, -1, \ldots, -1),$$

with π entries 1, r entries 0 and ν entries -1, where $\sigma = \pi - \nu$ is the signature of the quadratic form $q(x) = x^t Ax$.

5. Let $q(x_1, \ldots, x_n)$ be a quadratic form. Show that Steps (1) and (2) given in Lagrange's Method 1.30 may be written as

$$q(x) = \frac{1}{4a_{11}}\left(\frac{\partial q}{\partial x_1}\right)^2 + q_1(x),$$

$$q(x) = \frac{1}{8a_{12}}\left[\left(\frac{\partial q}{\partial x_1} + \frac{\partial q}{\partial x_2}\right)^2 - \left(\frac{\partial q}{\partial x_1} - \frac{\partial q}{\partial x_2}\right)^2\right] + q_2(x).$$

6. Apply Lagrange's Method 1.30 to reduce the following quadratic forms to a sum of squares:

   (i) $a_n(x_1, \ldots, x_n) = \sum_{i=1}^{n} x_i^2 - \sum_{i=1}^{n-1} x_i x_{i+1}$;
   (ii) $d_n(x_1, \ldots, x_n) = \sum_{i=1}^{n} x_i^2 - \sum_{i=1}^{n-2} x_i x_{i+1} - x_{n-2}x_n$, for $n \geq 3$;
   (iii) $e_6(x_1, \ldots, x_6) = \sum_{i=1}^{6} x_i^2 - \sum_{i=1}^{4} x_i x_{i+1} - x_3 x_6$;
   (iv) $q(x_1, \ldots, x_4) = 4x_1^2 + x_2^2 + x_4^2 - 3x_1x_2 - 4x_1x_3 + 4x_2x_4 - 3x_3x_4$;
   (v) $q_{2n-1}(x_1, x_2)$, as defined in Exercise 1.22.8.

7. Let $A = (a_{ij})$ be a real symmetric $n \times n$ matrix. Prove the following:

   (i) Let $r = \mathbf{rk}(A)$ be the rank of the matrix A. Then there are numbers $1 \leq i_1 \leq i_2 \leq \cdots \leq i_r \leq n$ such that $A(i_1) \neq 0$, $A(i_1, i_2) \neq 0, \ldots, A(i_1, i_2, \ldots, i_r) \neq 0$.
   (ii) Assume that $D_i = A(1, \ldots, i) \neq 0$ for $i = 1, \ldots, r$ and $r = \mathbf{rk}(A)$. Then $A = G^t DG$, where $D = \mathrm{diag}\,(D_1, \ldots, D_r, 0 \ldots, 0)$ and

$$G = \begin{pmatrix} g_{11} & g_{12} & \cdots & \cdots & g_{1n} \\ 0 & g_{22} & \cdots & \cdots & g_{2n} \\ \cdots & \cdots & \cdots & \cdots & \cdots \\ 0 & \cdots & g_{rr} & \cdots & g_{rn} \\ 0 & \cdots & 0 & \cdots & 0 \\ 0 & \cdots & 0 & \cdots & 0 \end{pmatrix},$$

where $g_{ij} = \dfrac{A\begin{pmatrix}1 \dots i-1 & i \\ 1 \dots i-1 & j\end{pmatrix}}{A\big(1 \dots i-1\big)}$ for $1 \le i \le j \le r$. Hence $g_{ii} = \frac{C_i}{D_{i-1}}$ for $i = 1, \dots, r$.

(iii) In the situation of point (ii) we have $A = T^t DT$ for an invertible upper triangular matrix.

1.7 Positive and Nonnegative Quadratic Forms

A real quadratic form $q(x_1, \dots, x_n)$ is called *positive* (resp. *nonnegative*) if for any nonzero vector $0 \neq v \in \mathbb{R}^n$ we have $q(v) > 0$ (resp. $q(v) \geq 0$). The same terminology will be used for quadratic forms over subrings of $\mathbb{R}$, in particular for integral quadratic forms. In this section we give classical characterizations of positive and nonnegative real quadratic forms (following Gantmacher [27, X§4]).

Proposition 1.32. *Let $q(x) = x^t A_q x$ be a real quadratic form and A_q its associated symmetric matrix. The following are equivalent:*

a) The form q is positive.
b) The form q is nonnegative and regular.
c) All eigenvalues of A_q are positive.
d) If $D_i = A_q(1, \dots, i)$ is the i-th principal minor of A_q, we have

$$D_1 > 0,\ D_2 > 0,\ \dots,\ D_n > 0.$$

Proof. That (a) implies (b) is obvious. To show that (b) implies (c) recall from Proposition 1.28(a) that any eigenvector $0 \neq v \in \mathbb{R}^n$ of A_q has real eigenvalue λ. Then

$$\lambda \|v\|^2 = v^t A_q v = q(v) \geq 0,$$

which implies that $\lambda \geq 0$. If $\lambda = 0$ we get $A_q v = \lambda v = 0$ and $\mathbf{rad}(q) \neq 0$.

Let us show now that (d) follows from (a). By Exercise 1.31.7 there is an invertible matrix T such that $A_q = T^t DT$, where $D = \operatorname{diag}\left(\frac{1}{D_1}, \frac{D_1}{D_2}, \dots, \frac{D_{r-1}}{D_r}, 0, \dots, 0\right)$ and $r = \mathbf{rk}(A_q)$. Then the product of the eigenvalues of A_q is $0 < \det(A_q) = (\det T)^2 \det D$. Hence $D_n > 0$. Consider the $(n-1) \times (n-1)$ symmetric matrix $A_q^{(n)}$ obtained from A_q by omitting the n-th row and column. Since the quadratic form $x^t A_q^{(n)} x$ is positive, $D_{n-1} > 0$. By induction we get (d).

We show now that (c) implies (a). Let $0 < \lambda_1 \leq \lambda_2 \leq \cdots \leq \lambda_n$ be the eigenvalues of A_q. For any $0 \neq v \in \mathbb{R}^n$, we get

$$q(v) = \|v\|^2 q\left(\frac{v}{\|v\|}\right) \geq \lambda_1 \|v\|^2 > 0.$$

Hence $q(x)$ is positive. Finally, that d implies (a) follows from Jacobi's Formula $q(x) = \sum_{i=1}^{r} \frac{D_{i-1}}{D_i} X_i^2$. □

For instance, the quadratic form $q(x) = \sum_{k=1}^{6} x_k^2 - \sum_{k=1}^{4} x_k x_{k+1} - x_3 x_6$ is positive. Indeed, $q(x) = x^t A_q x$ where

$$A_q = \frac{1}{2}\begin{pmatrix} 2 & -1 & 0 & 0 & 0 & 0 \\ -1 & 2 & -1 & 0 & 0 & 0 \\ 0 & -1 & 2 & -1 & 0 & -1 \\ 0 & 0 & -1 & 2 & -1 & 0 \\ 0 & 0 & 0 & -1 & 2 & -1 \\ 0 & 0 & -1 & 0 & -1 & 2 \end{pmatrix}; \qquad B_q =$$

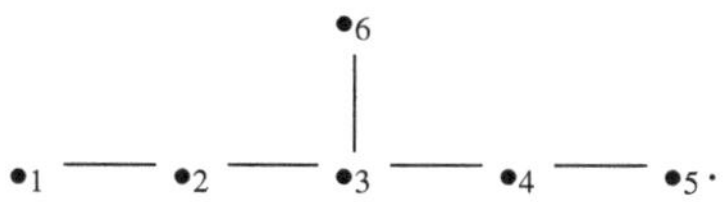

Taking $B = 2A_q$ we have the minors $B(1) = 2$, $B(1, 2) = 3$, $B(1, 2, 3) = 4$, $B(1, 2, 3, 4) = 5$, $B(1, 2, 3, 4, 5) = 6$ and $\det(B) = 3$, which are all positive. Then the positivity of q follows from Proposition 1.32.

Observe that from the fact $A(1) \geq 0$, $A(1, 2) \geq 0, \ldots, A(1, 2, \ldots, n) \geq 0$ it does not follow that A is nonnegative. Consider for instance the quadratic form $q(x_1, x_2) = -x_2^2$. Instead, we have the following appropriate generalization.

Proposition 1.33. *Let $q(x) = x^t A_q x$ be a real quadratic form with associated symmetric matrix A_q. The following are equivalent,*

a) The form q is nonnegative.
b) All eigenvalues of A_q are nonnegative.
c) Every principal minor of A_q is nonnegative (that is, $A_q(i_1, \ldots, i_s) \geq 0$ for $1 \leq i_1 < i_2 < \cdots < i_s \leq n$).

Proof. The equivalence of (a) and (b) follows as in the proof of Proposition 1.32. To show that (a) implies (c) observe that if q is nonnegative, then $\det(A_q) \geq 0$. The quadratic form $q^{(I)}$, obtained by making $x_i = 0$ for $i \in I = \{1, \ldots, n\} \setminus \{i_1, \ldots, i_s\}$, is also nonnegative. Hence $A_q(i_1, \ldots, i_s) \geq 0$.

We show now that (a) follows from (c). Take $\varepsilon > 0$ and consider the quadratic form

$$q_\varepsilon(x) = q(x) + \varepsilon \sum_{i=1}^{n} x_i^2 =: x^t A_\varepsilon x,$$

for an appropriate symmetric matrix A_ε. Observe that

$$A_\varepsilon(1, \ldots, j) = \varepsilon^j + \sum_{i=1}^{j} A_q(i)\varepsilon^{j-1} + \sum_{1 \leq i_1 < i_2 \leq j} A_q(i_1, i_2)\varepsilon^{j-2} + \cdots + A_q(1, \ldots, j)$$
$$\geq \varepsilon^j > 0.$$

Therefore by Proposition 1.32 the form q_ε is positive. Notice finally that $q = \lim_{\varepsilon \to 0} q_\varepsilon$ is nonnegative. □

For instance, for each $n \in \mathbb{N}$, the quadratic form $\sum_{i=1}^{n} x_i^2 - \sum_{i=1}^{n-1} x_i x_{i+1} - x_1 x_n$ is nonnegative, which can be shown using any of the equivalent conditions of Proposition 1.33.

Remark 1.34. Concerning the invariants rank $\mathbf{rk}(q)$ (given as the rank of the associated symmetric matrix A_q) and signature $\boldsymbol{\sigma}(q)$ (as introduced after Theorem 1.29) of a real quadratic form q, we have:

i) A quadratic form $q(x) = \sum_{i=1}^{n} q_{ij} x_i x_j$ is positive if and only if $\mathbf{rk}(q) = 0$ and $\boldsymbol{\sigma}(q) = n$;
ii) The quadratic form $q(x)$ is nonnegative if and only if $\boldsymbol{\sigma}(q) = n - \mathbf{rk}(q)$.

1.8 Cones in Real Vector Spaces

Following Vandergraft [50], we say that a closed subset K of the real vector space $\mathbb{R}^n$ is called a *cone* if the following conditions hold.

a) $K + K \subset K$ (that is, for v and w in K we have $v + w \in K$).
b) $\lambda K \subset K$ for any $\lambda \geq 0$ (that is, $\lambda v \in K$ for any v in K).

We further say that K is a *proper cone* if

c) $K \cap (-K) = \{0\}$.

We say that K is a *solid cone* if moreover

d) K generates the vector space $\mathbb{R}^n$.

Let us consider some examples. The positive cone V^+ in $\mathbb{R}^n$ is given by

$$V^+ = \{v = (v_1, \ldots, v_n) \in \mathbb{R}^n \mid v_i \geq 0 \text{ for } 1 \leq i \leq n\}.$$

Given a linear transformation $\alpha : \mathbb{R}^n \to \mathbb{R}^n$ and a cone K in $\mathbb{R}^n$, the preimage $\alpha^{-1}(K)$ is a cone. Taking $V = \mathbb{R}^n$, the dual space V^* is the set of linear transformations from V to $\mathbb{R}$, written $V^* = \mathbf{Hom}_{\mathbb{R}}(V, \mathbb{R})$. For $K \subset \mathbb{R}^n$ consider the set $K^{\perp} = \{f \in V^* \mid f(v) \geq 0, \text{ for } v \in K\}$. If K is a cone, then $K^{\perp}$ is a cone.

Consider the natural isomorphism

$$\phi : V \to V^{**}, \quad \phi(v)(g) = g(v), \quad \text{for } g \in V^* \text{ and } v \in V.$$

If K is a proper solid cone, then $K^{\perp\perp} = \phi(K)$ (see Exercise 7 below).

Let $V = \mathbb{R}^n$ and A be a real $n \times n$ matrix. We say that a cone K is *invariant under* A if $A(K) \subset K$. For instance, if A is a nonnegative matrix (that is, all entries of A are nonnegative real numbers) then the positive cone V^+ is invariant under A.

Let $q(x) = x^t Ax$ be a real positive quadratic form with $0 < \lambda_1 \le \lambda_2 \le \cdots \le \lambda_n$ the eigenvalues of A. Consider a set $v_1, \ldots, v_n$ of linearly independent eigenvectors of A with $Av_i = \lambda_i v_i$ for $1 \le i \le n$. Define the (proper solid) cone $K = \left\{ \sum_{i=1}^{n} \mu_i v_i \mid \mu_i \ge 0 \right\}$. Then clearly $A(K) \subset K$. As last example consider the matrix $A = \begin{pmatrix} 1 & a \\ 1 & -1 \end{pmatrix}$ with $a > 0$. Then A leaves invariant the cone $K = \{(x, y) \in \mathbb{R}^2 \mid 0 \le y \le x\}$. Observe that the maximal eigenvalue of A is $\rho(A) = \sqrt{a+1}$, with eigenvector $(1, \frac{1}{a}(\rho(A) - 1)) \in K$.

Let us briefly recall some classical definitions. For a real $n \times n$ matrix A, it is a fundamental fact that there exists an invertible matrix T such that

$$T^{-1}AT = \bigoplus_{\lambda \in \mathbf{Spec}(A)} \left(\bigoplus_{i=1}^{d_\lambda} J_i(\lambda)^{\nu_\lambda^i} \right)$$

is the *Jordan form* of A, where $\mathbf{Spec}(A)$, the *spectrum* of A, denotes the set of eigenvalues of A (counting repetitions). For each $\lambda \in \mathbf{Spec}(A)$, the *degree* d_λ is the maximal size of a Jordan block with eigenvalue λ appearing in the decomposition and $\nu_\lambda^i \ge 0$ is the number of blocks $J_i(\lambda)$ in such a decomposition (*multiplicity of the block*),

$$J_i(\lambda) = \begin{pmatrix} \lambda & 1 & & \\ & \lambda & \ddots & \\ & & \ddots & 1 \\ & & & \lambda \end{pmatrix}.$$

Therefore we have $0 \le \nu_\lambda^i \le n$ with $\nu_\lambda^{d_\lambda} > 0$ and

$$\sum_{\lambda \in \mathbf{Spec}(A)} \sum_{i=1}^{d_\lambda} i \nu_\lambda^i = n.$$

The number $\rho(A) = \max\{|\lambda| \mid \lambda \in \mathbf{Spec}(A)\}$ is called the *spectral radius* of the matrix A. Observe that $\mathbf{Spec}(A) \subset B_{\rho(A)}$, the ball of radius $\rho(A)$ and center at 0 in $\mathbb{C}^n$.

We will only indicate the main steps of the proof of the following important result, which goes back to Perron and Frobenius for the case of positive matrices, and to Birkhoff and Vandergraft [12, 50] in the general situation (for details, see [50] and Exercises 8, 9 and 12).

Theorem 1.35 (Birkhoff–Vandergraft). *Let K be a proper solid cone in $\mathbb{R}^n$ and A a real $n \times n$ matrix such that $A(K) \subseteq K$. The following hold:*

a) The spectral radius $\rho(A)$ is an eigenvalue of A.
b) We have $d_{\rho(A)} \geq d_\lambda$ for any eigenvalue λ of A with $|\lambda| = \rho(A)$.
c) There is a nonzero vector $v \in K$ with $Av = \rho(A)v$.

Sketch of Proof. Let $\varepsilon_1, \dots, \varepsilon_n$ be a Jordan basis of A ordered in the following way: $\lambda_1, \dots, \lambda_s$ are the eigenvalues of A with multiplicities and such that $|\lambda_1| \geq |\lambda_2| \geq \cdots \geq |\lambda_s|$. Vectors $\varepsilon_{m_1}, \dots, \varepsilon_{m_s}$ are all the eigenvectors of A (corresponding to the eigenvalues $\lambda_1, \dots, \lambda_s$) with $0 = m_0 < m_1 < m_2 < \cdots < m_s = n$. Moreover, for each $1 \leq h \leq n$ there is a unique $1 \leq j(h) \leq s$ such that $m_{j(h)-1} < h < m_{j(h)}$ and

$$\begin{aligned} A(\varepsilon_h) &= \lambda_{j(h)}\varepsilon_h + \varepsilon_{h+1}, \qquad \text{if } h < m_{j(h)}, \\ A(\varepsilon_{m_{j(h)}}) &= \lambda_{j(h)}\varepsilon_{m_{j(h)}}. \end{aligned}$$

The size b_j of the Jordan block $J_{b_j}(\lambda_j)$ is $m_j - m_{j-1}$. Take $M := \max\{b_j \mid |\lambda_j| = \rho(A)\}$. We may assume that $1, 2, \dots, t$ are the indices j with $|\lambda_j| = \rho(A)$ and $b_j = M$.

The case $\rho(A) = 0$ is clear, so we may assume $\rho(A) > 0$. Exercise 8 shows the existence of numbers $d_1, \dots, d_t \in \mathbb{C}$ such that

$$0 \neq y = \sum_{j=1}^{t} d_j \varepsilon_{m_j} \in K.$$

Assume that $\lambda_1 \notin \mathbb{R}$, (that is, $\lambda_1 \neq \rho(A)$). By Exercise 9 there are real numbers $0 \leq c_0, \dots, c_q$, not all zero, such that

$$\sum_{p=0}^{q} c_p \lambda_1^p = 0.$$

We get then $y^* = \sum_{p=0}^{q} c_p A^p(y) = \sum_{j=2}^{t} d_j \left(\sum_{p=0}^{q} c_p \lambda_j^p\right) \varepsilon_{m_j} \in K$, and since K is a proper cone, $0 \neq y^*$. After canceling all summands corresponding to $\lambda_i \notin \mathbb{R}$, $(1 \leq i \leq t)$, we get a vector $0 \neq \tilde{y} \in K$ which is a linear combination of $\{\varepsilon_{m_j} \mid \lambda_j = \rho(A) \text{ and } b_j = M\}$. Hence

$$\begin{aligned} A(\tilde{y}) &= \rho(A)\tilde{y}, \\ d_{\rho(A)} &= M \geq d_\lambda, \qquad \text{if } |\lambda| = \rho(A). \end{aligned}$$

Therefore (a), (b) and (c) hold. □

It will be useful to have a sharper version of Theorem 1.35 in case the matrix $A = (a_{ij})$ is *nonnegative* (written $A \geq 0$), that is, $a_{ij} \geq 0$ for all $1 \leq i, j \leq n$. We say that A is a *reducible matrix* if there is a permutation matrix P such that

$$PAP^t = \begin{pmatrix} B & 0 \\ C & D \end{pmatrix},$$

where B is a $m \times m$ matrix with $m < n$. In fact, if $0 \leq A$ is an irreducible $n \times n$-matrix, then $(\mathbf{Id}_n + A)^{n-1} > 0$, that is, all its entries are positive real numbers.

The following statement is part of the classical Perron–Frobenius Theorem.

Theorem 1.36. *Let A be an irreducible real matrix with nonnegative entries. Then*

a) The spectral radius $\rho(A)$ is a simple eigenvalue of A.
b) There exists a vector $v \in \mathbb{R}^n$ with positive coordinates such that $Av = \rho(A)v$.

Exercises 1.37.

1. Let A be a symmetric real $n \times n$ matrix with $q(x) = x^t Ax$ a positive form. Consider the set E_A of points $x \in \mathbb{R}^n$ such that $x^t Ax = 1$. Show the following:

   (i) E_A is an ellipsoid.
   (ii) The lengths of the principal semi-axes of E_A are $\frac{1}{\sqrt{\lambda_1}}, \ldots, \frac{1}{\sqrt{\lambda_n}}$, where $0 < \lambda_1 \leq \cdots \leq \lambda_n$ are the eigenvalues of A.
   (iii) If $B(\varepsilon)$ is a ball with center at $0 \in \mathbb{R}^n$ and radius ε, then

   $$B\left(\frac{1}{\sqrt{\lambda_n}}\right) \subset E_A \subset B\left(\frac{1}{\sqrt{\lambda_1}}\right).$$

   (iv) Let H be any hyperplane generated by vectors $w_1, \ldots, w_k$ in $\mathbb{R}^n$. Then the intersection E' of E_A and H is an ellipsoid in H. Let b_i be the length of the semi-axes of E' in the direction given by w_i. We may assume that $b_1 \geq b_2 \geq \cdots \geq b_k$. Show the following inequalities:

   $$\frac{1}{\sqrt{\lambda_1}} \geq b_1, \frac{1}{\sqrt{\lambda_2}} \geq b_2, \ldots, \frac{1}{\sqrt{\lambda_k}} \geq b_k.$$

   As an illustration of the case $n = 3$ and $k = 2$ see the following figure:

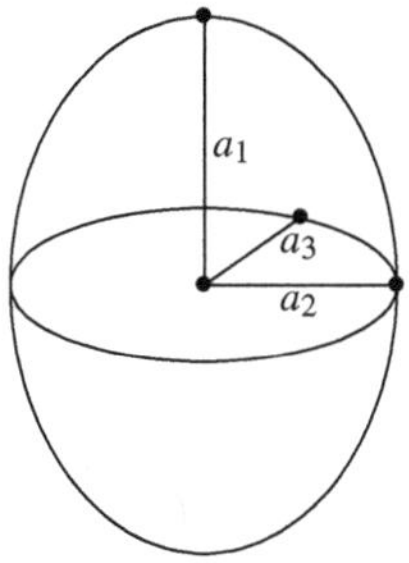

where $a_i = \frac{1}{\sqrt{\lambda_i}}$ for $i = 1, 2, 3$.

2. Let $q_n(x_1, x_2)$ be the forms associated to the Fibonacci numbers as introduced in Exercises 1.22.8 and 9. Prove the following:

   (i) For all $n \geq 1$, q_{2n} is positive.
   (ii) The form q_1 is positive, q_3 is nonnegative and q_{2n-1} is indefinite for $n \geq 3$.

3. Let $q(x_1, \ldots, x_n) = \sum_{1 \leq i \leq j \leq n} q_{ij} x_i x_j$ be a real quadratic form with $q_{ii} > 0$ for all i. Show that the following are equivalent:

   (i) The radical of q satisfies $\mathbf{rad}(q) \neq 0$.
   (ii) We have $0 \in \mathbf{D}_{\mathbb{R}}(q)$ and 0 is locally a minimal value for the function $q : \mathbb{R}^n \to \mathbb{R}$.

4. Let K be a cone in $\mathbb{R}^n$. Show that the following are equivalent:

   (i) K is a solid cone.
   (ii) The interior K^0 of K is nonempty.
   (iii) $K + (-K) = \mathbb{R}^n$.

5. Let K be a cone in $\mathbb{R}^n$. Take $x \in K^0$ and $y \in K$. Show that $x + y \in K^0$.
6. Given a cone $K \subset \mathbb{R}^n$ and a linear transformation $\alpha : \mathbb{R}^n \to \mathbb{R}^n$, find conditions for the cone $\alpha^{-1}(K)$ to be solid.
7. A cone $K \subset V$ is *generated* by a set of vectors $\{v_i\}_{i \in I}$ if

$$K = \left\{ \sum_{i \in I} \lambda_i v_i \mid 0 \leq \lambda_i \in \mathbb{R},\ i \in I \right\}.$$

   If I is finite, then K is called *polyhedral.*

   (i) Show that a polyhedral proper solid cone K has the shape

$$K = \bigcap_{i=1}^{s} H_{v_i}^{+},$$

   where $\{v_1, \ldots, v_s\}$ is a set in $\mathbb{R}^n$ and $H_{v_i}^{+} = \{x \in \mathbb{R}^n \mid x^t v_i \geq 0\}$ is a half-space determined by the vector v_i.
   (ii) Show that a polyhedral proper solid cone K satisfies $K^{\perp\perp} = \varphi(K)$, where $\varphi : V \to V^{**}$ is the evaluation map.
   (iii) Observe that any proper solid cone K is a limit of a sequence $(K_n)_{n \in \mathbb{N}}$ of polyhedral proper solid cones. Use (ii) to prove that $K^{\perp\perp} = \varphi(K)$.

8. Keeping the notation introduced in Theorem 1.35, show that there exists a nonzero

$$y = \sum_{j=1}^{t} d_j \varepsilon_{m_j} \in K,$$

for certain numbers $d_j \in \mathbb{C}$.
[Hint: Since K is a solid cone we may choose

$$0 \neq z \in \sum_{i=1}^{n} c_i \varepsilon_i \in K^0,$$

with $0 \neq c_i \in \mathbb{R}$. Then for $r \geq n$

$$A^r(z) = \sum_{i=1}^{n} c_i A^r(\varepsilon_i) = \sum_{h=1}^{n} \left(\sum_{m_{j(h)-1} < i < m_{j(h)}} c_i \lambda_{j(h)}^{r-(h-i)} \binom{r}{h-i} \right) \varepsilon_h,$$

where $\binom{r}{j}$ is the binomial coefficient.
It is not hard to see that

$$w = \lim_{r \to \infty} \frac{A^r(z)}{\rho(A)^r r^{(M-1)}}$$

is well-defined. Moreover, w is a linear combination of those ε_{m_j} with $1 \leq j \leq t$ and $w \in K$.]

9. For $\lambda \in \mathbb{C} \setminus \mathbb{R}$ show that there are nonnegative real numbers $c_0, \ldots, c_q$, not all of them zero, such that

$$\sum_{p=0}^{q} c_p \lambda^p = 0.$$

10. Assume that A is a real $n \times n$ matrix such that $A(K) \subset K$ for a cone $K \subset \mathbb{R}^n$. We say that K is *properly invariant under* A if $A^m(K - \{0\}) \subset K^0$ for some $m > 0$. In this case prove the following:

    (i) The spectral radius $\rho(A)$ is a simple eigenvalue of A, and for any $\rho(A) \neq \lambda \in \mathbf{Spec}A$ we have $|\lambda| < \rho(A)$.
    (ii) There is a vector $0 \neq y \in K^0$ with eigenvalue $\rho(A)$.

11. Let n be a natural number ≥ 2. Consider the matrices

$$P = \begin{pmatrix} 1 & 0 \\ n & 1 \end{pmatrix} \text{ and } C = -PP^t = \begin{pmatrix} -1 & n \\ -n & n^2 - 1 \end{pmatrix}.$$

Show the following:

    (i) The cone K generated by $C_{p_1}^m$ and $C_{p_2}^m$ for $m \geq 1$, where $p_1 = \binom{0}{1}$ and $p_2 = \binom{1}{n}$, is invariant under C. It is a solid cone in $(\mathbb{R}^2)^+$.

(ii) The spectral radius of C is $\rho = \frac{n^2-2+n\sqrt{n^n-4}}{2}$ and the eigenvector of C in K is $(1, (1+\rho)/n)$.

12. Give a proof of the Perron–Frobenius Theorem stated in 1.36. [Hint: Use Theorem 1.35.]

Chapter 2
Positive Quadratic Forms

In the previous chapter we reviewed the basic definitions and tools necessary to study properties of quadratic forms. Recall that an integral quadratic form $q(x) = \sum_{i \leq j} q_{ij} x_i x_j$ is *unitary* if all diagonal coefficients q_{ii} are equal to one. In this chapter we study *positive unit forms*, that is, those integral quadratic unit forms with $q(x) > 0$ for every nonzero vector x in $\mathbb{Z}^n$. In what follows the term *diagram* is synonymous with simple graph, and a vector $v = (v_1, \ldots, v_n)$ in $\mathbb{Z}^n$ is called *positive*, written $v > 0$, if v is nonzero and $v_i \geq 0$ for $i = 1, \ldots, n$.

2.1 Dynkin Graphs

Dynkin graphs (or *Dynkin diagrams*, see Table 2.1 below) appear in many places in mathematics: algebra, geometry, probability theory. In our context they are the main classification device for nonnegative quadratic unit forms, hence of great importance throughout the text. Here we are concerned with properties of integral quadratic forms associated to Dynkin diagrams.

Lemma 2.1. *Let Δ be one of the Dynkin graphs $\mathbb{A}_n$, $\mathbb{D}_m$ or $\mathbb{E}_p$ for $n \geq 1$, $m \geq 4$ and $p = 6, 7, 8$ shown in Table 2.1. Then q_Δ is a positive unit form.*

Proof. Let us consider first graphs $\mathbb{A}_n$ with $n \geq 1$. For $x = (x_1, \ldots, x_n)$ in $\mathbb{Z}^n$, observe that

$$\begin{aligned} q_{\mathbb{A}_n}(x) &= x_1^2 - x_1x_2 + x_2^2 - x_2x_3 + \ldots + x_{n-1}^2 - x_{n-1}x_n + x_n^2 \\ &= \frac{1}{2}\left[x_1^2 + (x_1 - x_2)^2 + (x_2 - x_3)^2 + \ldots + (x_{n-1} - x_n)^2 + x_n^2\right]. \end{aligned}$$

M. Barot et al., *Quadratic Forms*, Algebra and Applications 25,
https://doi.org/10.1007/978-3-030-05627-8_2

Table 2.1 Dynkin diagrams Δ_n with n vertices

Notation	Graph Δ	$\lvert R(q_\Delta)\rvert$
$\mathbb{A}_n$ $(n \geq 1)$		$n(n+1)$
$\mathbb{D}_m$ $(m \geq 4)$	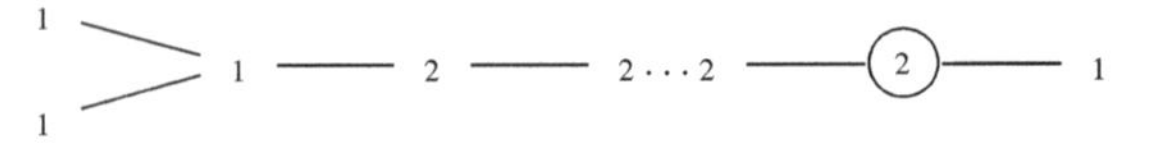	$2m(m-1)$
$\mathbb{E}_6$	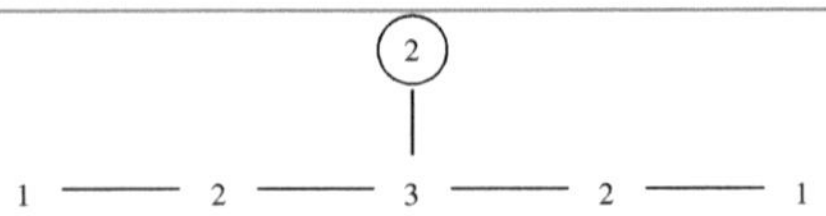	72
$\mathbb{E}_7$	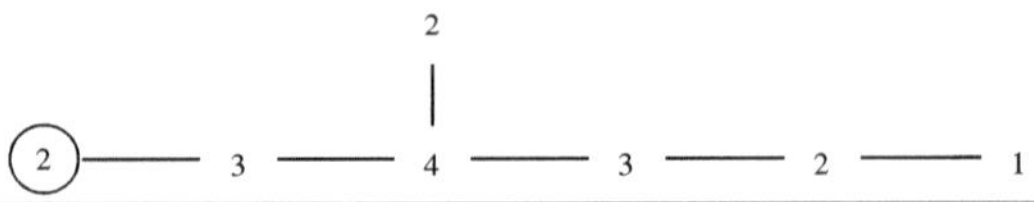	126
$\mathbb{E}_8$	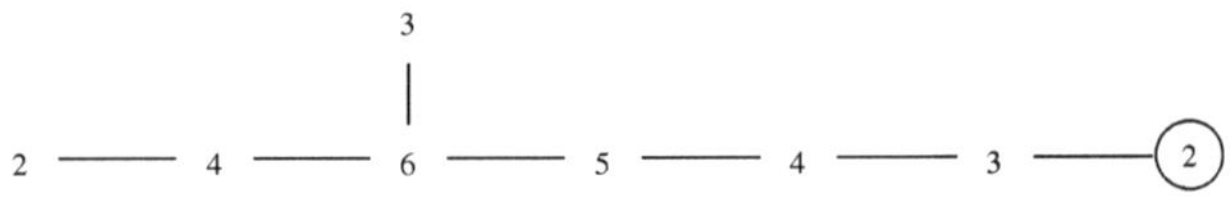	240

The vector p_{Δ_n} with entries given by the numbers in the vertices is the maximal root of q_{Δ_n}. Its largest entry is denoted by r_{Δ_n}

In particular, $q_{\mathbb{A}_n}(x) \geq 0$, and $q_{\mathbb{A}_n}(x) = 0$ if and only if $x_1 = 0$ and $x_{i+1} = x_i$ for $1 \leq i < n$. That is, $q_{\mathbb{A}_n}$ is a positive unit form. Notice also that if $x_1 \neq 0$ then $q_{\mathbb{A}_n}(x) > \frac{1}{2}x_1^2$.

Consider now the form $q_{\mathbb{D}_m}$ for $m \geq 4$, and observe that with an appropriate enumeration of the vertices in $\mathbb{D}_m$, for x in $\mathbb{Z}^m$ we have the equality

$$q_{\mathbb{D}_m}(x) = q_{\mathbb{A}_{m-1}}(x_1 + x_2, x_3, \dots, x_m) - 2x_1x_2.$$

If x in nonzero and $x_1 + x_2 = 0$ then we have $q_{\mathbb{D}_m}(x) = q_{\mathbb{A}_{m-1}}(x_1 + x_2, x_3, \dots, x_m) + 2x_1^2 > 0$. If $x_1 + x_2 \neq 0$, by the above we have

$$q_{\mathbb{D}_m}(x) > \frac{1}{2}(x_1 + x_2)^2 - 2x_1x_2 = \frac{1}{2}(x_1 - x_2)^2 \geq 0.$$

This shows that $q_{\mathbb{D}_m}$ is a positive unit form.

Alternatively we may use Lagrange's Method (Algorithm 1.30) to find that,

$$q_{\mathbb{D}_m}(x) = (x_1 - \frac{1}{2}x_3)^2 + (x_2 - \frac{1}{2}x_3)^2 + \frac{1}{2}\left[(x_3 - x_4)^2 + (x_4 - x_5)^2 + \ldots + (x_{m-1} - x_m)^2 + x_m^2\right].$$

The following expressions for exceptional cases $q_{\mathbb{E}_6}$, $q_{\mathbb{E}_7}$ and $q_{\mathbb{E}_8}$ and suitable ordering of vertices can be shown similarly,

$$q_{\mathbb{E}_6}(x) = (x_1 - \frac{1}{2}x_2)^2 + \frac{3}{4}(x_2 - \frac{2}{3}x_3)^2 + \frac{2}{3}(x_3 - \frac{3}{4}x_4 - \frac{3}{4}x_5)^2 + \frac{5}{8}(x_4 - \frac{3}{5}x_5)^2 + \frac{2}{5}(x_5 - \frac{5}{4}x_6)^2 + \frac{3}{8}x_6^2,$$

$$q_{\mathbb{E}_7}(x) = (x_1 - \frac{1}{2}x_2)^2 + \frac{3}{4}(x_2 - \frac{2}{3}x_3)^2 + \frac{2}{3}(x_3 - \frac{3}{4}x_4 - \frac{3}{4}x_5)^2 + \frac{5}{8}(x_4 - \frac{3}{5}x_5)^2 + \frac{2}{5}(x_5 - \frac{5}{4}x_6)^2 + \frac{3}{8}(x_6 - \frac{4}{3}x_7)^2 + \frac{1}{3}x_7^2,$$

$$q_{\mathbb{E}_8}(x) = (x_1 - \frac{1}{2}x_2)^2 + \frac{3}{4}(x_2 - \frac{2}{3}x_3)^2 + \frac{2}{3}(x_3 - \frac{3}{4}x_4 - \frac{3}{4}x_5)^2 + \frac{5}{8}(x_4 - \frac{3}{5}x_5)^2 + \frac{2}{5}(x_5 - \frac{5}{4}x_6)^2 + \frac{3}{8}(x_6 - \frac{4}{3}x_7)^2 + \frac{1}{3}(x_7 - \frac{3}{2}x_8)^2 + \frac{1}{4}x_8^2,$$

which completes the proof. □

It turns out that the forms q_Δ associated to Dynkin graphs are all the connected positive unit forms q with associated bigraph having no dotted edges (that is, those forms satisfying $q_{ij} \leq 0$ for all indices $i < j$).

Proposition 2.2. *Let G be a connected graph (without dotted edges). Then q_G is a positive unit form if and only if G is a Dynkin diagram (see Table 2.1).*

Proof. By Lemma 2.1, for a Dynkin diagram Δ the unit form q_Δ is positive.

For the converse take G a connected graph and observe that if a restriction q_G^I of q_G is nonpositive, then the form q_G itself is nonpositive. Indeed, if $q_G^I(x) \leq 0$ for a vector x in $\mathbb{Z}^{G_0 - I}$, then completing x with zeros to a vector x' in $\mathbb{Z}^{G_0}$, we have $q_G(x') = q_G^I(x) \leq 0$. In particular G is a simple graph (that is, G has no loop and no multiple edges).

The above implies that the graph G does not contain an extended Dynkin diagram (see Table 2.2) for their associated forms are nonpositive (as a direct calculation shows, cf. Exercise 4 below).

The result follows now from a combinatorial observation: *a simple graph G is a Dynkin diagram if and only if G does not contain as (full) subgraph any extended Dynkin diagram* (cf. Exercise 5). □

A vector x in $\mathbb{Z}^n$ is called *sincere* if all its entries x_i are nonzero. The set of roots (resp. positive roots) of a quadratic form q, that is, the set of (positive) vectors x in $\mathbb{Z}^n$ with $q(x) = 1$, is denoted by $R(q)$ (resp. by $R^+(q)$). Observe that in the following result we do not require the quadratic form to be unitary.

Table 2.2 Extended Dynkin diagrams $\widetilde{\Delta}_n$ with $n+1$ vertices

Notation	Graph
$\widetilde{\mathbb{A}}_n$ $(n \geq 2)$	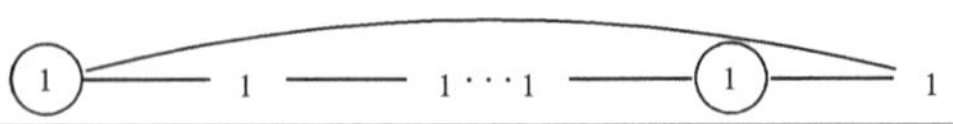
$\widetilde{\mathbb{D}}_n$ $(n \geq 4)$	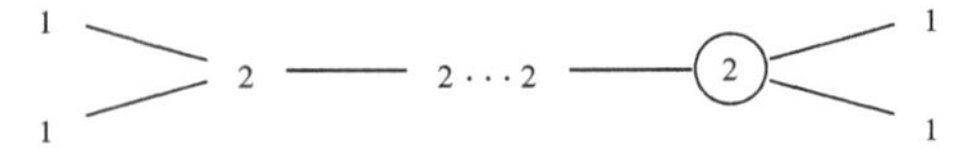
$\widetilde{\mathbb{E}}_6$	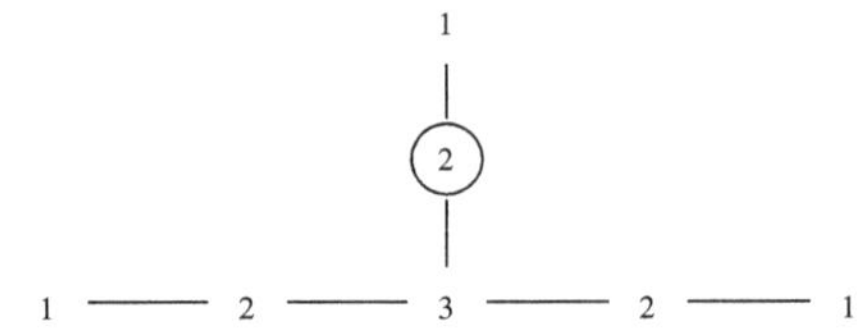
$\widetilde{\mathbb{E}}_7$	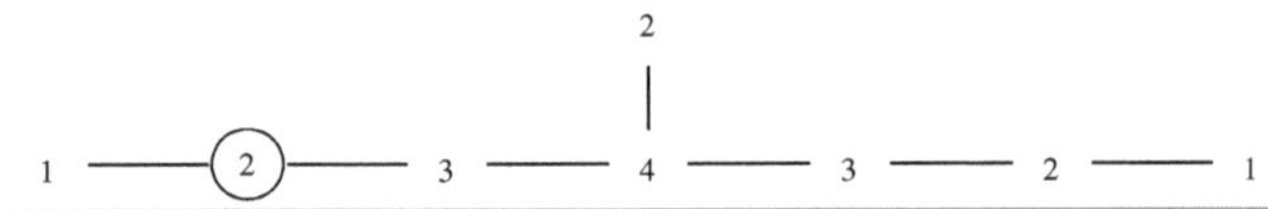
$\widetilde{\mathbb{E}}_8$	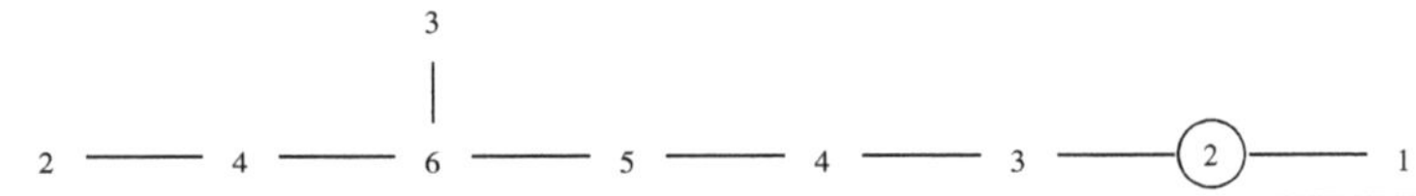

The vector $p_{\widetilde{\Delta}_n}$ with entries given by the numbers in the vertices is the generator of the radical of the quadratic form $q_{\widetilde{\Delta}_n}$

Proposition 2.3. *A positive integral quadratic form admits only finitely many roots.*

Proof. A polynomial q with integral coefficients may be considered as a function $q_{\mathbb{R}} : \mathbb{R}^n \to \mathbb{R}$. Notice that $q_{\mathbb{R}}(x) \geq 0$ for any x in $\mathbb{R}^n$. Indeed, for any y in $\mathbb{Q}^n$ there is a vector x in $\mathbb{Z}^n$ and $p \in \mathbb{Z}$ nonzero such that $y = x/p$. Hence $q_{\mathbb{R}}(y) = q(x)/p^2 \geq 0$, and the same holds for y in $\mathbb{R}^n$ by continuity.

We show that $q_{\mathbb{R}}(x) > 0$ for all nonzero x in $\mathbb{R}^n$. Let $x \in \mathbb{R}^n$ with $q_{\mathbb{R}}(x) = 0$. Then $q_{\mathbb{R}}$ has in x a local minimum, thus $0 = \frac{\partial q(x)}{\partial x_i} = 2q_{ii}x_i + \sum_{j \neq i} q_{ij}x_j$ for $i = 1, \ldots, n$ (see Lemma 1.1), that is, $x_1, \ldots, x_n$ satisfy a system of n linear equations. If the determinant of that system is zero, then it has a nonzero solution z/p with $z \in \mathbb{Z}^n$ and $0 \neq p \in \mathbb{Z}$. But then $q_{\mathbb{R}}(z/p)$ must be zero, in contradiction to $0 < q(z)/p^2 = q_{\mathbb{R}}(z)$. Hence the determinant of the above system is nonzero, forcing $x = 0$.

This shows that $q_{\mathbb{R}}$ restricted to the sphere $S = \{x \in \mathbb{R}^n \mid ||x|| = 1\}$ is a positive function. Since S is compact, $q_{\mathbb{R}}$ takes its minimum in some $x_0 \in S$. Then $q_{\mathbb{R}}(v) = ||v||^2 q_{\mathbb{R}}(\frac{v}{||v||}) \geq ||v||^2 q_{\mathbb{R}}(x_0)$ for all $v \neq 0$. Therefore any root v must satisfy $||v||^2 \leq \frac{1}{q_{\mathbb{R}}(x_0)}$, that is, any root of q is contained in $\{z \in \mathbb{Z}^n \mid ||z||^2 \leq \frac{1}{q_{\mathbb{R}}(x_0)}\}$, which is a finite set. $\square$

Remark 2.4. With the same proof we may actually show that, if q is a positive integral quadratic form, for any $c > 0$ the number of vectors y in $\mathbb{Z}^n$ with $q(y) = c$ is finite.

The number of roots $|R(q_\Delta)|$ of the quadratic form q_Δ associated to a Dynkin diagram Δ is shown in Table 2.1. These numbers will appear later in Chap. 4 in relation to the order of certain Coxeter matrices.

2.2 Roots and Reflections

Notice that Proposition 2.3 holds for all positive *integral* quadratic forms and provides in principle an algorithm for constructing their roots as far as the number $\min\{q_{\mathbb{R}}(x) \mid ||x|| = 1\}$ can be efficiently calculated. In the following we give a much more efficient way to inductively construct roots, which works nicely for *unitary* forms.

In Sect. 1.2 we have defined the i-th (simple) reflection associated to a unit form q as

$$\sigma_i : \mathbb{Z}^n \to \mathbb{Z}^n, \qquad \sigma_i(x) = x - q(x|e_i)e_i.$$

It was noted in Lemma 1.5(*c*) that $q(\sigma_i(x)) = q(x) - q(x|e_i)^2 + q(x|e_i)^2 q(e_i) = q(x)$, hence by applying simple reflections to already known roots we can hope to find new ones. The following result shows that this is in effect a powerful tool. Recall that there is a partial ordering in $\mathbb{Z}^n$, declaring $x < y$ whenever $0 < y - x$.

Proposition 2.5. *Let $x < y$ be positive roots of a positive unit form $q : \mathbb{Z}^n \to \mathbb{Z}$. Then there is a sequence of reflections $\sigma_{i_1}, \ldots, \sigma_{i_t}$ with $\sigma_{i_t} \cdots \sigma_{i_1}(x) = y$ and $\sigma_{i_s} \cdots \sigma_{i_1}(x) = x + e_{i_1} + \cdots + e_{i_s}$ for all $s \in \{1, \ldots, t\}$.*

Proof. Using the positivity of q and that $x \neq y$ are roots of q, we obtain from $q(x) = q(y-(y-x)) = q(y)+q(y-x)-q(y|y-x)$ that $0 < q(y-x) = q(y|y-x)$. Hence there is an index $i \in \{1, \ldots, n\}$ such that $x_i < y_i$ and $0 < q(y|e_i)$.

Notice that $y \neq e_i$, for x is a positive vector and $x < y$. Since $0 < q(y - e_i) = 2 - q(y|e_i)$ we get $q(y|e_i) = 1$ and therefore $x \leq \sigma_i(y) = y - e_i < y$. Set $i_1 = i$ and repeat the process with $\sigma_i(x)$ instead of x. □

Remark 2.6. Observe that in the proof of Proposition 2.5 we only require the quadratic form q to satisfy $q(x) > 0$ for positive vectors x. We dedicate Chap. 5 to the study of these kinds of quadratic forms, called *weakly positive unit forms.*

There seems to be one flaw in the above result, namely, it will produce only *positive* roots. But in the following result we will see that this is all we need for the forms associated to Dynkin graphs.

Lemma 2.7. *Let Δ be a Dynkin graph with associated unit form $q : \mathbb{Z}^n \to \mathbb{Z}$. If v is a root of q_Δ then either $v > 0$ or $-v > 0$. In particular q_Δ has a unique maximal positive root.*

Proof. Take $q = q_\Delta$ and let v^+ and v^- be defined by $v_i^+ = \max\{v_i, 0\}$ and $v_i^- = \max\{-v_i, 0\}$ for $i = 1, \ldots, n$. Then $v = v^+ - v^-$ and

$$1 = q(v) = q(v^+ - v^-) = q(v^+) + q(v^-) + \sum_{i,j=1}^{n} v_i^+ v_j^- (-q_{ij}).$$

Since q is a positive form by Lemma 2.1, the three summands on the right are nonnegative integers (for if $q_{ij} > 0$ then $i = j$ and $v_i^+ v_j^- = 0$). Since $v \neq 0$ we conclude that either $v^+ = 0$ or $v^- = 0$.

For the second claim consider a maximal positive root x of q. Since $\sigma_i(x) = x - q(x|e_i)e_i$ is again a root of q (Lemma 1.5(c)), by maximality we have $q(x|e_i) \geq 0$ for all $i = 1, \ldots, n$. Moreover, if x has a zero entry, since Δ is connected there are vertices i and j with $x_i = 0$ and $x_j > 0$ such that $q_{ij} < 0$. Then $q(x|e_i) = \sum_{j \neq i} q_{ij} x_j < 0$, which is impossible again by maximality of x. Hence x is a sincere vector, and the same holds for any other maximal positive root y. Since there is vertex i with $q(x|e_i) > 0$, and y is a positive sincere root, then

$$0 < q(x - y) = q(x) + q(y) - q(x|y) = 2 - \sum_{i=1}^{n} y_i q(x|e_i) < 2,$$

that is, $x - y$ is a root of q. Since both x and y are maximal positive roots, then $x - y$ has both negative and positive entries, which is impossible by the first claim of the lemma. □

By the above it suffices to know all positive roots of the quadratic form q_Δ associated to a Dynkin graph Δ. For instance, let $q = q_{\mathbb{D}_4}$ be the form associated to the Dynkin diagram $\mathbb{D}_4$ with the following enumeration of its vertices,

$\mathbb{D}_4$:

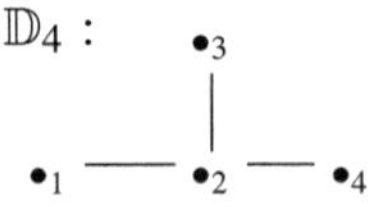

We already know that the canonical vectors e_1, e_2, e_3, e_4 of $\mathbb{Z}^n$ are roots of q_Δ. Reflecting these roots we get

$\sigma_1(e_1) = -e_1,$ (we discard this root, since it is negative)

$\sigma_2(e_1) = e_1 + e_2,$ (a new root)

$\sigma_3(e_1) = e_1 = \sigma_4(e_1).$

Since $\sigma_i(v) > v$ if and only if $q(v|e_i) < 0$, this last condition has to be investigated. We have from Lemma 1.1

$$q(v|e_i) = 2v_i + \sum_{j \neq i} q_{ij} v_j = 2v_i - \sum_{\substack{\text{there is}\\ \text{an edge}\\ i-j}} v_j .$$

Now the calculations become easy, since $\sigma_i(v) > v$ if and only if $\sum_{i-j} v_j > 2v_i$.

For instance we have

$$\begin{smallmatrix}0\\100\end{smallmatrix} \stackrel{\sigma_2}{\longmapsto} \begin{smallmatrix}0\\110\end{smallmatrix} \stackrel{\sigma_4}{\longmapsto} \begin{smallmatrix}0\\111\end{smallmatrix} \stackrel{\sigma_3}{\longmapsto} \begin{smallmatrix}1\\111\end{smallmatrix} \stackrel{\sigma_2}{\longmapsto} \begin{smallmatrix}1\\121\end{smallmatrix} =: \delta .$$

Observe now that $\sigma_i(\delta) = \delta$ for $i = 1, 3, 4$ and $\sigma_2(\delta) = \begin{smallmatrix}1\\111\end{smallmatrix}$, hence no new root is obtained, and the maximal root is found. The following picture exhibits all positive roots of the form $q_{\mathbb{D}_4}$.

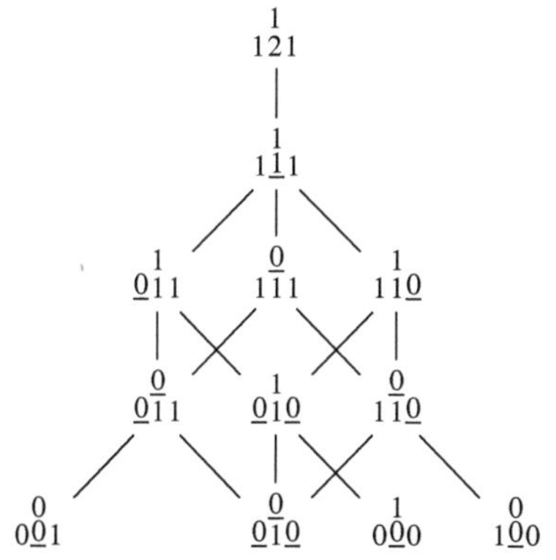

An edge indicates that the two roots at the end points are obtained by a reflection from each other (we omitted those reflections which leave a root unchanged or turn it negative). An entry v_i is depicted underlined if $\sigma_i(v) > v$. A patient reader may calculate the corresponding pictures for $q_{\mathbb{E}_6}, q_{\mathbb{E}_7}$ and $q_{\mathbb{E}_8}$ (Fig. 2.1). The least patient readers may take a look at the outcome on page 50, where we show all three in one (displaying only sincere roots). It thus remains to investigate the cases $q_{\mathbb{A}_n}$ and $q_{\mathbb{D}_n}$ in general. We leave the easier case to the reader and analyze the roots of $q_{\mathbb{D}_n}$ here.

First we construct all positive roots v in $\mathbb{Z}^n$ with $v_i = 0, 1$ and call such vectors *thin*. Denote by **supp**(v) the set of indices $i \in \{1, \ldots, n\}$ for which $v_i \neq 0$, and call it the *support* of v (thus a root v of q is *sincere* if **supp**$(v) = \{1, \ldots, n\}$). We say that a vector v is *connected* if so is the restricted form $q^{\mathbf{supp}(v)}$. Recall that a *tree graph* is a connected graph G such that $|G_0| = |G_1| + 1$ (for instance, all Dynkin graphs are trees).

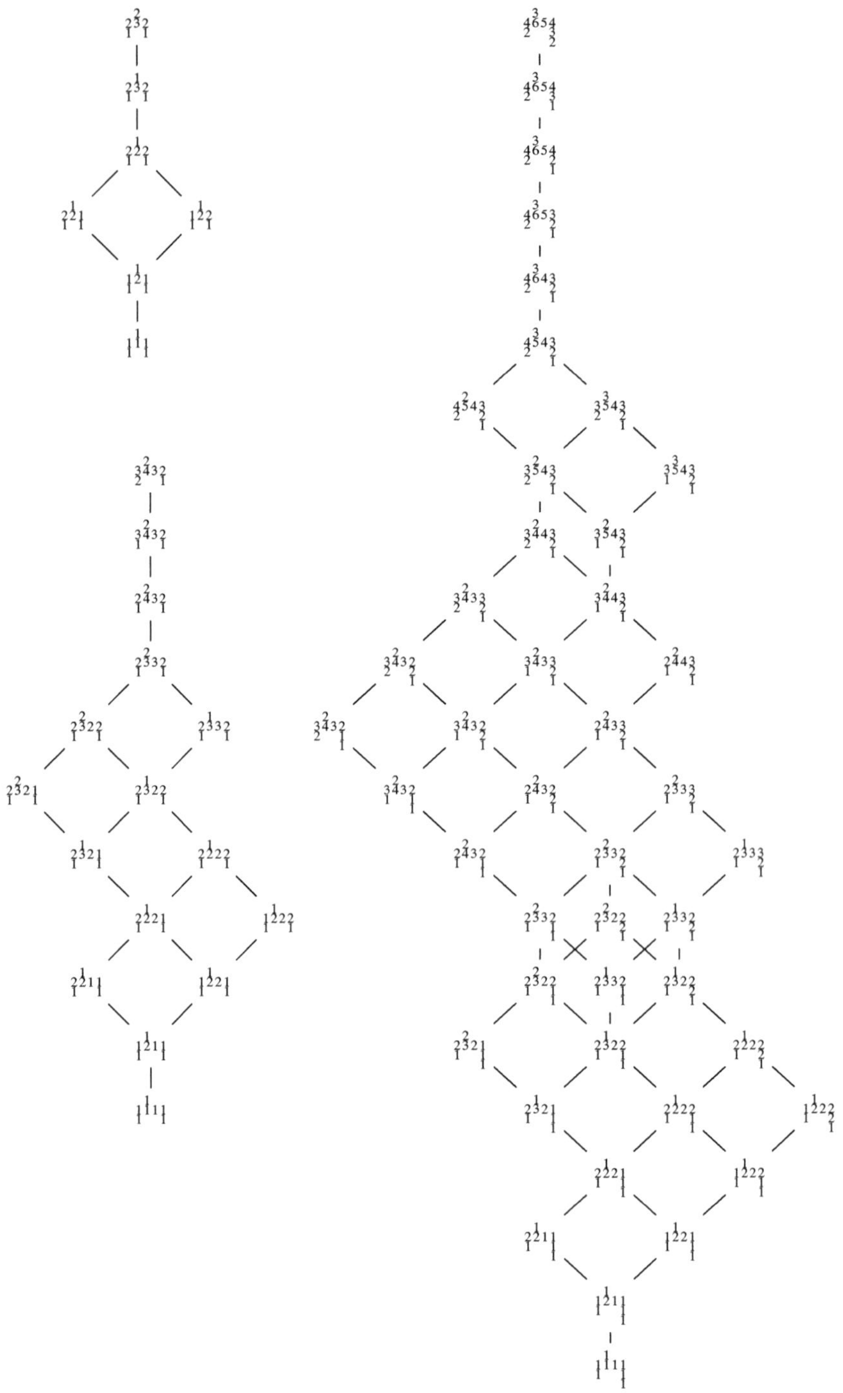

Fig. 2.1 Subposets of the root poset for $q_{\mathbb{E}_6}$ (left above), $q_{\mathbb{E}_7}$ (left below) and $q_{\mathbb{E}_8}$ (right) corresponding to sincere roots

Lemma 2.8. *Let G be a tree graph with n vertices, and v be a vector in $\mathbb{Z}^n$.*

a) If v is a thin vector, then v is a root of q_G if and only if v is a connected vector.
b) If G is a Dynkin graph and v is a q_G-root, then v is connected.

Proof. Take $q = q_G$. If v is a connected thin vector then the restriction of G to $\mathbf{supp}(v)$ is a tree, and

$$q(v) = \sum_{\substack{\text{vertices } i \\ \text{in } \mathbf{supp}(v)}} v_i^2 + \sum_{\substack{\text{edges } i-j \\ \text{in } \mathbf{supp}(v)}} q_{ij} v_i v_j$$
$$= \sharp\text{vertices} - \sharp\text{edges} = 1.$$

If v is not connected then $q(v) = \sharp\text{vertices} - \sharp\text{edges} > 1$, hence v is not a root. This shows (a).

For (b), assume G is a Dynkin graph. By Lemma 2.1 the form q is positive. Let us suppose that $v = v' + v''$ in such a form that $\mathbf{supp}(v') \cap \mathbf{supp}(v'') = \emptyset$, and $q_{ij} = 0$ for $i \in \mathbf{supp}(v')$ and $j \in \mathbf{supp}(v'')$. If v is a root of q, then

$$1 = q(v) = q(v' + v'') = q(v') + q(v''),$$

and by positivity either $v' = 0$ or $v'' = 0$, which shows that v is a connected vector. □

Consider a Dynkin graph $\mathbb{D}_n$ with the following ordering of vertices,

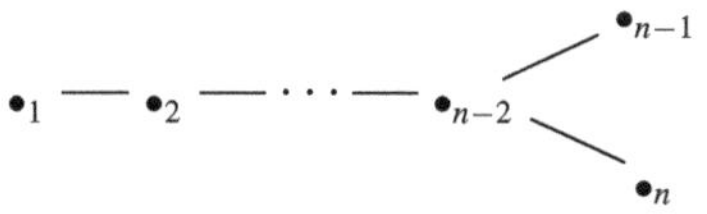

and for $1 \leq \ell < k \leq n-2$ take the vector $p_{\ell,k} = \sum_{i=\ell}^{n} e_i + \sum_{i=k}^{n-2} e_i$ in $\mathbb{Z}^n$ as depicted below,

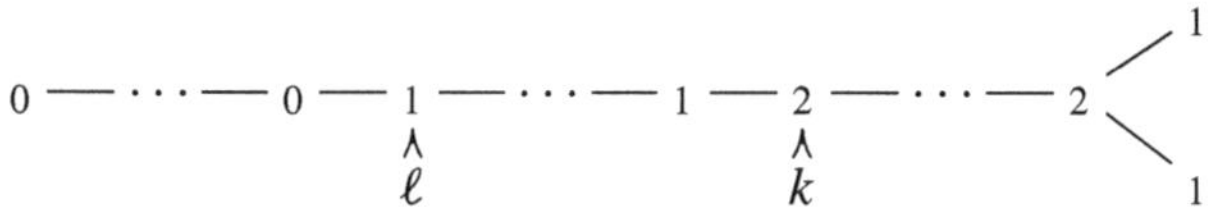

The following result completes the description of roots of the positive quadratic form $q_{\mathbb{D}_n}$.

Lemma 2.9. *If v is a positive nonthin root of $q = q_{\mathbb{D}_n}$, then $v = p_{\ell,k}$ for some $1 \leq \ell < k \leq n-2$.*

Proof. Let $w \in \mathbb{Z}^n$ be defined by $w_i = 1$ if $v_i > 0$ and $w_i = 0$ if $v_i = 0$. Since v is a root, by Lemma 2.8(b) the vector v is connected, and so is w for $\mathbf{supp}(w) = \mathbf{supp}(v)$. Hence by Lemma 2.8$(a)$, w is a thin root of q.

Since v is a positive nonthin vector we have $w < v$, and by Proposition 2.5 there exists a sequence of reflections $\sigma_{i_1}, \ldots, \sigma_{i_s}$ with $\sigma_{i_s} \cdots \sigma_{i_1}(w) = v$ such that $\sigma_{i_s} \cdots \sigma_{i_1}(w) = w + e_{i_1} + \cdots + e_{i_s}$. We want to show that $i_1 = n-2$ and that $i_{r-1} - r_r = 1$ for $1 < r \leq s$.

Recall from Lemma 1.1 that $q(v|e_i) = 2v_i - \sum_{i-j} v_j$, where the symbol $_{i \,-\, j}$ denotes the sum over all edges in $\mathbb{D}_n$ having i as an end-point. Since $\sigma_{i_1}(w) = w - q(w|e_{i_1})e_{i_1} > w$ if and only if $\sum_{i-j} w_j > 2w_i$ and since $\mathbf{supp}(w) = \mathbf{supp}(v)$, there must exist more than two edges ending in i_1, showing that $i_1 = n-2$ and that $w_n = w_{n-1} = w_{n-2} = w_{n-3} = 1$.

Then $\sigma_{i_1}(w) = p_{\ell,n-2}$ for some $1 \leq \ell < n-2$. Now, a direct calculation shows that, for $1 \leq \ell < k \leq n-2$, the inequality $q(p_{\ell,k}|e_i) < 0$ implies either $i = \ell - 1$ or $\ell < i = k-1$. In the first case we have $\mathbf{supp}(\sigma_i(p_{\ell,k})) \neq \mathbf{supp}(p_{\ell,k})$, which is unacceptable for our construction. Therefore we have $\ell < i = k-1$ and $\sigma_i(p_{\ell,k}) = p_{\ell,k-1}$. Proceeding inductively we get $v = \sigma_{i_s} \cdots \sigma_{i_2}(\sigma_{i_1}(w)) = p_{\ell,k}$ for some $\ell < k \leq n-2$, which completes the result. □

Exercises 2.10.

1. Let q be a positive unit form. Show that if v is a root of q then v is a connected vector. Is the same true for nonnegative unit forms?
2. Show that any positive root of $q_{\mathbb{A}_n}$ is thin. What is the shape of the picture of all positive roots if any two roots are connected by an edge if one is obtained by a reflection from the other?
3. Describe all positive roots of $\mathbb{E}_6$.
4. Let $\widetilde{\Delta}$ be an extended Dynkin diagram. Show that the vector $p_{\widetilde{\Delta}}$ described in Table 2.2 is a radical vector of the quadratic form $q_{\widetilde{\Delta}}$.
5. Suppose that G is a connected simple graph that does not contain as (full) subgraph any extended Dynkin diagram. Show that G is a Dynkin diagram. [Hint: define the degree of a vertex in G as the number of edges that contain it, and a ramification to be a vertex of degree greater than 2. Then notice that G must have at most one ramification vertex, with degree at most 3.]
6. Show that if G is a Dynkin graph, and i is any of its vertices, then the restriction $G^{(i)}$ is disjoint union of Dynkin graphs.

2.3 Criteria for Positivity

We call a quadratic unit form *critical nonpositive* if it is not positive but each proper restriction is. The critical nonpositive forms are, so to speak, the borderline which separates the positive forms from the rest. Since positive forms are always weakly positive (cf. Remark 2.6), we could compare critical nonpositive forms with critical nonweakly positive forms.

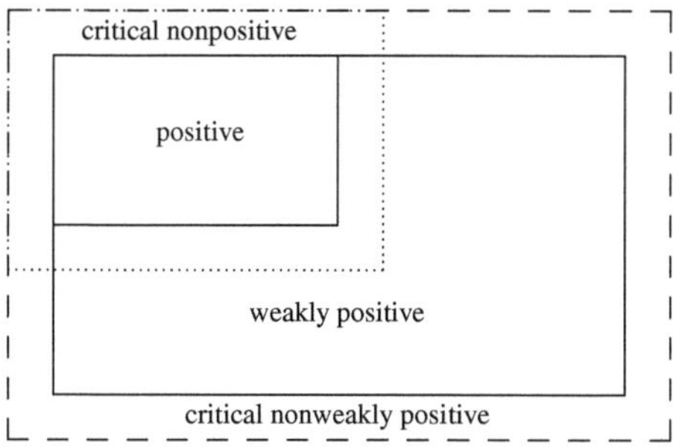

We will have to correct this picture (later in Sect. 5.1) when considering critical nonweakly positive forms. Now we restrict our discussion to unit forms.

First we need a simple result on nonnegative integral forms. Recall from Sect. 1.1 that a vector v in $\mathbb{Z}^n$ is called a *radical vector* for $q : \mathbb{Z}^n \to \mathbb{Z}$ if $q(v+u) = q(u)$ for all vectors u, or equivalently, if $q(v|e_i) = 0$ for $i = 1, \ldots, n$.

Lemma 2.11. *Let $q : \mathbb{Z}^n \to \mathbb{Z}$ be a nonnegative unit form and $p = q^I$ a restriction of q with $I \subset \{1, \ldots, n\}$. Then any radical vector of p, if extended by zeros to a vector in $\mathbb{Z}^n$, is a radical vector of q.*

Proof. Let $v \in \mathbf{rad}(p)$ and $\overline{v}$ be the extension of v by zeros in the missing coordinates. For $i \in I$ we have

$$q(\overline{v}|e_i) = q_{ii}\overline{v}_i + \sum_{j=1}^{n} \overline{v}_j q_{ij} = q_{ii}\overline{v}_i + \sum_{j\in I} \overline{v}_j q_{ij} = q^I(v|e_i) = 0.$$

Suppose that there is an $i \notin I$ with $q(\overline{v}|e_i) \neq 0$. Since $q(\overline{v}) = 0$, taking $\alpha = q_{ii} + 1$ we have

$$\begin{aligned} 0 &\leq q(\alpha\overline{v} - q(\overline{v}|e_i)e_i) = \alpha^2 q(\overline{v}) + q(\overline{v}|e_i)^2 q_{ii} - \alpha q(\overline{v}|e_i)^2 \\ &= q(\overline{v}|e_i)^2(q_{ii} - \alpha) \\ &< 0, \end{aligned}$$

a contradiction, hence the result. □

Consider the following example, which shows that the hypothesis of nonnegativity in the previous lemma is actually needed: Let $q = q_B$ be the quadratic form associated to the graph $B = 1 ═══ 2 ─── 3$. Then $\mathbf{rad}(q^{(3)}) = \mathbb{Z}(e_1 + e_2)$. Nevertheless, $e_1 + e_2$ is not a radical vector for q, since $q(e_1 + e_2|e_3) = -1$. We will come back to this situation later in Sect. 3.3.

Recall that the *corank* of an integral quadratic form q is given as the rank of its radical $\mathbf{rad}(q)$. The following characterization of critical nonpositive forms is an adaptation of a theorem of Ovsienko in the weakly positive setting (cf. Ringel [46] and Theorem 5.2 below).

Theorem 2.12. *Let q be a unit form. Then q is critical nonpositive if and only if either q is the Kronecker form q_m for some integer m with $|m| \geq 3$ or q is nonnegative of corank one with a sincere radical vector.*

Proof. Clearly, for $|m| \geq 3$ the m-Kronecker form

$$q_m(x_1, x_2) = x_1^2 - m x_1 x_2 + x_2^2,$$

is not positive since $q_m(1, \pm 1) = 2 \mp m$ is negative for the appropriate sign. Thus q_m is critical nonpositive.

If q is nonnegative of corank one with a sincere radical vector v, then for any vertex i the restriction $q^{(i)}$ is nonnegative. But if $w \in \mathbf{rad}(q^{(i)})$ then by Lemma 2.11 the extension by zeros $\overline{w}$ of w satisfies $\overline{w} \in \mathbf{rad}(q)$. Since $\mathbf{rad}(q)$ is generated by the sincere vector v and $\overline{w}_i = 0$, it follows that $w = 0$. In other words, q is critical nonpositive.

Assume now that $q : \mathbb{Z}^n \to \mathbb{Z}$ is critical nonpositive. Then there exists a nonzero vector $v \in \mathbb{Z}^n$ with $q(v) \leq 0$. Choose such a v with minimal weight $|v| = \sum_i |v_i|$. Since each restriction of q is positive, v is a sincere vector. Define $I = \{i \in \{1, \ldots, n\} \mid v_i > 0\}$ and $J = \{i \in \{1, \ldots, n\} \mid v_i < 0\}$.

For $i \in I$ we have $|v - e_i| < |v|$ and by minimality, $0 < q(v - e_i) = q(v) + 1 - q(v|e_i)$, hence $q(v|e_i) \leq q(v)$. Similarly, for each $i \in J$ we have $q(v|e_i) \geq -q(v)$, that is,

$$q(v|e_i) \begin{cases} \leq q(v), & \text{if } i \in I, \\ \geq -q(v), & \text{if } i \in J. \end{cases}$$

If $q(v) = 0$ we obtain $v_i q(v|e_i) \leq 0$ for all $i \in \{1, \ldots, n\}$, and thus from $0 = q(v) = \sum_i v_i q(v|e_i)$ we deduce that $q(v|e_i) = 0$, that is, that v is a radical vector.

For any $w \in \mathbb{Z}^n$ choose nonzero integers μ, λ such that $(\mu w - \lambda v)_i = 0$ for some index $i \in \{1, \ldots, n\}$. Then $\mu^2 q(w) = q(\mu w) = q(\mu w - \lambda v) = q^{(i)}(\mu w - \lambda v) \geq 0$ with equality if and only if $\mu w = \lambda v$. This shows that q is nonnegative. By minimality of v, we have that μ divides λ, therefore q has corank one with $\mathbf{rad}(q) = \mathbb{Z}v$.

If $q(v) < 0$, then we obtain from the values of $q(v|e_i)$ above that $v_i q(v|e_i) < 0$ for each index $i \in \{1, \ldots, n\}$. For $i \in I$, we deduce from $q(v|e_i) \leq q(v) = \frac{1}{2} \sum_j v_j q(v|e_j) < 0$ that

$$(v_i - 2) q(v|e_i) \geq - \sum_{i \neq j} v_j q(v|e_j) > 0,$$

where the last inequality is due to the fact that there must exist at least one other index apart from i. We obtain $v_i - 2 < 0$, thus $v_i = 1$. Similarly, for $i \in J$, we obtain $v_i = -1$. Choose a vertex x such that $|q(v|e_x)| \leq |q(v|e_i)|$ for all $i \in \{1, \ldots, n\}$, then

$$-\frac{n}{2}|q(v|e_x)| \geq \frac{1}{2}\sum_{i=1}^{n} -|q(v|e_i)| = q(v) \geq -|q(v|e_x)|,$$

which implies $n = 2$ and $|q(v)| = |q(v|e_x)|$, that is, q is a Kronecker form q_m. From $0 > q(v) = 2 - |q_{ij}| = 2 - |m|$ we get $|m| \geq 3$. □

We stress that a critical nonpositive form can very well be weakly positive. That happens precisely in the situation where q is the m-Kronecker form for some negative $m \leq -3$, or where the sincere radical vector is not positive.

Corollary 2.13. *Any critical nonpositive unit form q has infinitely many roots.*

Proof. By Proposition 1.23, Kronecker forms q_m with $|m| \geq 3$ have infinitely many roots. Now, if q is nonnegative with sincere radical vector v, then for any index i we have

$$1 = q(e_i) = q(e_i + mv),$$

for any $m \geq 0$, that is, $e_i + mv$ is a q-root. Thus the result is completed by the characterization given in Theorem 2.12. □

An integral quadratic form $q(x) = \sum_{i \leq j} q_{ij}x_i x_j$ is called *semi-unitary* or a *semi-unit form* given $q_{ii} \in \{0, 1\}$ for $i = 1, \ldots, n$. We need the following almost trivial observation.

Lemma 2.14. *Let $q : \mathbb{Z}^n \to \mathbb{Z}$ be a semi-unit form. If q is positive, then*

a) The form q is unitary.
b) We have $|q_{ij}| \leq 1$ for any indices $i, j \in \{1, \ldots, n\}$.

Proof. Since q is a positive semi-unit form we have $1 \geq q_{ii} = q(e_i) > 0$, thus q is unitary. Taking indices $i \neq j$ and a sign $\epsilon \in \{+1, -1\}$ such that $|q_{ij}| = \epsilon q_{ij}$, we have

$$0 < q(e_i - \epsilon e_j) = q(e_i) + q(e_j) - \epsilon q(e_i|e_j) = 2 - |q_{ij}|,$$

hence the result. □

The following handy **Positivity Criterion** [7, Theorem 2.2] will be useful in Chap. 3. By Proposition 1.20, a Kronecker form q_m with $|m| \geq 3$ satisfies $q_m^{-1}(0) = 0$, although q_m is (critical) nonpositive.

Theorem 2.15. *A semi-unit form $q : \mathbb{Z}^n \to \mathbb{Z}$ is positive if and only if the following conditions hold:*

P1) We have $|q_{ij}| \leq 2$ for $1 \leq i < j \leq n$.
P2) The form q is anisotropic (that is, $q(x) \neq 0$ for any nonzero vector x in $\mathbb{Z}^n$).

Proof. The necessity of conditions *(P1)* and *(P2)* follows by Lemma 2.14. Assume that q is not positive and satisfies these conditions. Take $0 \neq v \in \mathbb{Z}^n$ with $q(v) \leq 0$, and notice that q is a unit form (for if $q_{ii} = 0$ then $e_i \in q^{-1}(0)$), and that $|q_{ij}| \leq 1$ (for if $q_{ij} = \pm 2$ then $e_i \pm e_j \in q^{-1}(0)$).

We proceed by induction on n. For $n = 1$ we have $q(x_1) = x_1^2$, and for $n = 2$, $q(x_1, x_2) = x_1^2 + x_2^2 + ax_1x_2$ with $a \in \{\pm 1, 0\}$. All of these forms are positive. Assume $n \geq 3$, and observe that any restriction $q^{(i)}$ satisfies *(P1)* and *(P2)*. Therefore by induction we have that q is critical nonpositive. By Theorem 2.12, since $n \geq 3$ the form q is nonnegative with corank one, in contradiction to *(P2)*. □

We end this section with yet another characterization of positive unit forms.

Theorem 2.16. *Let $q : \mathbb{Z}^n \to \mathbb{Z}$ be a unit form. Then q is positive if and only if q has finitely many roots.*

Proof. If q is a nonpositive unit form, then there exists a restriction q^I of q which is critical nonpositive. By Corollary 2.13, q has infinitely many roots.

The converse was shown in Proposition 2.3. □

2.4 Inflations, Deflations and Dynkin Type

Gabrielov transformations have been used since the early seventies for the systematic study of quadratic unit forms (see for instance [22] and [23]). They are involutions that preserve unitary forms, and in many cases (for instance in the positive setting) their iterations generate all equivalences between such forms.

Denote by $\mathfrak{U}(n)$ the set of all unitary quadratic forms $q : \mathbb{Z}^n \to \mathbb{Z}$ in n variables. For indices $i, j \in \{1, \dots, n\}$ take E_{ij} to be the elementary $n \times n$ matrix having as unique nonzero entry a 1 at coordinates (i, j).

Proposition 2.17. *Let $\mathfrak{U}(n)$ be the set of unitary forms in n-variables. For each $i \neq j$ in $\{1, \dots, n\}$ define the* Gabrielov transformation

$$\mathcal{G}_{ij} : \mathfrak{U}(n) \longrightarrow \mathfrak{U}(n)$$

$$q \longmapsto \mathcal{G}_{ij}(q) = qG^q_{ij},$$

where G^q_{ij} is the linear transformation given by the matrix $G^q_{ij} = \mathbf{Id} - q_{ij}E_{ij}$. Then $\mathcal{G}_{ij}$ is an involution that preserves connected unit forms.

Proof. Notice that for $k \in \{1, \dots, n\}$ we have

$$G^q_{ij}(e_k) = \begin{cases} e_k, & \text{if } k \neq i, \\ e_i - q_{ij}e_j, & \text{if } k = i. \end{cases}$$

To show that $\mathscr{G}_{ij}(q)$ is a unitary form we observe that

$$q(G^q_{ij}(e_k)) = \begin{cases} q(e_k) = 1, & \text{if } k \neq i, \\ q(e_i - q_{ij}e_j) = q(e_i) + q^2_{ij}q(e_j) - q^2_{ij} = 1, & \text{if } k = i. \end{cases}$$

Therefore $\mathscr{G}_{ij} : \mathfrak{U}(n) \to \mathfrak{U}(n)$ is a well defined function.

We show now that $\mathscr{G}_{ij}$ is an involution. Take $q' = \mathscr{G}_{ij}(q)$ and observe that for $k \neq i$ and $\ell \neq i$ we have

$$q'_{k\ell} = q(G^q_{ij}(e_k)|G^q_{ij}(e_\ell)) = q(e_k|e_\ell) = q_{k\ell}.$$

On the other hand we have

$$q'_{i\ell} = \begin{cases} q(G^q_{ij}(e_i)|G^q_{ij}(e_\ell)) = q(e_i|e_\ell) - q_{ij}q(e_j|e_\ell) = q_{i\ell} - q_{ij}q_{j\ell}, & \text{if } \ell \neq j, \\ q(G^q_{ij}(e_i)|G^q_{ij}(e_j)) = q(e_i|e_j) - q_{ij}q(e_j|e_j) = -q_{ij}, & \text{if } \ell = j. \end{cases}$$

The same equations imply that, if $q'' = \mathscr{G}_{ij}(q')$, then $q''_{k\ell} = q_{k\ell}$ for $k, \ell \neq i$, and

$$q'_{i\ell} = \begin{cases} q'_{i\ell} - q'_{ij}q'_{j\ell} = (q_{i\ell} - q_{ij}q_{j\ell}) + q_{ij}q_{j\ell} = q_{i\ell}, & \text{if } \ell \neq j, \\ -q'_{ij} = q_{ij}, & \text{if } \ell = j. \end{cases}$$

Therefore $q'' = q$, that is, $\mathscr{G}_{ij}$ is an involution.

Assume now that q is a disconnected unit form. After a re-enumeration if necessary, we may take $1 < m < n$ such that $q_{ij} = 0$ if $1 \leq i \leq m$ and $m < j \leq n$. Then $q = q' \oplus q''$ for unitary forms $q' : \mathbb{Z}^m \to \mathbb{Z}$ and $q'' : \mathbb{Z}^{n-m} \to \mathbb{Z}$, and the symmetric matrix A_q associated to the unit form q is diagonal by blocks,

$$A_q = \begin{pmatrix} A_{q'} & 0 \\ 0 & A_{q''} \end{pmatrix} = A_{q'} \oplus A_{q''}.$$

If $1 \leq i \leq m$ and $m < j \leq n$ then $q_{ij} = 0$ and $\mathscr{G}_{ij}(q) = q = q' \oplus q''$. If $1 \leq i, j \leq m$ or $m < i, j \leq n$ we have respectively either

$$\mathscr{G}_{ij}(q) = \mathscr{G}_{ij}(q') \oplus q'' \qquad \text{or} \qquad \mathscr{G}_{ij}(q) = q' \oplus \mathscr{G}_{ij}(q''),$$

where in the right-hand side of the equalities the expression $\mathscr{G}_{ij}$ corresponds to a Gabrielov transformation of appropriate size. In any case $\mathscr{G}_{ij}(q)$ is a disconnected form. Hence, since Gabrielov transformations are involutions, they preserve connectedness. □

It is clear that $\mathscr{G}_{ij}(q) = q$ if and only if $q_{ij} = 0$. In what follows we take a slightly different approach, considering similar transformations called *inflations* and *deflations* of integral quadratic forms.

Take a sign $\varepsilon \in \{+, -\}$, and for different indices $1 \leq i, j \leq n$ define the linear transformation $T_{ij}^{\varepsilon} : \mathbb{Z}^n \to \mathbb{Z}^n$ by

$$T_{ij}^{\varepsilon} : v \mapsto v - \varepsilon v_i e_j.$$

Observe that T_{ij}^{+} is the inverse of T_{ij}^{-}, therefore the forms $q^{-} = qT_{ij}^{-}$ and $q^{+} = qT_{ij}^{+}$ are $\mathbb{Z}$-equivalent to q. In particular, q^{-} and q^{+} are positive if so is q.

We call the transformation T_{ij}^{-} a *deflation* for q if $q_{ij} < 0$, and the transformation T_{ij}^{+} is an *inflation* for q if $q_{ij} > 0$. Inflations and deflations are simply called *flations*, and a finite composition of flations is an *iterated flation*. To be precise, an iterated flation for q is a composition $T = T_{i_1 j_1}^{\epsilon_1} \cdots T_{i_r j_r}^{\epsilon_r}$ such that $T_{i_1 j_1}^{\epsilon_1}$ is a flation for q, and taking inductively $q_0 = q$ and $q_s = q_{s-1} T_{i_s j_s}^{\epsilon_s}$ for $s \geq 0$, $T_{i_s j_s}^{\epsilon_s}$ is a flation for q_{s-1}. Notice that if $|q_{ij}| = 1$ and $\epsilon \in \{+1, -1\}$ is such that $|q_{ij}| = \epsilon q_{ij}$, then the form qT_{ij}^{ϵ} coincides with the Gabrielov transformation $\mathscr{G}_{ij}(q)$.

Corollary 2.18. *For any $n \geq 1$ the set of positive unit forms in n variables, denoted $\mathfrak{U}^{>0}(n)$, is invariant under deflations and inflations. Moreover, these transformations are involutions that preserve connected positive unit forms.*

Proof. Flations preserve positivity since they are equivalences. Let $q : \mathbb{Z}^n \to \mathbb{Z}$ be a positive unit form. By Lemma 2.14(*b*) we have $|q_{ij}| \leq 1$ for $1 \leq i < j \leq n$. Therefore in the positive case, flations correspond to Gabrielov transformations, and the result follows from Proposition 2.17. □

If T_{ij}^{ϵ} is a flation for q and $q^{\epsilon} = qT_{ij}^{\epsilon}$, then there is a bijection

$$R(q^{\epsilon}) \longrightarrow R(q)$$

given by $x \mapsto T_{ij}^{\epsilon}(x)$. Perhaps the most important property of flations is that they allow us to control the number of positive roots of a unitary form, as we show next.

Lemma 2.19. *Let $q : \mathbb{Z}^n \to \mathbb{Z}$ be a unit form and take indices $1 \leq i \neq j \leq n$ and $1 \leq \ell \neq m \leq n$ with $q_{ij} < 0$ and $q_{\ell m} > 0$. Consider respectively quadratic forms $q^{-} = qT_{ij}^{-}$ and $q^{+} = qT_{\ell m}^{+}$. Then we have proper inclusions*

$$(T_{ij}^{-}) : R^{+}(q^{-}) \to R^{+}(q) \quad \textit{and} \quad (T_{\ell m}^{+})^{-1} : R^{+}(q) \to R^{+}(q^{+}).$$

Proof. For a positive root $x \in R^{+}(q^{-})$ we have $1 = q^{-}(x) = q(T_{ij}^{-}x)$. Thus $T_{ij}^{-}x \in R^{+}(q)$ and $e_i \in R^{+}(q)$ is a root not belonging to the image $T_{ij}^{-}(R^{+}(q^{-}))$. The second proper inclusion follows from the first since $(T_{\ell m}^{+})^{-1} = T_{\ell m}^{-}$ and $q = q^{+}T_{\ell m}^{-}$. □

We are now able to prove a central theorem in the theory of unitary forms, first shown by Ovsienko in the early seventies.

Theorem 2.20. *Let q be a positive unit form. Then there exists an iterated inflation T and a unique (up to permutation of components) disjoint union of Dynkin diagrams G such that $qT = q_G$.*

Proof. By Corollary 2.18 we may assume that q is a connected form. Consider the bigraph B^0 associated to q. If B^0 has no dotted edges, by Proposition 2.2 the graph B^0 is a Dynkin diagram and we are done. Assume B^0 has a dotted edge $\{i, j\}$ and take $T^1 = T_{ij}^+$. Again by Corollary 2.18 the quadratic form $q^1 = qT^1$ is also a positive unit form, having a connected associated bigraph B^1. By Lemma 2.19 and Theorem 2.16 we have $|R^+(q^0)| < |R^+(q^1)| < \infty$. Iterating this process we get a sequence of positive unit forms $q^0, q^1, q^2, \ldots$ in the same number of variables, and inequalities

$$|R^+(q^0)| < |R^+(q^1)| < |R^+(q^2)| < \ldots$$

We end the proof by showing that this process must stop, that is, that there must exist $i > 0$ such that the bigraph B^i associated to q^i has no dotted edges (and is therefore a Dynkin diagram). Indeed, since there is a finite number of unit positive forms with a fixed number of variables (for $|q_{ij}| \leq 1$ for all i, j, see Lemma 2.14), we conclude that the set of cardinalities $\{|R^+(q^i)|\}_{i \geq 0}$ is bounded. To complete the proof take T as the iterated flation $T^1 \cdots T^i$ so that $q_{B^i} = qT$ and B^i is a Dynkin graph.

For the uniqueness claim assume again that q is connected, and that there are iterated flations T and T' and Dynkin graphs G and G' with $qT = q_G$ and $qT' = q_{G'}$. Since the forms qT and qT' have the same number of roots, and $|G_0| = |G'_0|$, we conclude that $G = G'$ (cf. Table 2.1). □

By the Dynkin type of a positive unit form q we mean the disjoint union of Dynkin graphs $\mathbf{Dyn}(q) = G$ given in Theorem 2.20. That this is a well defined invariant of the quadratic form q is the content of our next result.

Corollary 2.21. *Two positive unit forms q and q' are equivalent if and only if* $\mathbf{Dyn}(q) = \mathbf{Dyn}(q')$.

Proof. If $\mathbf{Dyn}(q) = \mathbf{Dyn}(q') = G$, then there are iterated inflations T and T' such that $qT = q_G = q'T'$. Therefore $q = q'T'T^{-1}$, that is, q and q' are equivalent forms.

For the converse we may assume that q and q' are equivalent connected unit forms (see Corollary 2.18). Take iterated inflations T and T' such that $qT = q_G$ and $q'T' = q_{G'}$ for Dynkin diagrams G and G' as in Theorem 2.20. Since q_G is equivalent to $q_{G'}$, the graphs G and G' have the same number of vertices and their associated forms have the same number of roots. A direct inspection of Table 2.2 shows that $G = G'$, hence the result. □

Notice that each Dynkin diagram Δ has a unique *maximal root*, that is, a vector $p_\Delta \in R(q_\Delta)$ such that for all $v \in R(q_\Delta)$ we have $v \leq p_\Delta$ (the vector formed by the values in the vertices of Table 2.1). From this we obtain

$$|v_i| \leq r_\Delta, \quad \text{for all } i \text{ and all } v \in R(q_\Delta),$$

where $r_{\mathbb{A}_n} = 1$, $r_{\mathbb{D}_n} = 2$, $r_{\mathbb{E}_6} = 3$, $r_{\mathbb{E}_7} = 4$ and $r_{\mathbb{E}_8} = 6$. This remark can be extended to all connected positive unit forms.

Proposition 2.22. *If $v = (v_1, \dots, v_n)$ is an arbitrary root of a connected positive unit form $q : \mathbb{Z}^n \to \mathbb{Z}$ of Dynkin type Δ, then $|v_i| \leq r_\Delta$ for all $i \in \{1, \dots, n\}$.*

Proof. Let $q : \mathbb{Z}^n \to \mathbb{Z}$ be a connected positive unit form with $\mathbf{Dyn}(q) = \Delta$ and $v \in R(q)$. Let $C : \mathbb{Z}^n \to \mathbb{Z}^n$ be the transformation given by $C(e_i) = \varepsilon_i e_i$, where $\varepsilon_i = -1$ if $v_i \leq 0$ and $\varepsilon_i = 1$ otherwise. The form $q' = qC$ is a connected positive unit form and $v' = C^{-1}v$ is a root of q' with $v'_i = |v_i|$ for all $i \in \{1, \dots, n\}$. Moreover, by Corollary 2.21 we have $\mathbf{Dyn}(q') = \mathbf{Dyn}(q)$. If $q' = q_\Delta$, then $v' \leq p_\Delta$ by maximality of the root p_Δ of q_Δ. Otherwise we may apply an inflation T^+_{ij} to q' to obtain a connected positive unit form $q'' = q'T^+_{ij}$ and a positive root $v'' = (T^+_{ij})^{-1}v' = v' + v'_i e_j > v'$. Continuing with this process we get an iterated inflation T and a positive root $\widetilde{v}$ of $q'T$ such that $v' < \widetilde{v}$ and $qT = q_\Delta$. Again by maximality of the root p_Δ we have $\widetilde{v} < p_\Delta$, which completes the proof. □

As a direct consequence of Proposition 2.22 we have the following observation, which remarkably also holds in the context of weakly positive unit forms (see Ovsienko's Theorem 5.25).

Corollary 2.23. *Let $q : \mathbb{Z}^n \to \mathbb{Z}$ be a positive unit form. Then for any root $v = (v_1, \dots, v_n)$ of q we have $|v_i| \leq 6$ for $i = 1, \dots, n$.*

Proof. This follows from Proposition 2.22 since $r_\Delta \leq r_{\mathbb{E}_8} = 6$ for any Dynkin diagram Δ. □

Now we investigate how Dynkin diagrams behave under restriction of quadratic forms. For this purpose we introduce a partial ordering on Dynkin graphs by setting

$$\begin{aligned} \mathbb{A}_m &\leq \mathbb{A}_n, && \text{for } m \leq n; \\ \mathbb{A}_n < \mathbb{D}_n &\leq \mathbb{D}_p, && \text{for } 4 \leq n \leq p; \\ \mathbb{D}_p < \mathbb{E}_p &\leq \mathbb{E}_q, && \text{for } 6 \leq p \leq q \leq 8. \end{aligned}$$

The following easy observation will be used below.

Remark 2.24. For Dynkin graphs G and G' we have $G \leq G'$ if and only if $|G_0| \leq |G'_0|$ and $r_G \leq r_{G'}$ (cf. Table 2.1).

Recall that if I is a subset of indices $I \subseteq \{1, \dots, n\}$ then the restriction $q^I : \mathbb{Z}^I \to \mathbb{Z}$ of a quadratic form $q : \mathbb{Z}^n \to \mathbb{Z}$ is given by $q^I(x) = q(\sigma(x))$ for a vector

x in $\mathbb{Z}^I$, where $\sigma : \mathbb{Z}^I \to \mathbb{Z}^n$ is the linear transformation determined by $\sigma(e_i) = e_i$ for $i \in I$. As before, if $I = \{1, \dots, n\} - \{i\}$ for some index i then we use the notation $q^I = q^{(i)}$.

Proposition 2.25. *Let q be a connected positive unit form. Then for any connected restriction q^I of q we have* $\mathbf{Dyn}(q^I) \leq \mathbf{Dyn}(q)$.

Proof. We show that $\mathbf{Dyn}(q^{(i)}) \leq \mathbf{Dyn}(q)$ for any $i \in \{1, \dots, n\}$ with $q^{(i)}$ connected. By simplicity we may assume that $i = n$. Let $T : \mathbb{Z}^{n-1} \to \mathbb{Z}^{n-1}$ be an iterated inflation such that $q^{(n)}T = q_{\Delta'}$ given by Theorem 2.20.

Take $\widehat{T} = T \oplus [1]$ and $\widehat{q} = q\widehat{T}$. Since $\widehat{T}$ is $\mathbb{Z}$-invertible we have $\mathbf{Dyn}(\widehat{q}) = \mathbf{Dyn}(q)$ by Corollary 2.21. Extending the maximal root of $q_{\Delta'}$ by zeros we obtain a root v of $\widehat{q}$. By Proposition 2.22 we get

$$r_{\Delta'} \leq \max_i(|v_i|) \leq r_{\Delta},$$

where $\Delta = \mathbf{Dyn}(q)$. Clearly $|\Delta'_0| < |\Delta_0|$, thus the result follows from Remark 2.24. □

It might look a little odd at this point to define the order $\mathbb{A}_n < \mathbb{D}_n$ instead of $\mathbb{A}_n < \mathbb{D}_{n+1}$, but since this result will be generalized to nonnegative unit forms later in Sect. 3.6, we choose to introduce the final order at once.

Exercises 2.26.

1. Show that the number of roots of the quadratic form $q_{\mathbb{E}_p}$ associated to the Dynkin graph $\mathbb{E}_p$ for $p \in \{6, 7, 8\}$ is different from $|R(q_{\mathbb{A}_n})|$ and $|R(q_{\mathbb{D}_m})|$ for any $n \geq 1$ and $m \geq 4$.
2. Find two different Dynkin graphs Δ and Δ' such that the positive unit forms q_Δ and $q_{\Delta'}$ have the same number of roots. [Hint: Use Exercise 1 and the Pell equation $x_1^2 - 2x_2^2 = 1$ to find values for $a > 0$ such that there are integral solutions to the equations $n = m + a$ and $n(n+1) = 2m(m+1)$.]
3. Compute the Dynkin type of the following positive unit forms,

   a) $q(x) = x_1^2 + x_2^2 + x_3^2 + x_4^2 - (x_1 + x_2)(x_3 + x_4) + x_1x_2 + x_3x_4$.
   b) $q'(x) = x_1^2 + x_2^2 + x_3^2 + x_4^2 + x_1x_4 - x_1x_2 - x_2x_3 - x_3x_4$.

4. Show that any positive unit form q may be taken by deflations to q_B where B is a bigraph with no solid edges.
5. Give an example of a nonunitary positive integral quadratic form in at least three variables.
6. Determine which of the following unit forms is positive, and find the corresponding Dynkin type for those forms which are positive.

   a) $q(x) = x_1^2 + x_2^2 + x_3^2 + x_4^2 - x_2(x_1 - x_3) - x_4(x_1 + x_3) + x_1x_3$.
   b) $q'(x) = x_1^2 + x_2^2 + x_3^2 + x_4^2 + x_5^2 + x_6^2 - x_1(x_2 + x_4 - x_6) - x_2(x_3 + x_5 + x_6) + x_3(x_4 + x_5) - x_4x_6$.

7. Let $V : \mathbb{Z}^n \to \mathbb{Z}^n$ be a *point inversion*, that is, a linear transformation such that $V(e_i) = \pm e_i$ where $e_i, \dots, e_n$ are the canonical vector of $\mathbb{Z}^n$. Given a positive unit form q, is there an iterated flation T for q such that $V = T$?
8. Give an example of a unit form q and a flation T for q such that qT is no longer a unitary form.

2.5 Recognizing Positive Unit Forms

Let G be a simple graph (with only solid edges, no loop and no multiple edges) and for two different vertices r and s define $[r, s]_G$ to be the number of edges between vertices r and s (notice that $[r, s]_G = -(q_G)_{rs}$). For fixed vertices i, j we define a new graph G' with the same vertices as G and the same edges as G, except those containing vertex i for which we have

$$[r, i]_{G'} = \begin{cases} |[r, i]_G - [i, j]_G[r, j]_G|, & \text{if } r \neq j, \\ [j, i]_G, & \text{if } r = j. \end{cases}$$

We denote the new graph G' by GT_{ij} and say that G is *transformed by the graph flation* T_{ij} into G'. Observe that GT_{ij} is again a simple graph. Starting with a bigraph B we define a new graph **Frame**(B) (with only solid edges), referred to as the *frame* of bigraph B, by turning solid all dotted edges of B. For a unit form q we take **Frame**(q) to be the frame of the associated bigraph B_q of q.

Lemma 2.27. *Let q be a positive unit form. Take vertices $i \neq j$ such that $q_{ij} \neq 0$ and take $\epsilon \in \{+1, -1\}$ with $\epsilon q_{ij} > 0$. Then we have*

$$\textbf{Frame}(q)T_{ij} = \textbf{Frame}(qT_{ij}^{\varepsilon}).$$

Proof. By Lemma 2.14(*b*) we have $|q_{ij}| = 1$. Hence the claim follows from the expression

$$\begin{aligned} q(T_{ij}^{\varepsilon}x) &= q(x - q_{ij}x_ie_j) = q(x) + x_i^2 - q_{ij}x_iq(x|e_j) \\ &= q(x) + x_i^2 - q_{ij}x_i\left[2x_j + \sum_{k\neq j} q_{kj}x_k\right] \\ &= q(x) - q_{ij}x_i\left[2x_j + \sum_{k\neq i,j} q_{kj}x_k\right] \\ &= q(x) - q_{ij}\left[\sum_{k\neq i} q_{kj}x_kx_i + x_ix_j\right]. \end{aligned}$$

□

Notice that if the vertices i and j are not connected by an edge, then $GT_{ij} = G$.

Remark 2.28. Let G be a simple graph, and take vertices $i \neq j$. Then $G = (GT_{ij})T_{ij}$, that is, graph flations are involutions.

Proof. By the comment above we may assume that $[i, j]_G = 1$. Take $G' = GT_{ij}$ and $G'' = G'T_{ij}$. We only need to show that for $r \neq i, j$ we have $[r, i]_{G''} = [r, i]_G$. By definition,

$$[r,i]_{G''} = |[r,i]_{G'} - [i,j]_{G'}[r,j]_{G'}| = ||[r,i]_G - [i,j]_G[r,j]_G| - [i,j]_G[r,j]_G|.$$

If $[r, j]_G = 0$ then $[r, i]_{G''} = |[r, i]_G| = [r, i]_G$. Otherwise we have $[r, j]_G = 1 = [i, j]_G$ and therefore $|[r, i]_G - [i, j]_G[r, j]_G| = [i, j]_G[r, j]_G - [r, i]_G$, and in particular

$$||[r,i]_G - [i,j]_G[r,j]_G| - [i,j]_G[r,j]_G| = | - [r,i]_G| = [r,i]_G.$$

This completes the proof. □

We say that two simple graphs are *flation equivalent* if one is obtained from the other by a sequence of graph flations. By Remark 2.28 this is actually an equivalence relation. We call a graph G *positive admissible* if there exists a positive unit form q with **Frame**$(q) = G$.

Proposition 2.29. *The following statements hold.*

a) Let G and G' be flation equivalent simple graphs. Then G is positive admissible if and only if G' is positive admissible.
b) If q and q' are equivalent positive unit forms then their frames **Frame**(q) *and* **Frame**(q') *are flation equivalent.*

Proof. Assume G is a positive admissible graph and take a positive unit form q with $G = \mathbf{Frame}(q)$. Take an iterated graph flation $T = T_{i_1 j_1} \cdots T_{i_r j_r}$ for $r \geq 1$ such that $G' = GT$. Choose signs $\varepsilon_i, \ldots, \varepsilon_r$ such that $T^\varepsilon := T^{\varepsilon_1}_{i_1 j_1} \cdots T^{\varepsilon_r}_{i_r j_r}$ is an iterated flation for q. Taking $q' = qT^\varepsilon$ and using Lemma 2.27 observe that

$$\mathbf{Frame}(q') = \mathbf{Frame}(qT^\varepsilon) = \mathbf{Frame}(q)T = GT = G'.$$

Since q' is a positive unit form, then G' is a positive admissible graph, which shows (a).

For (b), since q and q' have the same Dynkin type, by Theorem 2.20 there are iterated inflations that take q and q' to q_Δ for some disjoint union of Dynkin diagrams Δ. In particular there is an iterated flation T with $q' = qT$. Then Lemma 2.27 implies that **Frame**(q) and **Frame**(q') are flation equivalent graphs. □

In the following results we describe bigraphs associated to positive unit forms. We start with some necessary conditions. A *(chordless) cycle* in a bigraph B is a full subbigraph which is connected and where every vertex has exactly two neighbors. We say that a bigraph B satisfies the *cycle condition* if every cycle in B has an odd number of dotted edges.

Proposition 2.30. *The bigraph B_q associated to a positive unit form q always satisfies the cycle condition.*

Proof. Suppose on the contrary that there exists a positive unit form q and a cycle C in $B = B_q$ which has an even number of dotted edges. Suppose that C has vertices $\{x_1, \ldots, x_n\}$ with $q_{x_i,x_{i+1}} \neq 0$ for $i = 1, \ldots, n-1$ and $q_{x_n,x_1} \neq 0$. Since q is a positive unit form we have $n > 2$. Define a vector v in $\mathbb{Z}^{B_0}$ by setting $v_y = 0$ for all $y \notin C$, $v_{x_1} = 1$ and take inductively $v_{x_{i+1}} = -q_{x_i,x_{i+1}} v_{x_i} = \pm 1$, using Lemma 2.14$(b)$. Then

$$\begin{aligned}
q(v) &= \sum_{i=1}^{n} v_{x_i}^2 + \sum_{i=1}^{n-1} q_{x_i x_{i+1}} v_{x_i} v_{x_{i+1}} + q_{x_1 x_n} v_{x_1} v_{x_n} \\
&= n - \sum_{i=1}^{n-1} q_{x_i x_{i+1}}^2 v_{x_i}^2 - (-q_{x_1 x_n}) \left(\prod_{i=1}^{n-1} (-q_{x_i x_{i+1}}) \right) v_{x_1}^2 \\
&= 0,
\end{aligned}$$

in contradiction to q being a positive unit form. □

By a *point inversion* we mean a linear transformation $V : \mathbb{Z}^n \to \mathbb{Z}^n$ such that $V(e_i) = \pm e_i$ for all $i \in \{1, \ldots, n\}$. If $q' = qV$ we also say that q' is a point inversion of q. Recall that a *walk* in a (bi)graph G is an alternating sequence of vertices and connecting edges,

$$w = (v_0, d_1, v_1, d_2, v_2 \ldots, v_{n-1}, d_n, v_n),$$

starting and ending in vertices v_0 and v_n (see for instance [18]). The number n is said to be the length of w. A walk w will be denoted by $w = (v_0|d_1 \cdots d_n|v_n)$, or simply by $w = d_1 \cdots d_n$. We say that a walk w is *reduced* if it does not contain any subsequence of the form (v_0, d, v_1, d, v_0) for an edge d joining vertices $v_0 \neq v_1$, and that w is *open* if it does not start and end at the same vertex. In what follows all walks are assumed to be reduced.

Proposition 2.31. *For two positive unit forms p and q the following are equivalent:*

a) There exists a point inversion V such that $q = pV$.
b) **Frame**(q) = **Frame**(p) *and for every cycle C in the bigraph B_q, the number of edges $\{i, j\}$ in C with $q_{ij} \neq p_{ij}$ is even.*

Proof. Let us first show that (a) implies (b). That **Frame**(q) = **Frame**(qV) for any point inversion V is clear. Assume that there is a j such that $V(e_j) = -e_j$ and

$V(e_i) = e_i$ for any other index i. If j is a vertex in cycle C, then there are exactly two edges in C containing j, namely $\alpha = \{i, j\}$ and $\alpha' = \{j, i'\}$ for some vertices i and i'. Then we have for $p = qV$ that $p_{ij} = -q_{ij}$ and $p_{ji'} = -q_{ji'}$. For the rest of the edges $\{a, b\}$ in B_q it is clear that $p_{ab} = q_{ab}$. Hence point (b) follows inductively for arbitrary point inversions.

For the converse choose a maximal subtree $\Sigma = (\Sigma_0, \Sigma_1)$ of $B = B_q$ (that is, a connected subgraph of B such that $|B_0| = |\Sigma_0| = |\Sigma_1| + 1$). Since $\mathbf{Frame}(q) = \mathbf{Frame}(p)$, there is a point inversion V such that $q' = pV$ satisfies $q_{ij} = q'_{ij}$ for all edges $\{i, j\}$ in Σ. We only need to verify that $q_{ij} = q'_{ij}$ for any other edge $\alpha = \{i, j\}$ in B. By maximality of Σ, both vertices i and j belong to the subtree Σ. In particular, there is a walk $w = (x_0, \alpha_1, x_1, \dots, x_{d-1}, \alpha_d, x_d)$ contained in Σ with $x_0 = i$ and $x_{d+1} = j$.

We proceed by induction on the (relative) distance $d_\Sigma(i, j)$ between vertices i and j along subtree Σ (the length d of the walk w above). If $d_\Sigma(i, j) = 1$ then the edge $\alpha = \{i, j\}$ actually belongs to Σ (for $\mathbf{Frame}(q)$ has no double edges by Lemma 2.14(b)), hence $q_{ij} = q'_{ij}$.

Assume now that $d_\Sigma(i, j) > 1$. If the full subgraph C of B consisting of vertices $x_0, \dots, x_d$ (which contains edges $\alpha = \{i, j\}, \alpha_1 = \{x_0, x_1\}, \dots, \alpha_d = \{x_{d-1}, x_d\}$) is a (chordless) cycle of B, then we have $q_{x_{r-1},x_r} = q'_{x_{r-1}x_r}$ for $r = 1, \dots, d$, which yields $q_{ij} = q'_{ij}$ by the hypothesis in (b). If C is not a cycle, then there are indices $1 \le r < s - 1 \le d$ such that $\beta = \{x_r, x_s\}$ is an edge of B different from α. Replacing any of the edges $\alpha_r, \dots, \alpha_s$ in Σ by the edge β, we get a maximal subtree Σ' of B where the relative distance $d_{\Sigma'}(i, j)$ is strictly smaller than $d_\Sigma(i, j)$. Apply induction in this case.

Altogether we have a point inversion V such that $q = q' = pV$, which completes the proof. □

Corollary 2.32. *If two positive unit forms p and q have the same frame then they are equivalent. In particular, they have same Dynkin type.*

Proof. By Proposition 2.30 both bigraphs B_p and B_q satisfy the cycle condition. Hence statement (b) in Proposition 2.31 is satisfied, and the forms p and q are equivalent. The second claim is consequence of Corollary 2.21. □

Therefore we might speak of the *Dynkin type* of a positive admissible graph G, which will be denoted by $\mathbf{Dyn}(G)$.

Proposition 2.33. *Let q be a unit form with positive admissible frame. Then q is positive if and only if B_q satisfies the cycle condition.*

Proof. Let p be a positive unit form with $\mathbf{Frame}(p) = \mathbf{Frame}(q)$. If q is positive then B_q satisfies the cycle condition by Proposition 2.30. On the other hand, if B_q satisfies the cycle condition then p and q satisfy condition (b) in Proposition 2.31, hence there is a point inversion V with $q = pV$. In particular q is a positive unit form. □

2.6 Assemblers

If Γ is a graph without loops, we will say that two edges a, b of Γ are *incident* if they have a single vertex in common. Define the *incidence graph* of Γ to be the graph G having as vertices the edges of Γ ($G_0 = \Gamma_1$), where two vertices in G are joined by a (single solid) edge if and only if they are incident edges in Γ. We denote the graph G by $\mathbf{Inc}(\Gamma)$ (compare with the definition of an *incidence (signed) graph* associated to a directed graph, as recently given in [35]). For the sake of clarity, graphs Γ with $\mathbf{Frame}(B) = \mathbf{Inc}(\Gamma)$ will be referred to as *assemblers* for the bigraph B, as well as for the unit form q in case $B = B_q$. Consider for instance the following assemblers Γ next to their corresponding incidence graphs $\mathbf{Inc}(\Gamma)$,

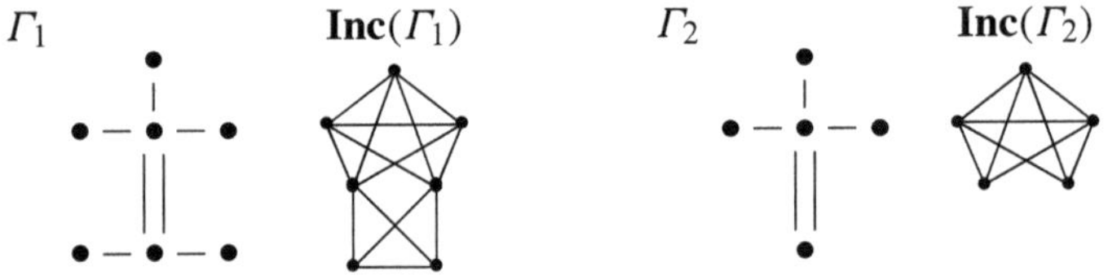

Keeping the notation introduced in Sect. 2.5, next we show that the class of incidence graphs is invariant under graph flations. For two incident edges $a \neq b$ in an assembler Γ, define the *flation for assemblers* of Γ to be the graph $\Gamma\mathscr{T}_{a,b}$ obtained from Γ by exchanging the edge a by the symmetric difference $a\Delta b$ (if $a = \{i, j\}$ and $b = \{i, k\}$ we have $a\Delta b = \{j, k\}$).

Proposition 2.34. *Let Γ be a graph without loop, and take a and b to be incident edges in Γ. Then*

$$\mathbf{Inc}(\Gamma)T_{ab} = \mathbf{Inc}(\Gamma\mathscr{T}_{ab}),$$

where T_{ab} is the graph flation as defined in Sect. 2.5.

Proof. Take $G = \mathbf{Inc}(\Gamma)$, $G' = \mathbf{Inc}(\Gamma\mathscr{T}_{ab})$ and $G'' = GT_{ab}$. It is clear that for vertices $c \neq a$ and $d \neq a$ in $G_0 = G'_0 = G''_0$ we get

$$[c, d]_G = [c, d]_{G'} = [c, d]_{G''}.$$

By definition we have $[a, c]_{G''} = |[a, c]_G - [a, b]_G[b, c]_G|$ for $c \neq b$, and $[a, b]_{G''} = [a, b]_G$. Let us assume that the edge c is incident to a in Γ (that is, $[a, c]_G = 1$). We distinguish two cases: In the first one we have that c is not incident to b in Γ (thus $[b, c]_G = 0$, and therefore $[a, c]_{G''} = 1$), where two subcases appear as shown below,

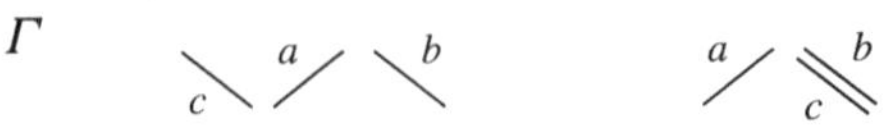

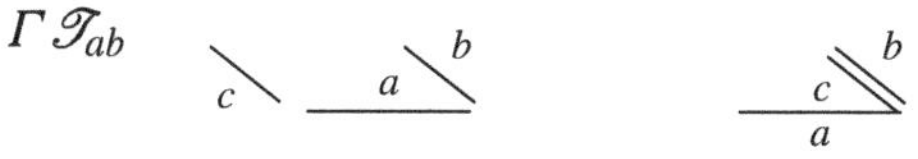

Notice that in these subcases the edges a and c remain incident in the graph $\Gamma\mathcal{T}_{ab}$ (that is, $[a, c]_{G'} = 1$). For the second case we assume that c is incident to b in Γ (that is, $[b, c]_G = 1$, and now $[a, c]_{G''} = 0$), where the possibilities are shown below,

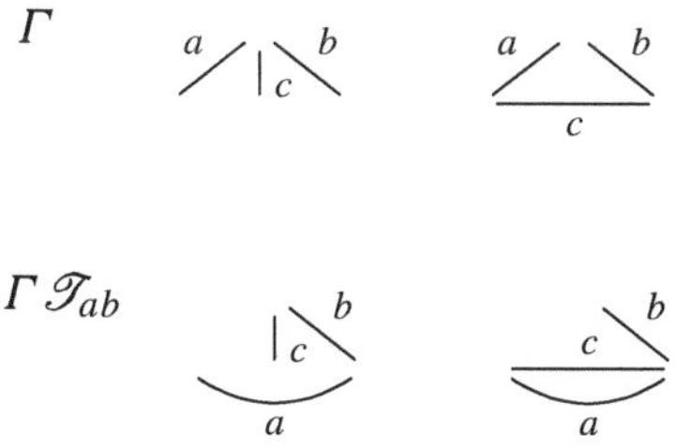

Observe now that edges a and c are not incident to each other in the graph $\Gamma\mathcal{T}_{ab}$, that is, we have $[a, c]_{G'} = 0$. In both cases we obtain $[a, c]_{G'} = [a, c]_{G''}$ whenever $[a, c]_G = 1$.

The case where edge c is not incident to edge a (that is, $[a, c]_G = 0$) can be treated in a similar way, and is left as exercise. Since we want to show that $[c, d]_{G'} = [c, d]_{G''}$ for all vertices c and d in G, by the above the proof is completed by noticing that $[a, b]_{G''} = [a, b]_G = 1 = [a, b]_{G'}$. □

We illustrate the previous proposition with some examples, showing on top of each graph G_i a corresponding assembler Γ_i with $\mathbf{Inc}(\Gamma_i) = G_i$, and below it a bigraph B_i with frame $\mathbf{Frame}(B_i) = G_i$. Notice that assemblers are not unique (see Exercise 2 below).

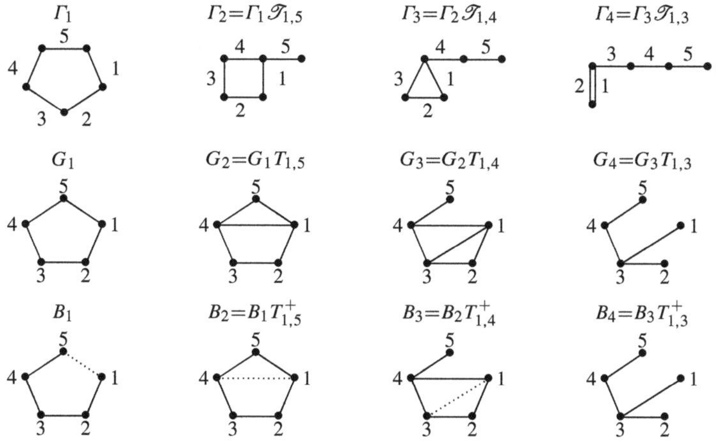

Assume now that Γ is a bigraph, and take $\overline{\Gamma} = \mathbf{Frame}(\Gamma)$. Define the *incidence graph of a bigraph* Γ, also denoted by $\mathbf{Inc}(\Gamma)$, to be the union of $\mathbf{Inc}(\overline{\Gamma})$ with a vertex ω joined exclusively to those vertices of $\mathbf{Inc}(\overline{\Gamma})$ corresponding to dotted edges of Γ. Flations for assemblers $\mathcal{T}_{ab}$ act in the same way on the frame of Γ, leave all edges different from a with the same type (solid or dotted) and change the type of a if and only if b is a dotted edge. We will also say that a bigraph Γ is an *assembler* of bigraph B (resp. of a unit form q) if $\mathbf{Inc}(\Gamma) = \mathbf{Frame}(B)$ (resp. $\mathbf{Inc}(\Gamma) = \mathbf{Frame}(q)$).

Corollary 2.35. *Let Γ be a bigraph and $a \neq b$ be incident edges in Γ. Then*

$$\mathbf{Inc}(\Gamma)T_{ab} = \mathbf{Inc}(\Gamma \mathcal{T}_{ab}).$$

Moreover, if Γ satisfies the cycle condition, so does $\Gamma \mathcal{T}_{ab}$.

Proof. Take $G = \mathbf{Inc}(\Gamma)$, $G' = \mathbf{Inc}(\Gamma \mathcal{T}_{ab})$ and $G'' = GT_{ab}$. To show that $[c,d]_{G'} = [c,d]_{G''}$ for all vertices c and d in $G_0 = G'_0 = G''_0$, by Proposition 2.34 we only need to assume that $c = a$ and $d = \omega$. Observe that $[a,\omega]_{G''} \neq [a,\omega]_G$ if and only if $[b,\omega]_G = 1$, that is, if and only if b is a dotted edge in Γ. On the other hand, by definition $[a,\omega]_{G'} \neq [a,\omega]_G$ if and only if b is a dotted edge in Γ, which shows the first claim.

For the second claim, assume that C is a cycle in Γ with $t \geq 2$ vertices, and that $\mathcal{T}_{ab}$ is a flation for assemblers that modifies C. Hence a must be an edge of C.

First, if b is also an edge of C, then $t > 2$ (for a and b are incident edges in Γ) and C is transformed by $\mathcal{T}_{ab}$ to a smaller cycle C' with the edge b (not belonging to C') attached next to a. Observe that b is dotted if and only if edge a changes its type, therefore $\Gamma \mathcal{T}_{ab}$ also satisfies the cycle condition.

Now, if b does not belong to the cycle C, then this cycle is transformed after $\mathcal{T}_{ab}$ to a larger cycle C' now including b. Again, edge b is dotted if and only if the type of a is changed, thus concluding that $\Gamma \mathcal{T}_{ij}$ satisfies the cycle condition. □

As for inflations and deflations (Corollary 2.18), and for graph flations (Remark 2.28), it is clear that flations for assemblers are involutions. In particular, any iterated flation for assemblers $\mathcal{T} = \mathcal{T}_{a_1 b_1} \cdots \mathcal{T}_{a_r b_r}$ is reversible, with inverse given by $\mathcal{T}^{-1} = \mathcal{T}_{a_r b_r} \cdots \mathcal{T}_{a_1 b_1}$.

Recall that a tree Γ is a connected graph satisfying $|\Gamma_0| = |\Gamma_1| + 1$. A connected graph Γ with $|\Gamma_0| = |\Gamma_1|$ is usually called a *pseudotree* (or 1-*tree*). Particular cases of the following result, corresponding to Dynkin types $\mathbb{A}$ and $\mathbb{D}$, were presented with different formulations respectively in [5] and [6]. The details for the proof of the following combinatorial observation, needed for the main result in this section, are left as exercise.

Remark 2.36. Every Dynkin graph Δ has an assembler Γ^{Δ} with the following shapes:

(i) $\Gamma^{\mathbb{A}_n} =$  for $n \geq 1$;

(ii) $\Gamma^{\mathbb{D}_m} =$  for $m \geq 4$;

(iii) $\Gamma^{\mathbb{E}_p} =$  for $6 \leq p \leq 8$.

Moreover, if Γ is respectively a solid tree with $n \geq 1$ edges, a solid pseudotree with $m \geq 4$ edges, or a pseudotree with the cycle condition and $p \in \{5, 6, 7\}$ edges, then there is a Dynkin graph Δ (respectively $\Delta = \mathbb{A}_n$, $\Delta = \mathbb{D}_m$ or $\Delta = \mathbb{E}_{p+1}$) and an iterated flation for assemblers $\mathscr{T}$ such that $\Gamma\mathscr{T} = \Gamma^{\Delta}$.

Sketch of the Proof. A direct observation yields $\Delta = \mathbf{Frame}(\Delta) = \mathbf{Inc}(\Gamma^{\Delta})$.

For the second claim, recall that a *pendant edge* in Γ is an edge containing an edge of degree one (that is, a vertex belonging to only one edge in Γ, cf. [18]). Assume that Γ is a solid tree with $n \geq 1$ edges, and observe that Γ is a linear graph (that is, $\Gamma = \mathbb{A}_{n+1}$) if and only if Γ has at most two pendant edges. It can be easily shown that if Γ has more than two pendant edges, then there is an iterated flation for assemblers $\mathscr{T}$ such that $\Gamma\mathscr{T}$ has fewer pendant edges than Γ, which shows the claim for the case $\mathbb{A}$.

Similarly, if Γ is a pseudotree with $m \geq 4$ edges then there is an iterated flation for assemblers $\mathscr{T}$ such that $\Gamma' = \Gamma\mathscr{T}$ has exactly one pendant edge (hence Γ' consists of a linear graph with a cycle attached at one end-point). Now, as illustrated in the figure before Corollary 2.35, such a graph Γ' may be taken by means of iterated flations for assemblers to a graph with shape $\Gamma^{\mathbb{D}_m}$. Finally, for the case $\mathbb{E}$ use the above and the fact that flations for assemblers preserve the cycle condition (Corollary 2.35). □

Theorem 2.37. *For a connected unit form q the following are equivalent:*

a) The form q is positive with Dynkin type Δ.
b) The associated bigraph B_q satisfies the cycle condition and there is an assembler Γ for q such that

i) Γ is a solid tree, for the case of Dynkin type $\mathbb{A}$.
ii) Γ is a solid pseudotree and $|\Gamma_1| \geq 4$, for the case of Dynkin type $\mathbb{D}$.
iii) Γ is a pseudotree with the cycle condition and $|\Gamma_1| + 1 \in \{6, 7, 8\}$, for the case of Dynkin type $\mathbb{E}$.

Proof. Assume first that q is a positive unit form. By Proposition 2.33 the bigraph B_q associated to form q satisfies the cycle condition. By Theorem 2.20 there is an iterated inflation $T^+ = T^+_{a_1 b_1} \cdots T^+_{a_r b_r}$ for q such that $qT^+ = q_{\Delta}$ for a Dynkin graph Δ (the Dynkin type of the connected form q). Observe that Dynkin graphs of type $\mathbb{A}_n$, $\mathbb{D}_m$ and $\mathbb{E}_p$ are respectively incidence graphs of a solid tree $\Gamma^{\mathbb{A}_n}$, a solid pseudotree $\Gamma^{\mathbb{D}_m}$ and a pseudotree with the cycle condition $\Gamma^{\mathbb{E}_p}$ (see Remark 2.36). Hence by Proposition 2.27 we have $\Delta = \mathbf{Inc}(\Gamma^{\Delta}) = \mathbf{Frame}(q_{\Delta}) = \mathbf{Frame}(qT^{+}) = \mathbf{Frame}(q)T$, where T is the iterated graph flation given by $T = T_{a_1b_1} \cdots T_{a_rb_r}$. Applying the reversed iterated graph flation T^{-1} we get by Proposition 2.34,

$$\mathbf{Frame}(q) = \mathbf{Inc}(\Gamma^{\Delta})T^{-1} = \mathbf{Inc}(\Gamma^{\Delta}\mathscr{T}^{-1}),$$

where $\mathscr{T}^{-1}$ is the corresponding iterated flation for graphs $\mathscr{T}^{-1} = \mathscr{T}_{a_rb_r} \cdots \mathscr{T}_{a_1b_1}$. Then (*b*) follows since (iterated) flations for assemblers preserve solid trees, solid pseudotrees and (by Corollary 2.35) pseudotrees with the cycle condition.

Assume now that (*b*) holds. By Remark 2.36 there is a Dynkin graph Δ and an iterated flation for assemblers $\mathscr{T}$ such that $\Gamma\mathscr{T} = \Gamma^{\Delta}$. Again by Proposition 2.34 and Corollary 2.35 we have

$$\Delta = \mathbf{Inc}(\Gamma^{\Delta}) = \mathbf{Inc}(\Gamma\mathscr{T}) = \mathbf{Inc}(\Gamma)T = \mathbf{Frame}(q)T,$$

where T is the iterated graph flation corresponding to $\mathscr{T}$, that is, $\Delta = \mathbf{Frame}(q_{\Delta})$ and $\mathbf{Frame}(q)$ are flation equivalent graphs. Since $\mathbf{Frame}(q_{\Delta})$ is positive admissible and B_q satisfies the cycle condition by hypothesis, we conclude by Proposition 2.33 that the unit form q is positive. By the above we also have $\mathbf{Dyn}(\mathbf{Frame}(q)) = \mathbf{Dyn}(\mathbf{Frame}(q_{\Delta})) = \Delta$, which implies that q has Dynkin type Δ. This completes the proof. □

We end this section with a description, up to point inversion, of the roots of connected positive unit forms of Dynkin type $\mathbb{A}_n$ and $\mathbb{D}_m$. In what follows we say that a (reduced) walk w is *minimal* if, for any cycle C in Γ, every edge in C appears at most once along the walk w. Consider a graph Γ, a walk w in Γ and the incidence graph $G = \mathbf{Inc}(\Gamma)$ of Γ. Define the *incidence vector* v^w of the walk w to be the vector in $\mathbb{Z}^{G_0} = \mathbb{Z}^{\Gamma_1}$ given as follows: for $a \in G_0 = \Gamma_1$ the entry v^w_a is the number of times that the edge a appears along the walk w.

Theorem 2.38. *Let Γ be a solid tree or a solid pseudotree which is assembler of a positive unit form q (of Dynkin type $\mathbb{A}$ or $\mathbb{D}$). Then for any open minimal walk w of Γ there exists a point inversion V such that $V(v^w)$ is a root of q. Moreover, every root of q can be found in this way.*

Proof. Take a minimal walk $w = (x|a_1 \cdots a_d|y)$ from vertex x to vertex y in Γ. We proceed by induction on d. Case $d = 1$ is obvious, since v^w is a canonical vector and q is unitary. Suppose that $d > 1$ and consider any flation for assemblers of the shape $\mathscr{T}_{a_i,a_{i+1}}$ for some $1 \leq i < d$. The walk w determines a smaller minimal open walk w' in $\Gamma' = \Gamma\mathscr{T}_{a_i,a_{i+1}}$ (also starting and ending at vertices x and y) by taking out the corresponding edge a_{i+1}. Since w' has smaller length than w, by induction there is a point inversion V' such that $v' := V'(v^{w'})$ is a root of the positive unit form $q' = qT^{\epsilon}_{a_i,a_{i+1}}$, for an appropriate sign ϵ. Notice that

$$T^{\epsilon}_{a_i,a_{i+1}}(v') = v' - \epsilon v'_{a_i} e_{a_{i+1}},$$

which is a root of q. By minimality and since Γ is a (pseudo)tree, a given edge will appear at most two times in the sequence of edges of the walk w, thus the walk w' contains the edge a_{i+1} at most once. Notice also that if v'_{a_i} is nonzero, then it has the same value of $\epsilon v'_{a_{i+1}}$. Hence, up to a change of sign in coordinate a_{i+1}, we deduce that the vector $T^{\epsilon}_{a_i,a_{i+1}}(v')$ is the incidence vector v^w of walk w, which shows the first claim.

For the second claim we count walks. Let us first assume that Γ is a solid tree with n edges (thus q has Dynkin type $\mathbb{A}_n$). Enumerating vertices in Γ observe that for any $x < y$ in Γ_0 there is a unique walk w starting at x and ending in y, thus producing $n(n+1)$ different roots. Indeed, a tree Γ with n edges has $n+1$ vertices, and for any walk w in Γ both v^w and $-v^w$ are roots of q up to a point inversion. This is the total number of q-roots of Dynkin type $\mathbb{A}_n$ (see Table 2.1).

Assume now that Γ is a pseudotree with $m \geq 4$ edges and notice that for any vertices $x < y$ in Γ_0 there are exactly two minimal walks starting at x and ending in y (either surrounding or not the unique cycle in Γ). Since pseudotrees with m edges have m vertices, as above we produce $2m(m-1)$ different q-roots, which is the total number of roots for positive unit forms of Dynkin type m (cf. Table 2.1). □

In what follows we will depict the shape of those bigraphs corresponding to positive unit forms of Dynkin type $\mathbb{A}$ or $\mathbb{D}$ having a sincere positive root (indicating the entries of such vectors in the vertices of the bigraph). Our pictures may contain some *snake edges* (a ~~~ a), each of them represents a *line* in B, that is, a full subgraph of type $\mathbb{A}_m$ with solid edges (and integer value a at each vertex).

Corollary 2.39. *Any sincere positive root of a connected positive unit form q of Dynkin type $\mathbb{A}_n$ is given by a line with values* 1 *at each entry, represented by the picture:*

(A)

Moreover, if v is a sincere positive root of a connected positive unit form q of Dynkin type $\mathbb{D}_n$, then v is shown in one of the illustrations in Table 2.3.

Proof. Case $\mathbb{A}$ is direct consequence of Theorem 2.38, since this is the shape of incidence vectors v^w for open walks in a tree Γ which is the assembler for q.

Assume now that q has Dynkin type $\mathbb{D}_m$ and that Γ is a pseudotree. Consider again Theorem 2.38 and observe that a walk in Γ which is not a line has the following form,

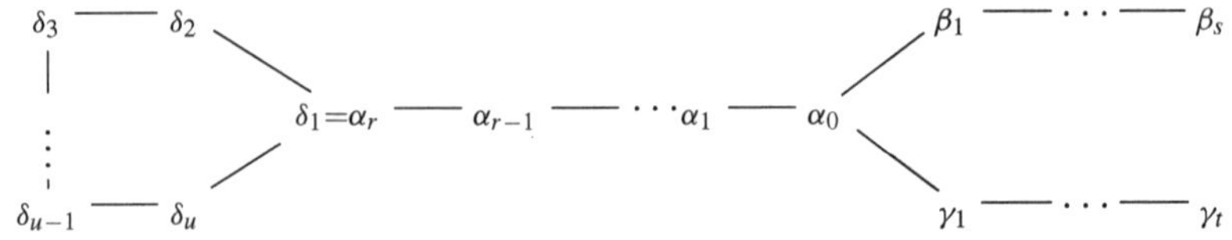

for integers $r, s \geq 0$, $t \geq 1$ and $u \geq 2$. When $u > 2$, cases $(D1–D4)$ in Table 2.3 correspond respectively to the cases $r = 0 = s$, $r > 0 = s$, $r > 0 < s$ and $r = 0 < s$. When $u = 2$, we have cases $(D1'–D4')$. □

Exercises 2.40.

1. Find an assembler for the positive unit forms given in Exercise 2.26.3.
2. Find a (solid) graph G with at least two assemblers (that is, different (bi)graphs Γ and Γ satisfying $\mathbf{Inc}(\Gamma) = G = \mathbf{Inc}(\Gamma'))$.
3. Let G be a complete solid graph with n vertices (that is, G has no loop and $[i, j]_G = 1$ for any pair of different vertices i and j). Find an iterated graph flation T such that $GT = \mathbb{A}_n$.
4. Let Γ be a pseudotree bigraph with the cycle condition such that $|\Gamma_1| + 1 \notin \{6, 7, 8\}$. Show that $\mathbf{Inc}(\Gamma)$ is a positive admissible graph, and determine its Dynkin type.
5. Give an example of a (solid) graph Γ such that no bigraph B with $\mathbf{Frame}(B) = \mathbf{Inc}(\Gamma)$ satisfies the cycle condition.

Table 2.3 Sincere positive roots of positive unit forms of Dynkin type $\mathbb{D}$

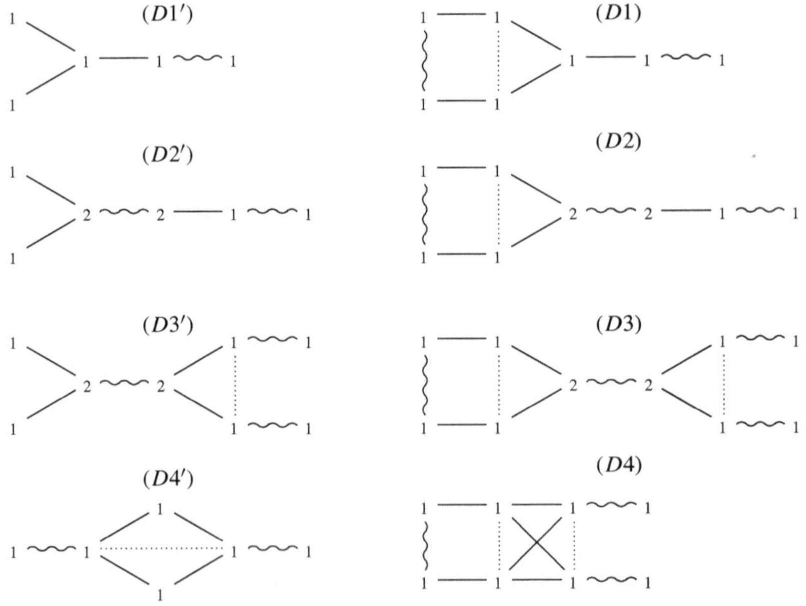

6. Show that a graph which is a solid cycle is positive admissible of Dynkin type $\mathbb{A}_3$ if it has three points, and of Dynkin type $\mathbb{D}_n$ if it has $n \geq 4$ points.

Recall that if $q : \mathbb{Z}^n \to \mathbb{Z}$ and $i \in \{1, \dots, n\}$, then the *restriction* of q to the set $\{1, \dots, i-1, i+1, \dots, n\}$ is denoted by $q^{(i)}$. Similarly we denote by $G^{(i)}$ the restriction to $\{1, \dots, i-1, i+1, \dots, n\}$ of a graph G with vertices $\{1, \dots, n\}$. Let x be a vertex in a connected graph G with more than one vertex. The *connecting valence* $\mathbf{v}_G(x)$ of x in G is the number of connected components of $G^{(x)}$. We say that G is a *block* if all its vertices have connecting valence one.

7. Let G be a connected graph. Show that G is a positive admissible block of Dynkin type $\mathbb{A}_n$ if and only if G is a complete graph.
8. Let G be a block of Dynkin type $\mathbb{D}_n$. Show that if $\mathbf{Dyn}(G^{(x)}) = \mathbb{D}_{n-1}$ for a vertex x, then $G^{(x)}$ is a block.
9. Describe all blocks of Dynkin type $\mathbb{D}_n$.
10. Describe all nonblocks of Dynkin type $\mathbb{E}_6$.
11. Find the Dynkin type and an assembler for the following positive unit forms.

    a) $q(x) = x_1^2 + \ldots + x_6^2 - x_1(x_2+x_3) + x_2(x_3+x_4-x_5) + x_3(x_4-x_6) - x_4x_5$.
    b) $p(x) = x_1^2 + \ldots + x_6^2 - x_1(x_2+x_3+x_4) + x_2(x_3+x_4) + x_3(x_4-x_5) - x_4x_6 - x_5x_6$.

12. Using inflations, show that the form q_B is positive of Dynkin type $\mathbb{E}_8$, where B is the following bigraph. How many positive roots does q_B have?

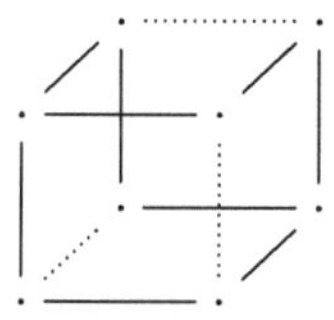

13. Show that if C is a (solid) cycle with $t > 1$ vertices and v is a sincere vector in $\mathbb{Z}^t$, then $q_C(v) > 1$.
14. We say that the roots of a positive unit form q of Dynkin type Δ attain their maximum if there exists a root v and a vertex i such that $|v_i| = r_\Delta$, see Proposition 2.22. Show that if $\mathbf{Dyn}(q) = \mathbb{D}_n$, then the roots of q attain their maximum if and only if $\mathbf{Frame}(q)$ is not a block.
15. Show that the following graphs contain a subgraph of Dynkin type $\mathbb{E}_6$:

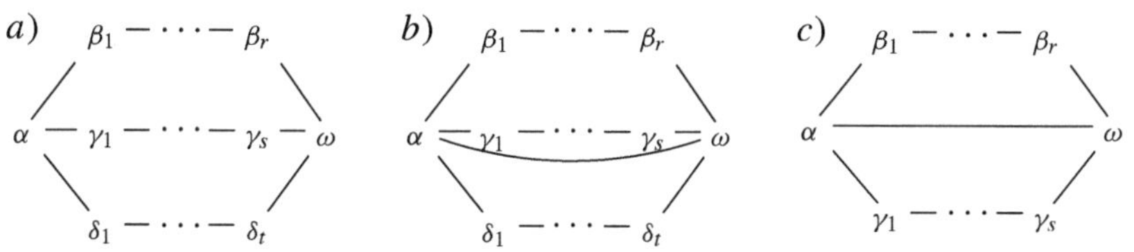

for $r, s, t \geq 1$ for $r, s \geq 2$.

16. For $i, j \geq 2$ consider the graph $H_{i,j}$ given by

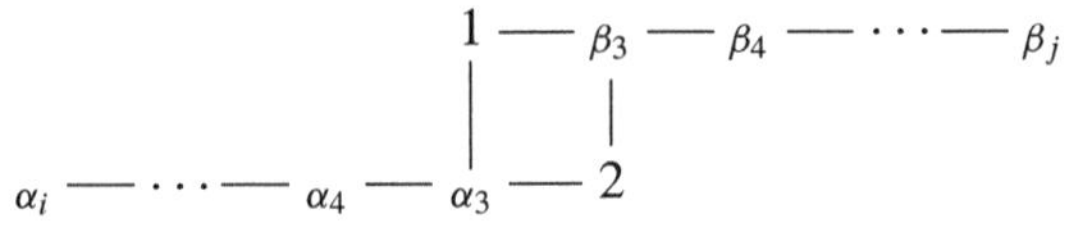

Show that if $H_{i,j}$ has $n \geq 4$ elements, then $\mathbf{Dyn}(H_{i,j}) = \mathbb{D}_n$.

17. For a positive admissible graph G of Dynkin type $\mathbb{D}_n$, what is the number of vertices x of G such that $\mathbf{Dyn}(G^{(x)}) = \mathbb{A}_{n-1}$?
18. Show that the following graphs are either of Dynkin type $\mathbb{E}$ or nonpositive admissible:

a)

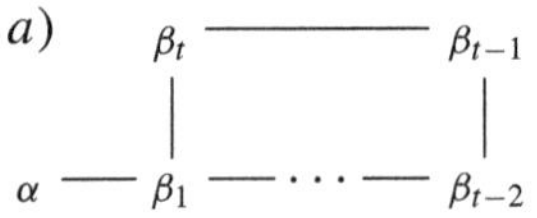

for $t \geq 5$

b)

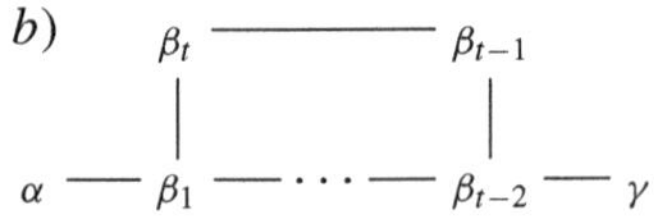

for $t \geq 4$.

19. Take a graph G of Dynkin type $\mathbb{D}$ and two proper subsets of vertices I, J with $I \cap J = \emptyset$. Assume that both restrictions G^I and G^J are connected. Show that if $\mathbf{Dyn}(G^I) = \mathbb{D}$, then $\mathbf{Dyn}(G^J) = \mathbb{A}$.
20. Assume that G is a connected graph of Dynkin type $\mathbb{D}_n$ with more than 5 vertices. Show that if x and y are vertices such that $\mathbf{Dyn}(G^{(x)}) = \mathbf{Dyn}(G^{(y)}) = \mathbb{D}_{n-1}$ and $G^{(x)(y)}$ is connected, then $\mathbf{Dyn}(G^{(x)(y)}) = \mathbb{D}_{n-2}$.

Chapter 3
Nonnegative Quadratic Forms

In this chapter we deal with *semi-unit forms* that are *nonnegative*, that is, integral quadratic forms $q(x_1, \dots, x_n) = \sum_{1 \leq i \leq j \leq n} q_{ij} x_i x_j$ with diagonal coefficients q_{ii} in the set $\{0, 1\}$ for $i = 1, \dots, n$ such that $q(x) \geq 0$ for any vector $x = (x_1, \dots, x_n)$ in $\mathbb{Z}^n$. As before it will be convenient to set $q_{ji} = q_{ij}$ for $i \neq j$. We begin by describing nonnegative forms related to (solid) graphs (those forms q satisfying $q_{ij} \leq 0$ for $i \neq j$).

3.1 Extended Dynkin Graphs

Recall that *extended Dynkin diagrams* (also known as *Euclidean graphs*) are obtained from Dynkin graphs Δ by adding a vertex ω and edges joining ω with certain *exceptional vertices* in Δ (cf. Tables 2.1 and 2.2 in Chap. 2 and Lemma 5.9 in Chap. 5). It was shown in Proposition 2.2 that the quadratic form q_G associated to a (solid) connected graph G is positive if and only if G is a Dynkin diagram. This result is generalized below to the nonnegative setting, by means of extended Dynkin diagrams (cf. [46]). Recall that for a set of indices G_0 the *support* of a vector $x \in \mathbb{Z}^{G_0}$ is given by $\mathbf{supp}(x) = \{i \in G_0 \mid x_i \neq 0\}$, and that the vector x is *positive* if $x \neq 0$ and $x_i \geq 0$ for all $i \in G_0$.

Proposition 3.1. *Let G be a connected (solid) graph. Then the associated quadratic form q_G is semi-unitary nonpositive and nonnegative if and only if G is a loop or an extended Dynkin diagram $\widetilde{\mathbb{A}}_n$, $\widetilde{\mathbb{D}}_m$ or $\widetilde{\mathbb{E}}_p$ for $n \geq 1$, $m \geq 4$ or $p = 6, 7, 8$ (see Table 2.2).*

Proof. Consider an extended Dynkin diagram $\widetilde{\Delta}$ and its associated quadratic form $q = q_{\widetilde{\Delta}}$. Observe that any proper subgraph of $\widetilde{\Delta}$ is union of Dynkin diagrams. Hence by Lemma 2.1 any proper restriction of q is positive. Now, it can be directly verified that the vector $p_{\widetilde{\Delta}}$ displayed as vertices in Table 2.2 is an isotropic vector

M. Barot et al., *Quadratic Forms*, Algebra and Applications 25,
https://doi.org/10.1007/978-3-030-05627-8_3

for q (see Exercise 2.10.4). Therefore q is a critical nonpositive unit form, and by Theorem 2.12 the form q is nonnegative (again since q has an isotropic vector and Kronecker forms q_m with $|m| > 2$ are anisotropic, see Proposition 1.20). If $\widetilde{\Delta}$ is a single loop, then $q_{\widetilde{\Delta}}$ is clearly a nonnegative semi-unit form (the zero form in one variable ξ).

Let now G be a connected graph with n vertices such that $q := q_G$ is nonpositive and nonnegative and take the canonical vectors $\{e_i\}_{i \in G_0}$ of $\mathbb{Z}^{G_0}$.

First notice that if $I \subset G_0$ is a subset of vertices such that q^I has a positive radical vector w, then $I = G_0$. Indeed, if $i \in G_0 - I$ we may complete w by zeros to a vector v in $\mathbb{Z}^{G_0}$, which is a (positive) radical vector of q by Lemma 2.11. Since $v_i = 0$, as shown in Lemma 1.1 we have

$$q(v|e_i) = 2q_{ii}v_i + \sum_{j \neq i} q_{ij}v_j = \sum_{j \neq i} q_{ij}v_j < 0,$$

where the last inequality is due to the connectedness of G, since $q_{ij} \leq 0$ for all $i \neq j$ and v is a positive vector. This is impossible since v is a radical vector of q.

Assume that G has a loop, say in vertex i (that is, $q_{ii} \leq 0$). By nonnegativity we have $0 \leq q(e_i) = q_{ii}$, that is, the vertex i has exactly one loop on it. Then e_i is a positive radical vector for the one-variable restriction $q^{\{i\}}$, and by the above we have $n = 1$ and G is a single loop (that is, q is the zero form ξ in one variable). Assume now that G has no loop, but has multiple edges (say $q_{ij} < -1$ for vertices $i \neq j$). Then $0 \leq q(e_i + e_j) = 2 + q_{ij}$, that is $q_{ij} = -2$, and in particular $e_i + e_j$ is a positive radical vector of the restriction $q^{\{i,j\}}$. Therefore $n = 2$ and q is the Kronecker form $q_2(x_1, x_2) = (x_1 - x_2)^2$.

Hence we may assume that G is a simple graph (with no loops nor multiple edges). By Proposition 2.2 the graph G is not a Dynkin diagram, for q is nonpositive. Recall that for any connected simple graph G that is not a Dynkin graph, there is a subset $E_0 \subset G_0$ such that the full subgraph E of G determined by E_0 is an extended Dynkin diagram (cf. Table 2.2 and Exercise 2.10.5). For any such diagram E, the restriction q^{E_0} of q has a positive radical vector p_E, the one exhibited in Table 2.2. Using again the above argument we have $E_0 = G_0$, that is, G is an extended Dynkin diagram which completes the proof. □

We give a useful result for nonnegative semi-unit forms which is analogous to Lemma 2.14, compare also with Lemma 2.11. We say that an integral quadratic form $q : \mathbb{Z}^n \to \mathbb{Z}$ is *pre-unitary* or a *pre-unit* form if $q(e_i) \leq 1$ for $i = 1, \ldots, n$, where $e_1, \ldots, e_n$ is the canonical basis for $\mathbb{Z}^n$. Recall that a vector z in $\mathbb{Z}^n$ is an *isotropic vector for* q if $q(z) = 0$.

Lemma 3.2. *Let* $q : \mathbb{Z}^n \to \mathbb{Z}$ *be a nonnegative pre-unit form. Then* q *is semi-unitary and the following hold.*

a) Any isotropic vector for q *is a radical vector.*
b) We have $|q_{ij}| \leq 2$ *for all indices* $i, j \in \{1, \ldots, n\}$.
c) If $q(e_i) = 0$ *for some* $i \in \{1, \ldots, n\}$, *then* $q_{ij} = 0$ *for all* $j \in \{1, \ldots, n\}$.

Proof. That q is semi-unitary is evident. For (a) consider an isotropic vector $x \in \mathbb{Z}^n$, an arbitrary integer m and an index $i \in \{1, \dots, n\}$. Then we have

$$0 \leq q(mx + e_i) = m^2 q(x) + q(e_i) + mq(x|e_i) \leq 1 + mq(x|e_i).$$

Since m is arbitrary the equality $q(x|e_i) = 0$ must hold, and since this is true for any index i, the vector x is radical for q.

Take now indices $i \neq j$ and observe that by nonnegativity, since $q_{ij} = q(e_i|e_j)$, we have

$$0 \leq q(e_i \pm e_j) = q(e_i) + q(e_j) \pm q_{ij} \leq 2 \pm q_{ij},$$

which shows (b). Assume finally that $q(e_i) = 0$ for some $i \in \{1, \dots, n\}$. By (a) the canonical vector e_i is radical for q, that is, for any $j \neq i$ we have

$$0 = q(e_i|e_j) = q_{ij},$$

thus (c) holds. □

Let q be an integral quadratic form. Recall form Sect. 2.4 that a *flation for* q is a linear transformation $T_{ij}^{\epsilon} : \mathbb{Z}^n \to \mathbb{Z}^n$ given by

$$T_{ij}^{\epsilon} : v \mapsto v - \epsilon v_i e_j,$$

where $\epsilon \in \{+, -\}$ is a sign such that $\epsilon q_{ij} = |q_{ij}|$. When $q_{ij} > 0$ we say that T_{ij}^{+} is an *inflation* for q, and when $q_{ij} < 0$ the transformation T_{ij}^{-} is called a *deflation*. A finite composition of flations is called an *iterated flation*.

In contrast to the positive case, nonnegative unit forms are not preserved under flations, as the following example shows. Let ξ denote the zero quadratic form in one variable and take $q_{\widetilde{\mathbb{A}}_2}$ to be the form associated to the extended Dynkin diagram $\widetilde{\mathbb{A}}_2$. Then $q := q_{\widetilde{\mathbb{A}}_2} T = \xi \oplus q_{\mathbb{A}_2}$ is not unitary, where the iterated flation T is the composition $T_{12}^{-} T_{13}^{-}$.

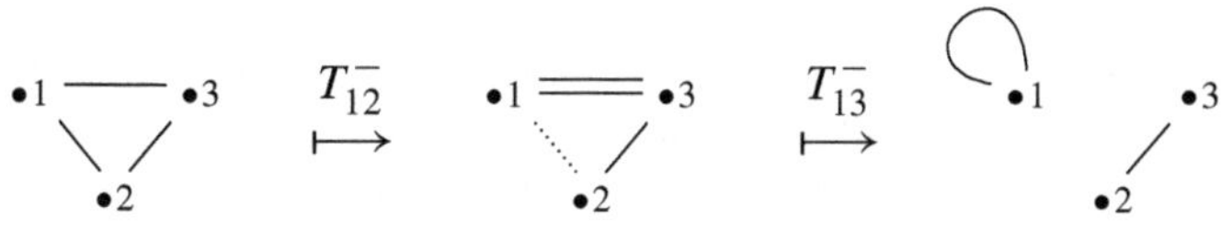

Notice that if q' denotes the quadratic form associated to the bigraph in the middle, then the vector $e_1 + e_3$ generates the radical of q', and $T_{12}^{-}(e_1 + e_3) = e_1 + e_2 + e_3$ is a generator of the radical of $q_{\widetilde{\mathbb{A}}_2}$. Notice also that T_{13}^{-} is not a Gabrielov transformation, and that its inverse T_{13}^{+} is neither an inflation nor a deflation for q (see Proposition 2.17). However, we show next that *semi-unitary forms* are actually preserved under flations.

Lemma 3.3. *For $n \geq 1$, the set of nonnegative semi-unit forms with n variables (denoted $\mathfrak{SU}^{\geq 0}(n)$) is invariant under inflations and deflations.*

Proof. Since flations are equivalences, we only need to show that they preserve semi-unitary forms in the nonnegative case.

Let us assume that T_{ij}^{ϵ} is a flation for a nonnegative semi-unit form $q : \mathbb{Z}^n \to \mathbb{Z}$ with $q_{ij} \neq 0$. By Lemma 3.2 we have $|q_{ij}| \in \{1, 2\}$ and $q(e_i) = 1 = q(e_j)$. Notice that

$$q(e_i - \epsilon e_j) = q(e_i) + q(e_j) - \epsilon q_{ij} = 2 - |q_{ij}| \in \{0, 1\}.$$

Take $q^{\epsilon} = qT_{ij}^{\epsilon}$. Since $q^{\epsilon}(e_i) = q(e_i - \epsilon e_j)$ and $q^{\epsilon}(e_k) = q(e_k)$ for $k \neq i$, by the above we conclude that q^{ϵ} is semi-unitary. □

The fact that flations do not necessarily preserve connectedness of nonnegative semi-unit forms (as exhibited in the example above) was used in [7] to classify those forms.

Lemma 3.4. *If q is a nonzero nonnegative connected semi-unit form, then q is unitary. Moreover, if T_{ij}^{ϵ} is a flation for q such that $q^{\epsilon} = qT_{ij}^{\epsilon}$ is not connected, then $|q_{ij}| = 2$ and there is a nonnegative connected unit form q' such that*

$$q^{\epsilon} = q' \oplus \xi,$$

where ξ is the zero form in one variable.

Proof. Assume first that q is connected but nonunitary, say $q(e_1) = 0$. If $n > 1$ then for any other index $1 < i \leq n$ we have $q_{1i} = 0$ by Lemma 3.2(c), which is impossible since q is connected. Then $n = 1$ and q is the zero form.

For the second claim let us assume that $q^{\epsilon} = qT_{ij}^{\epsilon}$ is not connected. Observe that we must have $|q_{ij}| = 2$ (for if $|q_{ij}| = 1$ then T_{ij}^{ϵ} is a Gabrielov transformation, hence q^{ϵ} is connected by Proposition 2.17). By Lemma 3.2(c) we have $q(e_i) = 1 = q(e_j)$, and

$$q^{\epsilon}(e_i) = q(e_i - \epsilon e_j) = 2 - |q_{ij}| = 0.$$

Again by Lemma 3.2(c) the bigraph B^{ϵ} associated to q^{ϵ} has an isolated loop at vertex i. The result will follow by showing that B^{ϵ} has exactly two connected components, that is, we will show that for any vertex $k \neq i$, if $k \neq j$ then k and j belong to the same connected component of B^{ϵ}. Let B be the bigraph associated to q. Since for $k \neq i$ we have $q_{k,i}^{\epsilon} = 0 = q^{\epsilon}(e_i)$, and considering that q is unitary, then

$$q^{\epsilon}(e_k + e_i) = q^{\epsilon}(e_k) + q^{\epsilon}(e_i) + q_{ki}^{\epsilon} = q(e_k) = 1,$$

and therefore, if moreover $k \neq j$,

$$\begin{aligned} 1 &= q^{\epsilon}(e_k + e_i) = q(e_k + e_i - \epsilon e_j) \\ &= q(e_k) + q(e_i) + q(e_j) - |q_{ij}| + q_{ki} - \epsilon q_{kj} \\ &= 1 + q_{ki} - \epsilon q_{kj}, \end{aligned}$$

that is, $q_{ki} = \epsilon q_{kj}$. Hence, since B is connected, for every $k \neq i, j$ there exists a walk w in B joining k and j and not containing vertex i. The same is true in B^{ϵ} since the bigraph B^{ϵ} differs from B only on edges containing vertex i (for clearly $q^{(i)} = (q^{\epsilon})^{(i)}$). □

Recall that a vector x in $\mathbb{Z}^n$ is called *sincere* if $\mathbf{supp}(x) = \{1, \dots, n\}$. The proof of the following result, analogous to Theorem 2.20 and due originally to Ovsienko [43], is based on an argument given by von Höhne in [52].

Theorem 3.5. *Let $q : \mathbb{Z}^n \to \mathbb{Z}$ be a connected nonnegative unit form with $\mathbf{rad}(q) = \mathbb{Z}v$ for a sincere positive vector v. Then there exists an iterated inflation T and an extended Dynkin graph $\widetilde{\Delta}$ such that $qT = q_{\widetilde{\Delta}}$.*

Proof. Let B^0 be the bigraph associated to $q = q_0$. If B^0 has no dotted edges, by Proposition 3.1 the connected graph B^0 is an extended Dynkin graph and we are done. Assume otherwise that $(q_0)_{ij} > 0$ for some $i \neq j$ and consider the inflation $T_0 = T_{ij}^{+}$ and $q_1 = q_0 T_0$ with associated bigraph B^1. Notice that if $(q_0)_{ij} > 1$, then $q_0(e_i - e_j) = 2 - (q_0)_{ij} \leq 0$. By nonnegativity and Lemma 3.2(a), the vector $e_i - e_j$ is radical for q_0, contradicting the hypothesis on the generator of $\mathbf{rad}(q_0)$. Therefore $(q_0)_{ij} = 1$ and by Lemma 3.4 the unit form q_1 is connected.

Observe that the radical of q_1 is generated by the positive sincere vector $v^1 := (T_{ij}^{+})^{-1} v = v + v_i e_j$. Iterating this process we find a sequence of connected nonnegative forms $q_0, q_1, q_2, \dots$ with associated connected bigraphs $B^0, B^1, B^2, \dots$ and an inflation T_r for q_r such that $q_{r+1} = q_r T_r$. Moreover, the radical of each form q_r is generated by a sincere positive vector v^r. We show that this process is finite, arriving in this way at a quadratic form q_r with B^r having no dotted edge, therefore B^r is an extended Dynkin graph again by Proposition 3.1.

For $r \geq 0$ consider the set

$$C_{q_r} = \{x \in \mathbb{Z}^n \mid q_r(x) = 1 \text{ and there are indices } i, j \text{ with } x_i > 0 \text{ and } x_j < 0\}.$$

We divide the proof into two steps.

Step 1. *The set C_{q_r} is finite for each $r \geq 0$.*

Let us assume that C_{q_r} is an infinite set. Notice that for any of its elements x and an arbitrary index $i \in \{1, \dots, n\}$ we have

$$0 \leq q(x \pm e_i) = q(x) + q(e_i) \pm q(x|e_i) = 2 \pm q(x|e_i),$$

therefore $|q(x|e_i)| \leq 2$. Consequently we may find a sequence $\{a^0, a^1, a^2, \ldots, \}$ of different vectors in C_{q_r} such that for any $i \in \{1, \ldots, n\}$ and any $k \geq 0$ we have $q(a^0|e_i) = q(a^k|e_i)$. By construction, for any $k \neq \ell$ the difference $a^k - a^\ell$ is a radical vector for q_r, therefore a nonzero integral multiple of v^r. In particular, $a^k - a^\ell$ is a sincere vector, hence $a_i^k \neq a_i^\ell$ for any index i. This implies that for any integer $m \geq 1$ there is an integer $M > 0$ such that for any $k \geq M$ none of the entries a_i^k of the vector a^k belongs to the interval $[-m, m]$. Therefore we may find $k < \ell$ such that

$$\min_{i=1,\ldots,n}(a_i^\ell) < \min_{i=1,\ldots,n}(a_i^k) < 0 < \max_{i=1,\ldots,n}(a_i^k) < \max_{i=1,\ldots,n}(a_i^\ell).$$

Then the difference $a^k - a^\ell$ is a radical vector for q_r with a negative entry as well as a positive entry. This is impossible since the radical of q_r is generated by a positive vector v^r. Thus C_{q_r} is a finite set for $r \geq 0$.

Step 2. *For $r \geq 0$ the inflation T_r determines a proper inclusion $C_{q_{r+1}} \to C_{q_r}$.* First assume that for a q_{r+1}-root x the vector $T_r(x)$ does not belong to C_{q_r}, and assume that $T_r = T_{ij}^+$ for indices $i \neq j$. Multiplying by (-1) if necessary, we may assume that $T_r(x) = x - x_i e_j$ is a positive vector, that is, that $x_k \geq 0$ for $k \neq j$ and $x_j - x_i \geq 0$. Since $x_i \geq 0$ we must have $x_j \geq 0$, that is, the vector x itself is positive. This shows that $T_r(x) \in C_{q_r}$ for any vector x in $C_{q_{r+1}}$. Thus $T_r : C_{q_{r+1}} \to C_{q_r}$ is an inclusion (for T_r is $\mathbb{Z}$-invertible) which is proper since $T_r(e_i) = e_i - e_j \in C_{q_r} - T_r(C_{q_{r+1}})$.

Using Steps 1 and 2 we get a sequence of proper inclusions between finite sets

$$C_{q_r} \xrightarrow{T_{r-1}} C_{q_{r-1}} \xrightarrow{T_{r-2}} \cdots \longrightarrow C_{q_2} \xrightarrow{T_1} C_{q_1} \xrightarrow{T_0} C_{q_0},$$

hence the iterative process must stop, which completes the proof. □

In the last result of this section we reformulate Vinberg's characterization of extended Dynkin diagrams (presented originally in the context of Cartan matrices, see [51] and [32]) to the setting of integral quadratic forms (adapting the short presentation given in [3]). For a unit form $q : \mathbb{Z}^n \to \mathbb{Z}$ denote by $\mathbf{rad}^+(q)$ the subset of $\mathbf{rad}(q)$ consisting of positive vectors.

Theorem 3.6 (Vinberg). *Let G be a connected (solid) graph without loops. The following are equivalent:*

a) The graph G is an extended Dynkin diagram (see Table 2.2).
b) The associated unit form q_G satisfies $\mathbf{rad}^+(q_G) \neq \emptyset$.

Proof. That (a) implies (b) is clear, since for an extended Dynkin graph $\widetilde{\Delta}$, the vector $p_{\widetilde{\Delta}}$ given in Table 2.2 belongs to $\mathbf{rad}^+(q_{\widetilde{\Delta}})$.

For the converse consider the following classical terminology (cf. [3] or [32]). For a unit form $q : \mathbb{Z}^n \to \mathbb{Z}$, a vector x in $\mathbb{Z}^n$ is said to be *subadditive* if x is positive and $q(x|e_i) \geq 0$ for $i = 1, \ldots, n$. If moreover $q(x|e_i) = 0$ for all i then x

is said to be an *additive* vector for q (observe that x is additive for q if and only if $x \in \mathbf{rad}^+(q)$). Next we divide the proof into several steps.

Step 1. *The Kronecker form $q_m(x_1, x_2) = x_1^2 + x_2^2 - mx_1x_2$ admits no subadditive vector for $m > 2$.* A direct calculation shows that for $x = (x_1, x_2)$ we have

$$q_m(x|e_1) = 2x_1 - mx_2, \quad \text{and} \quad q_m(x|e_2) = 2x_2 - mx_1.$$

This shows that for $m > 2$ and x a positive vector we have either $q_m(x|e_1) < 0$ or $q_m(x|e_2) < 0$.

Step 2. *If G is an extended Dynkin diagram then any subadditive vector for q_G is additive.* Let x be a subadditive vector for q_G and take $y = p_G$ where p_G is the positive vector given in Table 2.2. Since y is a radical vector for q_G, then

$$0 = q_G(x|y) = \sum_{i=1}^{n} y_i q_G(x|e_i).$$

This implies that $q_G(x|e_i) = 0$ since $q_G(x|e_i) \geq 0$ and $y_i > 0$ for $i = 1, \dots, n$, that is, x is an additive vector for q_G.

Step 3. *Let G be a connected (solid) graph without loops. If q_G admits a subadditive vector x, then for any proper restriction q_G^I of q_G, the restriction x' of x to the coordinates of I is a subadditive vector for q_G^I which is not additive.* First notice that x must be a sincere vector (otherwise, by connectedness there are vertices $i \in \mathbf{supp}(x)$ and $j \notin \mathbf{supp}(x)$ with $(q_G)_{ij} < 0$, and therefore $q_G(x|e_j) < 0$). In particular the restriction x' is also a positive vector. Then for $i \in I$, using that x is a positive vector and that $(q_G)_{ij} \leq 0$ for $j \neq i$, we have

$$0 \leq q_G(x|e_i) = 2x_i + \sum_{j \neq i} (q_G)_{ij} x_j \leq 2x_i + \sum_{j \in I,\ j \neq i} (q_G)_{ij} x_j = q_G^I(x'|e_i),$$

which shows that x' is a subadditive vector for q_G^I. To show that x' is not additive observe that, since G is connected and I is a proper subset of vertices, we may find vertices $i \in I$ and $j \notin I$ such that $(q_G)_{ij} < 0$. Since x is sincere, this shows that the second inequality in the expression above is strict, therefore $q_G^I(x'|e_i) > 0$ for such $i \in I$.

We are able now to complete the proof. Take a graph G as in the hypothesis, and assume that $x \in \mathbf{rad}^+(q_G)$, that is, x is an additive vector for q_G. Steps 3 and 1 imply that q_G has no Kronecker restriction of the shape q_m for $m > 2$ (that is, G has at most double edges).

Now, if G has as full subgraph an extended Dynkin graph G', then Step 3 implies that the restriction x' is a subadditive vector for $q_{G'}$, which is not additive. This contradicts Step 2, therefore G admits no extended Dynkin diagram as proper full subgraph. Since G is not a Dynkin diagram (for $\mathbf{rad}(q_G) \neq 0$), then G is an extended Dynkin diagram (see Exercise 2.10.5). □

The following examples show that, in the Theorem above, condition $\mathbf{rad}^+(q_G) \neq \emptyset$ cannot be replaced by $\mathbf{rad}(q_G) \neq \emptyset$.

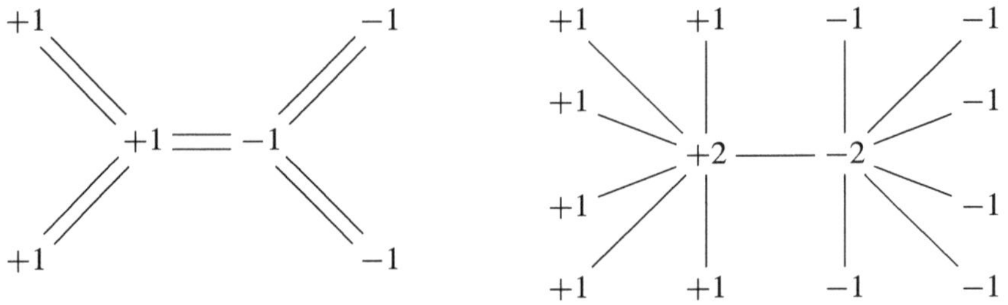

Indeed, the depicted connected graphs G are not extended Dynkin diagrams, but the vector with entries as displayed in the figures is a radical vector of q_G. Subadditive roots of q, also called *locally maximal roots*, will be studied later in Sect. 6.2.

3.2 Dynkin Type and Corank

An integral quadratic form q is said to be *balanced* if $q^{-1}(0) = \mathbf{rad}(q)$, that is, if the linear form $q(x|-)$ vanishes for every $x \in \mathbb{Z}^n$ with $q(x) = 0$.

Let $q : \mathbb{Z}^n \to \mathbb{Z}$ be a nonnegative quadratic semi-unit form. It was shown in Lemma 3.2(a) and (b) that q is a balanced form and that $|q_{ij}| \leq 2$ for all $1 \leq i < j \leq n$. We show next that these conditions characterize all nonnegative forms. This **Nonnegativity Criterion**, given in [7], (see also [8]), will be useful in subsequent chapters. Observe that m-Kronecker forms $q_m(x_1, x_2) = x_1^2 - mx_1x_2 + x_2^2$ are balanced (cf. Proposition 1.20), but for $|m| \geq 3$ they fail to be nonnegative forms.

Theorem 3.7. *A semi-unit form* $q : \mathbb{Z}^n \to \mathbb{Z}$ *is nonnegative if and only if the following conditions hold:*

N1) For $1 \leq i < j \leq n$ *we have* $|q_{ij}| \leq 2$.
N2) The form q *is balanced, that is, every isotropic vector for* q *is a radical vector.*

Proof. The necessity was shown in Lemma 3.2(a) and (b).

Assume now that q satisfies conditions ($N1$) and ($N2$), and let us also assume that n is minimal such that there is a vector $v \in \mathbb{Z}^n$ with $q(v) < 0$. Notice in particular that $n > 2$, since Kronecker forms satisfying condition ($N1$) are nonnegative. For any vertex $i \in \{1, \ldots, n\}$ the restriction $q^{(i)}$ satisfies condition ($N1$), which is condition ($P1$) in Theorem 2.15. If $q^{(i)}$ is anisotropic then it is positive by Theorem 2.15. Hence there exists an index i and an isotropic vector z for $q^{(i)}$, for otherwise q would be critical nonpositive, thus nonnegative by Theorem 2.12. Viewing z as a vector in $\mathbb{Z}^n$ by setting $z_i = 0$, notice that there are nonzero integers a and b such that $av + bz$ is nonsincere. But since q is balanced and $q(bz) = 0$, the vector bz is radical for q, therefore

$$q(av + bz) = a^2 q(v) < 0.$$

This is impossible by minimality of n, for clearly any restriction of q satisfies conditions $(N1)$ and $(N2)$ (see Exercise 1 below). □

The following reduction theorem, given in [7], is the main tool for the classification of nonnegative semi-unit forms in terms of Dynkin diagrams. For $\varepsilon \in \{+, -\}$ and a vector x in $\mathbb{Z}^n$ define the vectors x^ε by taking $x_i^\varepsilon = \max(\varepsilon x_i, 0)$ for $i = 1, \dots, n$, so we have $x = x^+ - x^-$ (recall that x is positive if $x \neq 0$ and $x = x^+$). Consider also the *weight* of a vector x in $\mathbb{Z}^n$ given by $|x| = \sum_{i=1}^n |x_i|$. Recall that the *corank* of a semi-unit form q is the rank of its radical.

Theorem 3.8. *Let q be a connected nonnegative semi-unit form with corank c. Then there exists an iterated flation T such that $qT = p \oplus \xi^c$, where ξ^c is the zero quadratic form in c variables and p is a connected positive unit form.*

Proof. Notice first that by connectedness and Lemma 3.2(c), we may assume that q is a unitary form. We proceed by induction on the corank c of q. If $c = 0$ then q is positive and there is nothing to show. For $c > 0$ the proof is divided into two steps:

Step 1. *There is an iterated inflation T such that qT has a positive radical vector.*
For a nonzero radical vector v assume that there are vertices $i \in \mathbf{supp}(v^+)$ and $j \in \mathbf{supp}(v^-)$ with $q_{ij} > 0$ and $|v_i| \leq |v_j|$ (exchange the roles of i and j otherwise). Define $q' = qT_{ij}^+$ and $v' = (T_{ij}^+)^{-1}v = v + v_i e_j$, and observe that since v_i and v_j have opposite sign we have $|v_i + v_j| < |v_j|$. Since $|v'| < |v|$ this process must stop, getting an iterated inflation T, a quadratic semi-unit form $\widehat{q} = qT$ and a vector $\widehat{v} = T^{-1}v$ satisfying

$$0 = \widehat{q}(\widehat{v}) = \widehat{q}(\widehat{v}^+ - \widehat{v}^-) = \widehat{q}(\widehat{v}^+) + \widehat{q}(\widehat{v}^-) + \sum_{(i,j)} \widehat{q}_{ij}\widehat{v}_i\widehat{v}_j,$$

where the sum runs over the set $\mathbf{supp}(\widehat{v}^+) \times \mathbf{supp}(\widehat{v}^-)$. Since every summand on the right side of the equation is nonnegative, all of them are equal to zero (for $(\widehat{q})_{ij} \leq 0$ if $(i, j) \in \mathbf{supp}(\widehat{v}^+) \times \mathbf{supp}(\widehat{v}^-)$). By Lemma 3.2$(a)$ all three vectors $\widehat{v}^+$, $\widehat{v}^-$ and $\widehat{v}^+ + \widehat{v}^-$ are positive radical vectors of $\widehat{q}$. Notice that by Lemma 3.4, if the form $\widehat{q}$ is not connected then there is a connected unit form $\widehat{q}'$ and an integer c' with $\widehat{q} = \widehat{q}' \oplus \xi^{c'}$, for $0 \leq c' \leq c$. Thus by induction we may assume that $\widehat{q}$ is connected.

Step 2. *If q has a positive radical vector, there exists an iterated deflation T' such that qT' is the direct sum of a zero form in k variables (for $1 \leq k \leq c$) and a connected nonnegative unit form with corank $c - k$.*
Assume that v is a positive radical vector of q and that there exist $i, j \in \mathbf{supp}(v)$ with $q_{ij} < 0$ and $v_i \leq v_j$.

Take $q' = qT_{ij}^-$ and $v' = (T_{ij}^-)^{-1}v = v - v_i e_j$, and observe that v' is a positive radical vector for q' with $|v'| < |v|$. Repeating this procedure as long as possible we end up with a quadratic form $\widetilde{q}$ and a positive radical vector $\widetilde{v}$ such that

$$0 = \widetilde{q}(\widetilde{v}) = \sum_{i=1}^{n} \widetilde{q}_{ii}\widetilde{v}_i^2 + \sum_{1 \leq i < j \leq n} \widetilde{q}_{ij}\widetilde{v}_i\widetilde{v}_j.$$

Again, both summands on the right side are nonnegative, hence zero. Then $\widetilde{q}_{ii} = 0$ for any i in the support of $\widetilde{v}$, and the claim follows from Lemma 3.4.

We conclude the proof of the theorem by induction, using Steps 1 and 2 above, and Lemma 3.4 for connectivity. □

Considering Theorems 3.8 and 2.20, for a nonnegative semi-unit form q there is an iterated flation T such that $qT = q_\Delta \oplus \xi^c$, where Δ is a disjoint union of Dynkin diagrams, q_Δ is its associated (positive) unit form, and c is the corank of q. Notice that if there are iterated inflations T and T' for q such that $qT = p \oplus \xi^c$ and $qT' = p' \oplus \xi^c$, then p and p' are equivalent positive unit forms, therefore by Theorem 2.20 the disjoint union of Dynkin graphs Δ related to q is unique up to a permutation of its components. This disjoint union Δ is referred to as the *Dynkin type* of q, written $\mathbf{Dyn}(q) = \Delta$. We now show that the Dynkin type of a nonnegative semi-unit form, together with its corank, determine the equivalence class of such forms. Here and in what follows, the zero quadratic form in $c \geq 1$ variables will be denoted by ξ^c.

Corollary 3.9. *Let q and q' be nonnegative semi-unit forms. Then q and q' are equivalent forms if and only if they have the same Dynkin type and the same corank.*

Proof. Assume first that q and q' are equivalent forms. Then $\mathbf{cork}(q) = \mathbf{cork}(q') =: c$. Take iterated flations T and T' such that $qT = p \oplus \xi^c$ and $q'T' = p' \oplus \xi^c$, where p and p' are positive unit forms. By transitivity observe that p and p' are equivalent forms, hence using Corollary 2.21 we have

$$\mathbf{Dyn}(q) = \mathbf{Dyn}(p) = \mathbf{Dyn}(p') = \mathbf{Dyn}(q').$$

For the converse assume there is a disjoint union of Dynkin graphs Δ with $\mathbf{Dyn}(q) = \Delta = \mathbf{Dyn}(q')$, and an integer c with $\mathbf{cork}(q) = c = \mathbf{cork}(q')$. Then there are iterated flations T and T' with

$$qT = q_\Delta \oplus \xi^c = q'T'.$$

In particular, we have $q' = qT(T')^{-1}$, that is, q and q' are equivalent forms. □

Example 3.10. The quadratic form q associated to the following bigraph is nonnegative with Dynkin type $\mathbf{Dyn}(q) = \mathbb{D}_4$ and corank one.

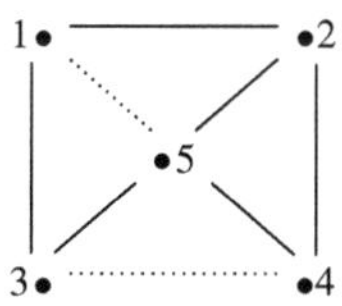

Its radical is generated by the vector $e_2+e_4+e_5$. Moreover, the restricted forms $q^{(2)}$, $q^{(4)}$ and $q^{(5)}$ are positive with Dynkin type $\mathbb{D}_4$, while $\mathbf{Dyn}(q^{(1)}) = \mathbf{Dyn}(q^{(3)}) = \mathbb{A}_3$.

Exercises 3.11. 1. Let $q : \mathbb{Z}^n \to \mathbb{Z}$ be a balanced semi-unit form, and take a subset of indices $I \subset \{1, \dots, n\}$. Show that the restricted form q^I is also balanced.

2. Show that if $q : \mathbb{Z}^n \to \mathbb{Z}$ is a nonnegative unit form of Dynkin type $\mathbb{A}_n$ with radical generated by a single positive sincere vector then $q = q_{\widetilde{\mathbb{A}}_n}$ for the extended Dynkin diagram $\widetilde{\mathbb{A}}_n$.
3. Find a connected bigraph B with at least three dotted edges such that q_B is a nonnegative unit form with Dynkin type $\mathbb{E}_6$ and radical generated by a sincere positive vector.
4. Which of the following unit forms is nonnegative?

   a) $q(x) = x_1^2 + x_2^2 + x_3^2 + x_4^2 + x_1x_2 - x_3(x_1 - x_2 + x_4)$.
   b) $q(x) = x_1^2 + \dots + x_5^2 - x_1(x_2 + x_3 + x_4 + x_5) + x_4x_5$.
   c) $q(x) = x_1^2 + \dots + x_5^2 - x_1(x_2 + x_3 + x_4) + x_5(x_2 - x_3 + x_4) - x_3(x_2 - x_4)$.

5. Show that for any integer $c \geq 0$ and any Dynkin graph G there is a connected nonnegative unit form q with $\mathbf{Dyn}(q) = G$ and corank c.
6. Prove that the following unit forms are nonnegative and determine their Dynkin type and corank.

   a) $x_1^2 + \dots + x_5^2 - x_2(x_1 - x_5) + x_3(x_1 - x_2 + x_4 - x_5) + x_4(x_1 + x_5)$.
   b) $x_1^2+\dots+x_6^2+x_1(x_2-x_3-x_5+x_6)-x_4(x_2-x_3+x_5-x_6)+(x_2+x_3)(x_5-x_6)$.
   c) $x_1^2+\dots+x_7^2-x_1(x_2+x_3+x_4)+x_2x_3+x_4(x_2+2x_3)+x_5(x_6-x_7)+x_6x_7$.

3.3 Radicals and Their Extensions

Recall that a quadratic form $q : \mathbb{Z}^I \to \mathbb{Z}$ is said to be *regular* if $\mathbf{rad}(q) = 0$. For a subset of indices $J \subset \{1, \dots, n\}$ consider the inclusion $\sigma : \mathbb{Z}^J \to \mathbb{Z}^n$ determined by $e_j \mapsto e_j$. The restriction $q^J : \mathbb{Z}^J \to Z$ of q is given by $q^J(x) = q(\sigma(x))$ for $x \in \mathbb{Z}^J$. In that situation we say $\mathbf{rad}(q^J) \subseteq \mathbf{rad}(q)$ if the restriction of σ to the radical of q^J determines an injective map $\sigma : \mathbf{rad}(q^J) \to \mathbf{rad}(q)$. As mentioned in Lemma 2.11 and its following example, it is not always true that $\mathbf{rad}(q') \subset \mathbf{rad}(q)$ for a unit form q. Our purpose here is to show that this property characterizes nonnegativity.

Instead of using the somehow clumsy term "critical not nonnegative form", we say that a quadratic form q is *hypercritical nonnegative* if any proper restriction q'

of q is nonnegative, but q itself is not. Notice, for instance, that a Kronecker form q_m is hypercritical nonnegative if and only if $|m| \geq 3$. The following graph $\tilde{\tilde{\mathbb{E}}}_8$ with 10 vertices,

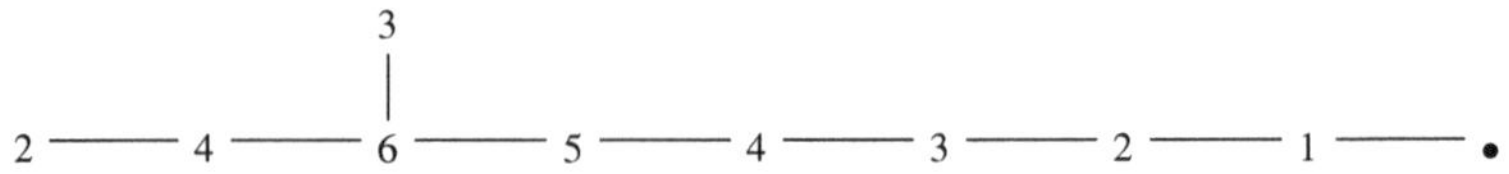

has hypercritical nonnegative associated form $q = q_{\tilde{\tilde{\mathbb{E}}}_8}$, where the bullet • is the unique vertex in $\tilde{\tilde{\mathbb{E}}}_8$ satisfying that the restriction $q^{(\bullet)}$ is nonpositive (thus critical nonpositive.). The vector in $\mathbb{Z}^9$ indicated by the numbers at the vertices is the generator of the radical of $q^{(\bullet)}$. It is convenient to point out that if q is simultaneously a critical nonpositive and hypercritical nonnegative form then q is a Kronecker form q_m with $|m| \geq 3$. We say that a nonzero vector z in $\mathbb{Z}^n$ is called a *critical vector* for a critical nonpositive form $q : \mathbb{Z}^n \to \mathbb{Z}$ if z generates the radical of q (cf. Theorem 2.12).

Proposition 3.12. *Let $q : \mathbb{Z}^n \to \mathbb{Z}$ be a unit form with $n \geq 3$. Then q is hypercritical nonnegative if and only if q is not nonnegative and for any critical nonpositive restriction q^I of q, there exists an index i such that $I = \{1, \dots, n\} - \{i\}$ and a critical vector z' of q^I such that $q(z|e_i) < 0$ where z is the vector in $\mathbb{Z}^n$ obtained by extending z' by zeros.*

Proof. First let q be a hypercritical nonnegative unit form, and take $v \in \mathbb{Z}^n$ with $q(v) < 0$. Since any proper restriction of q is nonnegative, v is a sincere vector. Assume q^I is a critical nonpositive form. Since $n \geq 3$, the form q^I is a proper restriction of q. Further, q^I is not the Kronecker form q_m with $|m| \geq 3$, for q is hypercritical nonnegative. Therefore q^I has a critical vector z' (see Theorem 2.12). Complete z' with zeros to a vector z in $\mathbb{Z}^n$.

Take integers m, k and a vertex $j \in I$ such that $(kv + mz)_j = 0$. Since

$$0 \leq q^{(j)}(kv + mz) = k^2 q(v) + m^2 q(z) + kmq(z|v) < km \sum_{i=1}^{n} v_i q(z|e_i),$$

there must exist a vertex $i \in \{1, \dots, n\}$ satisfying $q(z|e_i) \neq 0$ (hence $i \notin I$). Multiplying z by (-1) if necessary we may assume that $q(z|e_i) < 0$. Moreover,

$$q(2z + e_i) = 4q(z) + 1 + 2q(z|e_i) = 1 + 2q(z|e_i) < 0,$$

therefore q hypercritical implies that $2z + e_i$ is a sincere vector, that is, $I = \{1, \dots, n\} - \{i\}$ (and z' is a critical vector for $q^{(i)}$).

For the converse we need to show that $q^{(i)}$ is nonnegative for any $i = 1, \dots, n$. If $q^{(i)}$ is not nonnegative for some $i \in \{1, \dots, n\}$ then there is a critical nonpositive restriction q^I of $q^{(i)}$ (that is, $I \subset \{1, \dots, \widehat{i}, \dots, n\}$). By hypothesis, $I = \{1, \dots, \widehat{i}, \dots, n\}$ and $q^{(i)}$ is the m-Kronecker form for some m with $|m| \geq 3$

(since $q^{(i)}$ is not nonnegative, cf. Theorem 2.12). This contradicts the existence of a critical vector for q^I (cf. Proposition 1.20). □

Corollary 3.13. *Every hypercritical nonnegative unit form q is regular (that is, $\mathbf{rad}(q) = 0$).*

Proof. If q is a binary form, then q is the Kronecker form q_m with $|m| \geq 3$, and by Proposition 1.20 the form q is anisotropic, in particular regular.

Let v be a radical vector of a hypercritical nonnegative unit form $q : \mathbb{Z}^n \to \mathbb{Z}$ with $n > 2$. Consider a vertex i such that q^I is a critical nonpositive restriction of q for $I = \{1, \dots, n\} - \{i\}$, with critical vector z' whose extension by zeros z to $\mathbb{Z}^n$ satisfies $q(z|e_i) < 0$. If $v_i = 0$ then v is an integral multiple of z (for the restriction of v to a vector in $\mathbb{Z}^I$ is a radical vector for q^I), which is impossible since $z \notin \mathbf{rad}(q)$). Suppose now that $v_i \neq 0$ and consider the vector v' in $\mathbb{Z}^I$ such that $v = v' + v_i e_i$. Then

$$0 = q(v|z) = q(v'|z) + v_i q(e_i|z) = q^I(v'|z) + v_i q(z|e_i) = v_i q(z|e_i) \neq 0,$$

again a contradiction. □

As an illustration consider the (solid) *r-pointed star graph* $\mathbb{S}_r$ with $r + 1$ vertices and r edges

$\mathbb{S}_r =$

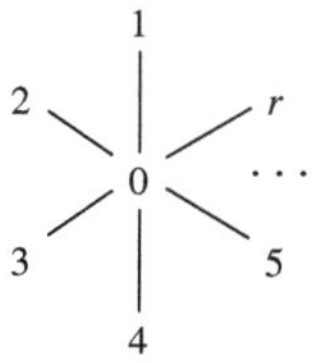

for $r \geq 1$. Observe that $q_{\mathbb{S}_r}$ is nonnegative if and only if $r \leq 4$, and is regular if and only if $r \neq 4$. The first assertion is consequence of $q_{\mathbb{S}_5}$ being hypercritical nonnegative. For the second claim, take $q = q_{\mathbb{S}_r}$ and $x = (x_0, x_1, \dots, x_r)$ in $\mathbb{Z}^{r+1}$ such that $q(x|e_i) = 0$ for $i = 0, \dots, r$. These equations can be written as

$$\begin{aligned} 2x_0 &= x_1 + \ldots + x_r, \\ 2x_1 &= x_0, \\ \ldots &\quad \ldots \\ 2x_r &= x_0, \end{aligned}$$

and in particular $4x_0 = rx_0$. Therefore there exists such nonzero x if and only if $r = 4$.

Theorem 3.14. *For a semi-unit form q the following are equivalent:*

a) *The form q is nonnegative.*
b) *For any restriction q' of q we have $\mathbf{rad}(q') \subseteq \mathbf{rad}(q)$.*

Proof. That (a) implies (b) was shown in Lemma 3.2 (see also Lemma 2.11). Assume that q is not nonnegative and take a hypercritical nonnegative restriction q^I of q (with $I \subset \{1, \dots, n\}$) and a vertex $i \in I$ such that $q' = (q^I)^{(i)}$ is a critical nonpositive restriction of q^I (see Proposition 3.12). Then there is a critical vector $z \in \mathbf{rad}(q')$ but its extension by zeros $\sigma(z) \in \mathbb{Z}^I$ is not radical for q^I by Corollary 3.13, in particular not a radical vector for q. □

We say that a semi-unit form $q' : \mathbb{Z}^n \to \mathbb{Z}$ is a *radical extension* of a semi-unit form $q : \mathbb{Z}^m \to \mathbb{Z}$ (with $m \leq n$) if there is a subgroup U of $\mathbb{Z}^n$ and a subgroup U' of $\mathbf{rad}(q')$ such that $\mathbb{Z}^n = U \oplus U'$ and $q = q'|_U$. In other words, q' is radical extension of q if there is a $\mathbb{Z}$-invertible transformation $S : \mathbb{Z}^n \to \mathbb{Z}^n$ such that

$$q'S = q \oplus \xi^{n-m}.$$

In particular the columns of S consists of roots or isotropic vectors of q'. Throughout the text we will find many instances of radical extensions: Theorem 3.8 implies that every nonnegative semi-unit form with Dynkin type Δ is radical extension of the positive unit form q_Δ (see details below in Theorem 3.15). In Sect. 3.5 we will consider *one-point extensions*, one of the main tools in the construction of unitary forms. *Radical explosions* are defined in Sect. 5.5, with a particular case known as *doubling of vertices* used in Sect. 6.5 for the construction of graphical forms.

Let $q : \mathbb{Z}^n \to \mathbb{Z}$ be a nonnegative semi-unit form. Observe that $\mathbf{rad}(q)$ is a *pure subgroup* of $\mathbb{Z}^n$ (that is, if $0 \neq n \in \mathbb{Z}$ with $nv \in \mathbf{rad}(q)$, then $v \in \mathbf{rad}(q)$), hence there is an isomorphism $\mathbb{Z}^n/\mathbf{rad}(q) \to \mathbb{Z}^{n-c}$ where $c = \mathbf{cork}(q)$ is the corank of q. Recall that for $v \in \mathbf{rad}(q)$ we have $q(w+v) = q(w)$ for any $w \in \mathbb{Z}^n$, thus we may consider a well-defined induced mapping

$$\overline{q} : \mathbb{Z}^n/\mathbf{rad}(q) \longrightarrow \mathbb{Z}, \qquad w + \mathbf{rad}(q) \longmapsto q(w).$$

We show that there is a basis in $\mathbb{Z}^n/\mathbf{rad}(q)$ which makes $\overline{q}$ a positive unit form.

Theorem 3.15. *Let $q : \mathbb{Z}^n \to \mathbb{Z}$ be a nonnegative semi-unit form. Then the induced mapping $\overline{q} : \mathbb{Z}^n/\mathbf{rad}(q) \to \mathbb{Z}$ is $\mathbb{Z}$-equivalent to q_Δ, where Δ is the Dynkin type of q. In particular, q is radical extension of q_Δ.*

Proof. Applying Theorems 3.8 and 2.20 to each connected component of q, there is a $\mathbb{Z}$-invertible linear transformation $T : \mathbb{Z}^n \to \mathbb{Z}^c \oplus \mathbb{Z}^{n-c}$ such that

$$qT^{-1} = \xi^c \oplus q_\Delta : \mathbb{Z}^c \oplus \mathbb{Z}^{n-c} \to \mathbb{Z},$$

where Δ is the Dynkin type of q and $c = \mathbf{cork}(q)$. Notice that $\mathbb{Z}^c = T(\mathbf{rad}(q)) = \mathbf{rad}(qT^{-1})$, thus we have an induced isomorphism $\overline{T} : \mathbb{Z}^n/\mathbf{rad}(q) \to \mathbb{Z}^{n-c}$ which makes the following diagram commutative,

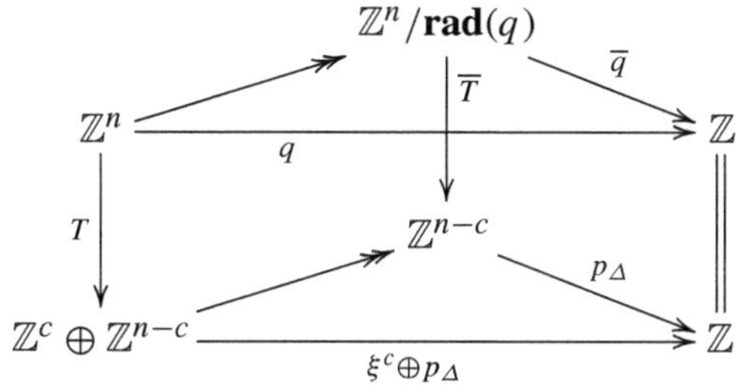

where $\twoheadrightarrow$ denotes canonical projections. Hence $\overline{T}$ is the desired equivalence, since we have $\overline{q} = p_\Delta \overline{T}$. Taking $U = T^{-1}(Z^{n-c})$ and $U' = T^{-1}(\mathbb{Z}^c)$ as in the definition of radical extension above, it is clear that q is a radical extension of q_Δ. □

Exercises 3.16.

1. Show that the quadratic forms associated to the following bigraphs are nonnegative, and find their radicals and Dynkin type.

2. Prove that a semi-unit form q is positive if and only if $\mathbf{rad}(q') = 0$ for any restriction q' of q.
3. Give an example of a nonregular unit form q which fails to be nonnegative.
4. Determine all hypercritical nonnegative unit forms in 5 variables.

3.4 Omissible Variables

In this section we analyze, following [7] and [9], how Dynkin type and corank change under restrictions of nonnegative semi-unit forms.

Lemma 3.17. *Let $q : \mathbb{Z}^n \to \mathbb{Z}$ be a nonnegative semi-unit form. For any vertex $i \in \{1, \ldots, n\}$ we have*

$$0 \leq \mathbf{cork}(q) - \mathbf{cork}(q^{(i)}) \leq 1.$$

Proof. Take $I = \{1, \ldots, n\} - \{i\}$ and consider the canonical inclusion $\sigma : \mathbb{Z}^I \to \mathbb{Z}^n$. By Theorem 3.14 we have $\sigma(\mathbf{rad}(q^{(i)})) \to \mathbf{rad}(q)$, which shows the first inequality. Observe now that $\mathbf{rad}(q)$ is a pure subgroup of $\mathbb{Z}^n$, and that $\sigma(\mathbf{rad}(q^{(i)}))$ is a pure subgroup of $\mathbf{rad}(q)$ (cf. Remark 4.2). Hence, if $v^1, \ldots, v^r$ is a $\mathbb{Z}$-basis of $\mathbf{rad}(q^{(i)})$, then their image $w^i := \sigma(v^i)$ may be completed to a basis $w^1, \ldots, w^r, w^{r+1}, \ldots, w^c$ of $\mathbf{rad}(q)$ (where $c = \mathbf{cork}(q)$ and $r = \mathbf{cork}(q^{(i)})$, see Proposition 4.1). If $r < c-1$ then there are non-zero integers a and b with $(aw^{c-1} + bw^c)_i = 0$, which means that $aw^{c-1} + bw^c$ is a radical vector of q belonging to $\sigma(\mathbf{rad}(q^{(i)}))$. This is impossible since $w^1, \ldots, w^r, w^{r+1}, \ldots, w^c$ is linearly independent and $w^1, \ldots, w^r$ generate $\sigma(\mathbf{rad}(q^{(i)}))$. Therefore $c - r \leq 1$, which completes the proof. □

We now generalize Proposition 2.25 to the nonnegative setting. The following partial ordering of Dynkin graphs was introduced in Sect. 2.4.

$$\begin{aligned} \mathbb{A}_m &\leq \mathbb{A}_n, \quad \text{for } m \leq n; \\ \mathbb{A}_n < \mathbb{D}_n &\leq \mathbb{D}_p, \quad \text{for } 4 \leq n \leq p; \\ \mathbb{D}_p < \mathbb{E}_p &\leq \mathbb{E}_q, \quad \text{for } 6 \leq p \leq q \leq 8. \end{aligned}$$

As before we take $r_{\mathbb{A}_n} = 1$, $r_{\mathbb{D}_m} = 2$, $r_{\mathbb{E}_6} = 3$, $r_{\mathbb{E}_7} = 4$ and $r_{\mathbb{E}_7} = 6$ to be the maximal value the entries of a maximal positive root of q_Δ may attain, where Δ is a Dynkin graph (cf. Table 2.1 and Remark 2.24).

Proposition 3.18. *Let q be a connected nonnegative unit form. Then for any connected restriction q' of q we have* $\mathbf{Dyn}(q') \leq \mathbf{Dyn}(q)$.

Proof. We will show that $\mathbf{Dyn}(q^{(i)}) \leq \mathbf{Dyn}(q)$ for any $1 \leq i \leq n$ such that $q^{(i)}$ is still connected (see Exercise 6 below). For simplicity we take $i = n$.

Suppose first that $\mathbf{cork}(q^{(n)}) = c = \mathbf{cork}(q)$. Using Theorems 3.8 and 2.20 there is an iterated flation $T : \mathbb{Z}^{n-1} \to \mathbb{Z}^{n-1}$ such that $q^{(n)}T = \xi^c \oplus q_{\Delta'}$ where $\Delta' = \mathbf{Dyn}(q^{(n)})$. Consider the linear transformation $\widetilde{T} = T \oplus [1] : \mathbb{Z}^n \to \mathbb{Z}^n$, and observe that $q\widetilde{T} = \xi^c \oplus \widetilde{q}$ with $\widetilde{q}$ a positive unit form (for $\mathbf{cork}(q\widetilde{T}) = \mathbf{cork}(q) = c$). Let $v = p_{\Delta'}$ be the maximal positive root of $q_{\Delta'}$ (see Table 2.1) and take $v_n = 0$, so that we may view v as a positive root of $\widetilde{q}$. If $\mathbf{Dyn}(q) = \Delta$, then $\mathbf{Dyn}(\widetilde{q}) = \Delta$ and we have

$$r_{\Delta'} = \max_{i=1,\ldots,n} (|v_i|) \leq r_\Delta,$$

where the last inequality follows from Proposition 2.22. Since the number of vertices of Δ' is $n-1-c$, and that of Δ is $n-c$, we get $|\Delta'_0| < |\Delta_0|$ and $r_{\Delta'} \leq r_\Delta$. Thus by Remark 2.24 we have $\Delta' \leq \Delta$.

Suppose now that $\mathbf{cork}(q^{(n)}) \neq \mathbf{cork}(q)$ hence by Lemma 3.17 we have $\mathbf{cork}(q^{(n)}) = \mathbf{cork}(q) - 1$. Taking $\mathbf{Dyn}(q^{(n)}) = \Delta'$ and $\mathbf{Dyn}(q) = \Delta$, we notice as above that $|\Delta'_0| = |\Delta_0|$. As in the proof of Theorem 3.15, the inclusion $\sigma : \mathbb{Z}^{n-1} \to \mathbb{Z}^n$ induces an injection $\overline{\sigma} : \mathbb{Z}^{n-1}/\mathbf{rad}(q^{(n)}) \to \mathbb{Z}^n/\mathbf{rad}(q)$. If $\overline{q}$ and $\overline{q^{(n)}}$ are the induced positive unit forms of Theorem 3.15, then $\overline{\sigma}$ determines an inclusion

$R(\overline{q^{(n)}}) \to R(\overline{q})$. Observe finally that if $|\Delta'_0| = |\Delta_0|$ and $|R(q_{\Delta'})| \leq |R(q_\Delta)|$, then $\Delta' \leq \Delta$, which completes the proof. □

In what follows we give conditions on an index $1 \leq i \leq n$ and a nonnegative semi-unit form q ensuring that the restriction q^i and q have same Dynkin type. We say that an index $i \in \{1, \ldots, n\}$ is an *omissible point* (or an *omissible variable*) for a nonnegative semi-unit form $q : \mathbb{Z}^n \to \mathbb{Z}$ if $q(e_i) = 1$ and there is a radical vector v of q with $v_i = 1$. In Example 3.10, for instance, indices 2, 4 and 5 are omissible points.

Example 3.19. Let q be the quadratic form associated to the following bigraph:

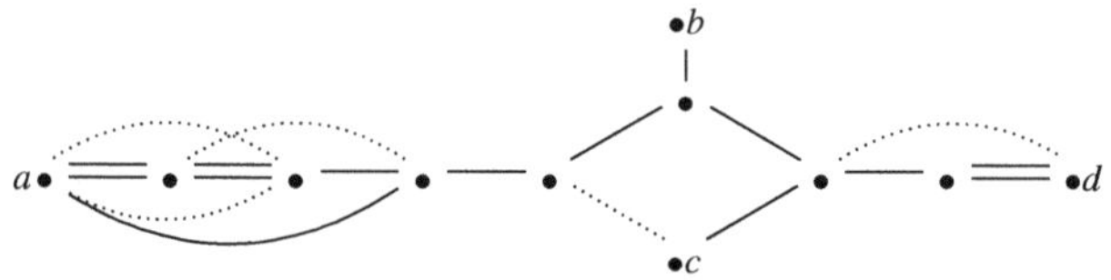

Then q is nonnegative, $\mathbf{Dyn}(q) = \mathbb{E}_8$ and $\mathbf{cork}(q) = 3$. Moreover, vertices a and d are omissible points, $\mathbf{Dyn}(q^{(b)}) = \mathbb{D}_7$ and $\mathbf{Dyn}(q^{(c)}) = \mathbb{E}_7$ with $\mathbf{cork}(q^{(b)}) = \mathbf{cork}(q^{(c)}) = 3$.

Proposition 3.20. *Let $q : \mathbb{Z}^n \to \mathbb{Z}$ be a connected nonnegative semi-unit form.*

a) For any omissible variable i for q, the restriction $q^{(i)}$ is connected and satisfies $\mathbf{Dyn}(q^{(i)}) = \mathbf{Dyn}(q)$.
b) If q is unitary and $\mathbf{cork}(q) > 0$, then q admits an omissible variable.

Proof. Let i be an omissible point of q and $v \in \mathbf{rad}(q)$ with $v_i = 1$. Consider $\overline{x} = \{\overline{x}^1, \ldots, \overline{x}^\ell\}$ a $\mathbb{Z}$-basis of $\mathbb{Z}^n/\mathbf{rad}(q)$ (recall that $\mathbf{rad}(q)$ is a pure subgroup of $\mathbb{Z}^n$) and take a representative $x^j \in \mathbb{Z}^n$ of $\overline{x}^j$ with $x^j_i = 0$ for $j = 1, \ldots, \ell$ (which is possible since $v_i = 1$). Denote by $\sigma : \mathbb{Z}^{n-1} \to \mathbb{Z}^n$ the canonical inclusion with $q^{(i)} = q\sigma$, and take $y^j \in \mathbb{Z}^{n-1}$ with $x^j = \sigma(y^j)$. Since $\mathbf{rad}(q^{(i)}) \subset \mathbf{rad}(q)$ (Theorem 3.14), the set $\overline{y} = \{\pi(y^1), \ldots, \pi(y^\ell)\}$ is linearly independent, where $\pi : \mathbb{Z}^{n-1} \to \mathbb{Z}^{n-1}/\mathbf{rad}(q^{(i)})$ is the canonical projection. Since $\mathbf{cork}(q^{(i)}) = \mathbf{cork}(q) - 1$ (Lemma 3.17), the rank of $\mathbb{Z}^{n-1}/\mathbf{rad}(q^{(i)})$ is ℓ, thus $\overline{y}$ is a $\mathbb{Z}$-basis of $\mathbb{Z}^{n-1}/\mathbf{rad}(q^{(i)})$. If T denotes the change of basis transformation between $\overline{x}$ and $\overline{y}$, we have $\overline{q} = \overline{q^{(i)}}T$.

Since $\overline{q}$ is connected, this implies first that $\overline{q^{(i)}}$ is connected, thus $q^{(i)}$ is also connected. Moreover, we have $\mathbf{Dyn}(q) = \mathbf{Dyn}(q^{(i)})$, which shows (a).

Assume now that $\mathbf{cork}(q) > 0$ and that q is unitary. By Lemma 3.17, and restricting q to a subset of vertices if necessary, we may assume that $\mathbf{cork}(q) = 1$ and that $\mathbf{rad}(q)$ is generated by a sincere vector v. Moreover, composing with a point inversion S we get a nonnegative quadratic form $q' = qS$ with radical generated by a positive sincere vector $v' = Sv$. By Theorem 3.5, there is an iterated flation T such that $q'T$ is the quadratic form associated to an extended Dynkin graph. All these forms have an omissible point (cf. Table 2.2), hence the same is true for q'. Since $|v_i| = |v_i'|$ for $i = 1, \ldots, n$, the form q admits an omissible variable. □

Remark 3.21. As a consequence of Proposition 3.20, for any nonnegative semi-unit quadratic form q and any $c \leq \mathbf{cork}(q)$, there exists a restriction q' of q such that $\mathbf{cork}(q') = c$ and $\mathbf{Dyn}(q') = \mathbf{Dyn}(q)$.

In particular, taking $c = 0$ in the last remark, there is a positive restriction q' of q with $\mathbf{Dyn}(q') = \mathbf{Dyn}(q)$, called a *core* of q. The form q in Example 3.10 has exactly three cores, namely $q^{(2)}$, $q^{(4)}$ and $q^{(5)}$.

Exercises 3.22.

1. Give an example of a semi-unit form q and a flation T for q such that qT is no longer semi-unitary.
2. Determine which of the following quadratic forms are nonnegative:

   i) $q_1 = x_1^2 + x_2^2 + x_3^2 + x_4^2 - x_1x_2 + x_1x_3 - x_1x_4 + x_2x_3 + x_2x_4 - 2x_3x_4$.
   ii) $q_2 = x_1^2 + \ldots + x_6^2 - (x_1 + x_4)(x_2 + x_3) - x_3x_5 - x_4x_6 + x_2x_3 + x_4x_5$.
   iii) $q_3 = x_1^2 + \ldots + x_4^2 - x_1(x_2 + x_3 + x_4) - x_2x_4 + x_3x_4$.

3. Prove that if q is a nonnegative unit form then q is connected if and only if the induced quadratic form $\overline{q}$ given in Theorem 3.15 is connected.
4. Show that if the quadratic form q associated to a complete bigraph with at least four vertices is nonnegative, then q is a positive form.
5. Let q be a nonnegative unit form such that $\mathbf{Dyn}(q) = \mathbf{Dyn}(q^{(i)})$. Is i necessarily is an omissible variable for q?
6. Show that if q is a connected nonnegative unit form and q' is a connected restriction of q, then there is a sequence of indices $i_1, \ldots, i_r$ with $q' = q^{(i_1)\cdots(i_r)}$ and such that $q^{(i_1)\cdots(i_s)}$ is connected for any $s = 1, \ldots, r$.

3.5 Root Induction and One-Point Extensions

Let $q : \mathbb{Z}^n \to \mathbb{Z}$ be a semi-unit form. Following [9], for a finite collection of q-roots $r = (r^j)_{j \in J}$ define the *quadratic form induced by r*, denoted by $q_r : \mathbb{Z}^J \to \mathbb{Z}$, to be the form

$$q_r(y) = q\left(\sum_{j \in J} y_j r^j\right).$$

Notice that $q_r(e_j) = q(r^j) = 1$ for any $j \in J$, that is, q_r is a unit form. Moreover, if q is nonnegative then q_r is again nonnegative. Observe also that if $I \subset \{1, \ldots, n\}$ and $r = (e_i)_{i \in I}$, then the root induction q_r is precisely the restriction q^I. We say that two unit forms q and q' are *root equivalent* if q is root induced from q', and q'

is root induced from q. First we show that root equivalence is indeed an equivalence relation in the set of unit quadratic forms. Clearly we only need to prove transitivity.

Lemma 3.23. *Let $r = (r^i)_{i \in I}$ and $s = (s^j)_{j \in J}$ be finite collections of q-roots and q_r-roots respectively, where q is a unit form. Then there exists a collection t of q-roots such that $(q_r)_s = q_t$.*

Proof. Take $t^j = \sum_{i \in I} s_i^j r^i$ for $j \in J$. Observe that $q(t^j) = q\left(\sum_{i \in I} s_i^j r^i\right) = q_r(s^j) = 1$, thus $t = (t^j)_{j \in J}$ is a collection of q-roots. Then we have

$$(q_r)_s(x) = q_r\left(\sum_{j \in J} x_j s^j\right) = q\left(\sum_{i \in I}\left(\sum_{j \in J} x_j s^j\right)_i r^i\right)$$
$$= q\left(\sum_{j \in J} x_j \left(\sum_{i \in I} s_i^j r^i\right)\right) = q\left(\sum_{j \in J} x_j t^j\right) = q_t(x). \qquad \square$$

We now show how root induction behaves with respect to connectivity. For convenience, for an empty collection of q-roots r we denote by q_r the trivial quadratic form in zero variables. Let us first analyze the positive case.

Remark 3.24. Consider a positive unit form q that decomposes as $q = q^1 \oplus q^2$, and take a finite collection $r = (r^j)_{j \in J}$ of q-roots. There is a partition $J = J^1 \cup J^2$ such that $q^1(\overline{r}^j) = 1$ for $j \in J^1$ and $q^2(\overline{r}^j) = 1$ for $j \in J^2$, where r^j is obtained from $\overline{r}^j$ by extending by zeros. Take collections $r' = (\overline{r}^j)_{j \in J^1}$ and $r'' = (\overline{r}^j)_{j \in J^2}$. Then

$$q_r(x) = q\left(\sum_{j \in J} x_j r^j\right) = q^1\left(\sum_{j \in J^1} x_j \overline{r}^j\right) + q^2\left(\sum_{j \in J^2} x_j \overline{r}^j\right) = q^1_{r'}(x') \oplus q^2_{r''}(x''),$$

where x' and x'' are the restrictions of x to the entries indexed by J^1 and J^2 respectively.

Lemma 3.25. *Let $p : \mathbb{Z}^J \to \mathbb{Z}$ and $q : \mathbb{Z}^I \to \mathbb{Z}$ be root equivalent positive unit forms. If $p = p^1 \oplus \cdots \oplus p^m$ and $q = q^1 \oplus \cdots \oplus q^n$ are decompositions with p^a and q^b connected for $a = 1, \ldots, m$ and $b = 1, \ldots, n$, then $m = n$ and there is a permutation π such that p^k is root equivalent to $q^{\pi(k)}$ for $k = 1, \ldots, n$.*

Proof. Let $r = (r^j)_{j \in J}$ and $s = (s^i)_{i \in I}$ be finite collections of q-roots and p-roots respectively with $p = q_r$ and $q = p_s$. Using Remark 3.24, we have a partition $I = \bigcup_{a=1}^m I_a$ such that $\overline{s}^i$ is a p^a-root for $i \in I_a$, where s^i is obtained from $\overline{s}^i$ by extending by zeros. Take $s^{(a)} = (\overline{s}^i)_{i \in I_a}$ for $a = 1, \ldots, m$.

Observe first that $|I| = |J|$. Indeed, if there is an integral linear dependence in the collection s, say $\sum x_j s^j = 0$, then taking $x = (x_j)$ we have $q(x) = p_s(x) =$

$p(\sum_{i\in I} x_j s^j) = p(0) = 0$, contradicting the positivity of q. Then s is a linearly independent set, and $|I| \leq |J|$. Exchanging positions of p and q we get $|J| \leq |I|$.

Now we have

$$q = p_s = p^1_{s(1)} \oplus \ldots \oplus p^a_{s(a)} \oplus \ldots \oplus p^m_{s(m)},$$

and from $|I| = |J|$ it follows that $s^{(a)}$ is a $\mathbb{Q}$-basis in the domain of p^a. In particular, $m \leq n$. Exchanging the roles of p and q we get $n \leq m$. Hence $m = n$ and there is a permutation π of the set $\{1, \ldots, n\}$ such that p^a is root equivalent to $q^{\pi(a)}$ using the Remark above. □

Assume that $q : \mathbb{Z}^n \to \mathbb{Z}$ is a nonnegative semi-unit form for which the last variable n is omissible, and take a radical vector $\overline{v}$ of q with $\overline{v}_n = 1$. As shown before, the restriction $q^{(n)}$ has the same Dynkin type as q. We want to recover q from its restriction $q^{(n)}$. With that purpose define the *one-point extension* $p[v] : \mathbb{Z}^n \to \mathbb{Z}$ of a semi-unit form $p : \mathbb{Z}^{n-1} \to \mathbb{Z}$ with respect to a p-root v as

$$p[v] = p_{e(v)}, \qquad \text{where } e(v) = (e_1, \ldots, e_{n-1}, -v).$$

Lemma 3.26. *Let $q : \mathbb{Z}^n \to \mathbb{Z}$ be a nonnegative semi-unit form such that n is an omissible point for q, and take $p = q^{(n)}$. Then there exists a p-root $v \in \mathbb{Z}^{n-1}$ such that $q = p[v]$ and such that $v + e_n$ is a radical vector for q.*

Proof. Take $\overline{v}$ a radical vector for q such that $\overline{v}_n = 1$, and let $v \in \mathbb{Z}^{n-1}$ with $\overline{v} = v + e_n$. Then

$$p(v) = q(\overline{v} - e_n) = q(\overline{v}) + q(e_n) + q(\overline{v}|e_n) = 1,$$

since $q(e_n) = 1$. Observe that the coefficients of $p[v]$ are given as follows,

$$\begin{aligned} p[v](e_i|e_n) &= p[v](e_i + e_n) - p[v](e_i) - p[v](e_n) \\ &= p(e_i - v) - p(e_i) - p(-v) \\ &= -p(e_i|v). \end{aligned}$$

Notice now that $q(e_i|e_n) = q^{(n)}[v](e_i|e_n)$ for $i = 1, \ldots, n-1$. Since $\overline{v}$ is a radical vector for q we have

$$0 = q(e_i|\overline{v}) = q(e_i|v + e_n) = q^{(n)}(e_i|v) + q(e_i|e_n) = q(e_i|e_n) - q^{(n)}[v](e_i|e_n),$$

which completes the proof. □

Proposition 3.27. *Let q be a nonnegative quadratic unit form, and consider a core p of q. Then q and p are root equivalent forms.*

Proof. Being a restriction of q, the core p is the root induced from q. For the converse we proceed by induction on $c = \mathbf{cork}(q)$. If $c = 0$ then $p = q$ and there is nothing to prove. Assume that $c > 0$ and take an omissible variable i (using Proposition 3.20(b)) such that p is a restriction of $q^{(i)}$ (written $p \subset q^{(i)}$). By Proposition 3.20 we have $\mathbf{Dyn}(q^{(i)}) = \mathbf{Dyn}(p)$, thus p is a core of $q^{(i)}$. By induction there is a collection r of p-roots such that $q^{(i)} = p_r$. By Lemma 3.26 there is a $q^{(i)}$-root v such that $q = q^{(i)}[v] = (p_r)_{e(v)}$, and by the Transitivity Lemma 3.23, q is root induced from p. □

We proceed now to prove the main result of this section, as provided in [9].

Theorem 3.28 (Barot-de la Peña). *Two nonnegative unit forms have the same Dynkin type if and only if they are root equivalent forms.*

Proof. Assume first that $p : \mathbb{Z}^I \to \mathbb{Z}$ and $q : \mathbb{Z}^J \to \mathbb{Z}$ are root equivalent forms. By Proposition 3.27, p and any of its cores are root equivalent, as well as q and any of its cores. Thus we may assume that both p and q are positive unit forms. In this case we have shown that root induction preserves connected components (Lemma 3.25), therefore we may also assume that p and q are connected.

Take a collection of q-roots $r = (r^i)_{i \in I}$ with $p = q_r$ and a collection of p-roots $s = (s^j)_{j \in J}$ with $q = p_s$. Consider the linear maps

$$\begin{array}{ccc} \mathbb{Z}^I \xrightarrow{\ \varphi\ } \mathbb{Z}^J & \text{and} & \mathbb{Z}^J \xrightarrow{\ \psi\ } \mathbb{Z}^I \\ x \longmapsto \sum_{i \in I} x_i r^i & & y \longmapsto \sum_{j \in J} y_j s^j. \end{array}$$

Since $p(x) = q_r(x) = q(\varphi(x))$ and $q(y) = p_s(y) = p(\psi(y))$ and p, q are positive unit forms, both φ and ψ are injective maps, which implies $|I| = |J|$. Moreover, φ and ψ induce respectively injective functions $p^{-1}(1) \to q^{-1}(1)$ and $q^{-1}(1) \to p^{-1}(1)$, and by Proposition 2.3 both sets $p^{-1}(1)$ and $q^{-1}(1)$ are finite. Hence p and q are connected positive unit forms in the same number of variables and with the same number of roots. This implies that p and q have the same Dynkin type (cf. Table 2.1).

Assume now that $\mathbf{Dyn}(p) = \mathbf{Dyn}(q)$, and take cores p' and q' of p and q respectively. By Proposition 3.20 we have $\mathbf{Dyn}(p') = \mathbf{Dyn}(q')$. Since p' and q' are positive unit forms, they are equivalent by Corollary 2.21. Take a matrix T with columns $r^1, \dots, r^m$ such that $p' = q'T$. Then r^i is a q'-root for $i = 1, \dots, m$ (for p' is unitary) and the collection $r = (r^1, \dots, r^m)$ of q'-roots clearly satisfies $p' = q'_r$, that is, p' and q' are root equivalent unit forms. By Proposition 3.27, p is root equivalent to p' and q is root equivalent to q', hence by transitivity we conclude that p and q are root equivalent forms. □

As an interesting consequence of the result above, we show that the number of nonnegative unit forms q without double edges (that is, such that $|q_{ij}| < 2$ for all $i < j$) of fixed Dynkin type is bounded.

Proposition 3.29. *Let q be a nonnegative unit form of Dynkin type Δ, and take $p = q_\Delta$. Then q has no double edge if and only if there exists a collection of p-roots r such that $r \cap -r = \emptyset$ and $q = p_r$.*

Proof. From Theorem 3.28 we know that there is a finite collection r of p-roots such that $q = p_r$. Assume first that there are $r^i = \varepsilon r^j$ in the collection r with $i \neq j$ and $\varepsilon = \pm 1$. Then

$$\varepsilon q_{ij} = p_r(e_i|\varepsilon e_j) = p_r(e_i + \varepsilon e_j) - p_r(e_i) - p_r(e_j) = p(r^i + \varepsilon r^j) - 2 = 2,$$

that is, q has a double edge. On the other hand, if $r^i \neq r^j$ and $r^i \neq -r^j$ for any $i \neq j$, then

$$0 < p(r^i \pm r^j) = p_r(e_i \pm e_j) = 2 \pm q(e_i|e_j) = 2 \pm q_{ij},$$

that is, q has no double edge. □

Proposition 3.29 yields the following immediate consequence.

Corollary 3.30. *There are only finitely many nonnegative unit forms without double edges of a given Dynkin type.*

We end this section with a result necessary for Chaps. 5 and 6.

Corollary 3.31. *Let $q : \mathbb{Z}^n \to \mathbb{Z}$ be a nonnegative unit form with radical generated by a vector $v \in \mathbb{Z}^n$. Then $|v_i| \leq 6$ for $i = 1, \ldots, n$.*

Proof. Consider a core $p : \mathbb{Z}^{n-1} \to \mathbb{Z}$ of q and take a p-root w such that $q = p[w]$. By Proposition 2.22 we have $|w_i| \leq 6$ for $i = 1, \ldots, n-1$, and the result follows since $v = \pm(w + e_n)$, see Lemma 3.26. □

Exercises 3.32.

1. Find all nonnegative unit forms of Dynkin type $\mathbb{A}_3$ without double edges.
2. Find a bound for the number of connected nonnegative unit forms of a given Dynkin type Δ without double edges.
3. Let $q : \mathbb{Z}^n \to \mathbb{Z}$ be a unit form with a root u, and consider a flation $T : \mathbb{Z}^n \to \mathbb{Z}^n$ for q. Show that $T^{-1}u$ is a qT-root, and that $q[u]\overline{T} = (qT)[T^{-1}u]$, where $\overline{T} = T \oplus [1]$.
4. **Doubling vertices.** Let $q : \mathbb{Z}^n \to \mathbb{Z}$ be a unit form. For an index $i \in \{1, \ldots, n\}$ the one-point extension $q[i] = q[-e_i] = q_{v_i}$, where $v_i = (e_1, \ldots, e_n, e_i)$, is called the *doubling of the vertex i*. Consider the morphism $\pi : \mathbb{Z}^{n+1} \to \mathbb{Z}^n$ given by $\pi(e_k) = e_k$ if $k \leq n$ and $\pi(e_{n+1}) = e_i$. Show that:

   a) The mapping π is order preserving.
   b) For all $x, y \in \mathbb{Z}^{n+1}$ we have $q[i](x, y) = q(\pi(x), \pi(y))$.
   c) $\mathbf{rad}(q[i]) = \mathbf{rad}(q) \oplus \mathbb{Z}(e_{n+1} - e_i)$.
   d) A vector $x \in \mathbb{N}^{n+1}$ is a maximal positive root of $q[i]$ if and only if $\pi(x)$ is a maximal positive root of q.

5. Considering Exercise 4 above, show that for any pair of vertices $i \neq j$ in I and $q : \mathbb{Z}^I \to \mathbb{Z}$ a unit form we have $q[i][j] = q[j][i]$.
6. Show that if $q : \mathbb{Z}^{n+1} \to \mathbb{Z}$ is a semi-unit form such that $e_{n+1} - e_i \in \mathbf{rad}(q)$, then $q = q^{(n+1)}[i]$.

3.6 Order of Dynkin Types

With the above analysis of root induction we may generalize the partial order within Dynkin diagrams studied in Proposition 3.18. As defined in [9], for two Dynkin types Γ and Δ we set $\Gamma \leq \Delta$ if there is a nonnegative unit form q such that $\mathbf{Dyn}(q) = \Delta$ and a q-root induced form p with $\mathbf{Dyn}(p) = \Gamma$. In what follows an empty graph will be considered as a Dynkin type, corresponding to the form q_r for an empty set of q-roots r.

Lemma 3.33. *Let Γ and Δ be Dynkin types such that Γ is an immediate predecessor of Δ. Then either $\Delta = \Gamma \sqcup \mathbb{A}_1$ or there is a Dynkin type Θ with $\Gamma = \Theta \sqcup \Gamma'$ and $\Delta = \Theta \sqcup \Delta'$, where Δ' is connected and Γ' is an immediate predecessor of Δ'.*

Proof. Notice first that for Dynkin types Δ_1 and Δ_2, any predecessor of $\Delta_1 \sqcup \Delta_2$ has the shape $\Gamma_1 \sqcup \Gamma_2$ where $\Gamma_i \leq \Delta_i$ for $i = 1, 2$. Indeed, suppose that q is a nonnegative unit form with Dynkin type $\Delta := \Delta_1 \sqcup \Delta_2$. By Theorem 3.28 there is a collection of q_Δ-roots s such that $q = (q_\Delta)_s$. Thus for any q-root induced form q_r (where r is a collection of q-roots), using Lemma 3.23 we have q_Δ-roots t with $q_r = ((q_\Delta)_s)_r = (q_\Delta)_t$. Since $q_\Delta = q_{\Delta_1} \oplus q_{\Delta_2}$ and q_Δ is a positive unit form, by Remark 3.24 we have

$$q_r = (q_{\Delta_1} \oplus q_{\Delta_2})_t = (q_{\Delta_1})_{t'} \oplus (q_{\Delta_2})_{t''},$$

for appropriate collections of q_{Δ_i}-roots t' and t''. This shows that any predecessor of $\Delta_1 \sqcup \Delta_2$ has the shape $\mathbf{Dyn}(q_r) = \Gamma_1 \sqcup \Gamma_2$ with $\Gamma_i \leq \Delta_i$ for $i = 1, 2$.

Now, if Δ is connected, taking $\Theta = \emptyset$ there is nothing to prove. Otherwise there is a Dynkin type Θ such that $\Delta = \Theta \sqcup \Delta'$ with Δ' connected. By the above we have $\Gamma = \Theta' \sqcup \Gamma'$ with $\Theta' \leq \Theta$ and $\Gamma' \leq \Delta'$, with exactly one strict inequality since Γ is an immediate predecessor of Δ (see Exercise 1 below). If $\Gamma' = \Delta'$ then we apply the result to Θ (using induction on the number of connected components) and rearrange components. If $\Theta' = \Theta$ then Γ' is an immediate predecessor of Δ', which completes the proof (observe that Γ' is empty if and only if $\Delta' = \mathbb{A}_1$). □

In order to understand the partial relation in Dynkin types determined by root induction, using Lemma 3.33 it is sufficient to determine the immediate predecessors of all (connected) Dynkin diagrams. This is done in the following result, given in [9], which is used below in Table 3.1 to compute immediate predecessors of Dynkin graphs.

Table 3.1 Immediate predecessors of a Dynkin diagram Δ

Dynkin diagram Δ	Immediate predecessors Γ of Δ with $\lvert\Gamma\rvert = \lvert\Delta\rvert$	Immediate predecessors Γ of Δ with $\lvert\Gamma\rvert < \lvert\Delta\rvert$
$\mathbb{A}_n$ $(n \geq 1)$		$\mathbb{A}_{n-1}$
		$\mathbb{A}_i \sqcup \mathbb{A}_{n-i-1}$ $(i = 1, \ldots, n-2)$
$\mathbb{D}_4$	$\mathbb{A}_1^4$	$\mathbb{A}_3$
$\mathbb{D}_5$	$\mathbb{A}_1^2 \sqcup \mathbb{A}_3$	$\mathbb{A}_4$
		$\mathbb{D}_4$
$\mathbb{D}_6$	$\mathbb{A}_1^2 \sqcup \mathbb{D}_4$	$\mathbb{A}_5$
	$\mathbb{A}_2^3$	$\mathbb{D}_5$
$\mathbb{D}_m$ $(m > 6)$	$\mathbb{A}_1^2 \sqcup \mathbb{D}_{m-2}$	$\mathbb{A}_{m-1}$
	$\mathbb{A}_3 \sqcup \mathbb{D}_{m-3}$	$\mathbb{D}_{m-1}$
	$\mathbb{D}_i \sqcup \mathbb{D}_{m-i}$ $(i = 4, \ldots, m-4)$	
$\mathbb{E}_6$	$\mathbb{A}_1 \sqcup \mathbb{A}_5$	$\mathbb{D}_5$
	$\mathbb{A}_2^3$	
$\mathbb{E}_7$	$\mathbb{A}_7$	$\mathbb{D}_6$
	$\mathbb{A}_1 \sqcup \mathbb{D}_6$	
	$\mathbb{A}_2 \sqcup \mathbb{A}_5$	
$\mathbb{E}_8$	$\mathbb{A}_8$	
	$\mathbb{D}_8$	
	$\mathbb{A}_1 \sqcup \mathbb{E}_7$	
	$\mathbb{A}_2 \sqcup \mathbb{E}_6$	
	$\mathbb{A}_3 \sqcup \mathbb{D}_5$	
	$\mathbb{A}_4^2$	

The notation Σ^m indicates the disjoint union of m copies of a Dynkin type Σ

Theorem 3.34. *Let Γ be an immediate predecessor of a Dynkin diagram Δ. Then Γ is a restriction (by either one or two points) of the extended Dynkin diagram $\widetilde{\Delta}$.*

Proof. Take a nonnegative unit form $q : \mathbb{Z}^J \to \mathbb{Z}$ with $\mathbf{Dyn}(q) = \Delta$, and consider a collection $r = (r^i)_{i \in I}$ of q-roots such that $p = q_r$ has Dynkin type $\mathbf{Dyn}(p) = \Gamma$. The *multi-point* extension $q[r]$, defined by the root induction $q_{e(r)}$ where $e(r) = (e_j)_{j \in J} \sqcup (-r^i)_{i \in I}$ (see Exercise 4 below), also satisfies $\mathbf{Dyn}(q) = \Delta$, and clearly p is equal to the restriction $q[r]^I$. Thus substituting q by $q[r]$ we may assume that $p = q^I$ for some subset of indices $I \subset J$.

Take $j_1, \ldots, j_t$ such that $J = I \sqcup \{j_1, \ldots, j_t\}$, and for $0 \leq a \leq t$ define $I_a = I \sqcup \{j_1, \ldots, j_a\}$. Then we have

$$\Gamma = \mathbf{Dyn}(q^{I_0}) \leq \mathbf{Dyn}(q^{I_1}) \leq \ldots \leq \mathbf{Dyn}(q^{I_{t-1}}) \leq \mathbf{Dyn}(q^{I_t}) = \Delta,$$

and since Γ is an immediate predecessor of Δ, there is exactly one a for which $\mathbf{Dyn}(q^{I_{a-1}}) \neq \mathbf{Dyn}(q^{I_a})$. Of course we may substitute q by q^{I_a} and p by $q^{I_{a-1}}$, so that there is a vertex $i \in J$ such that $p = q^{(i)}$. Observe that Lemma 3.2(a) implies that any omissible vertex j for p is also omissible for q, therefore if the restriction $p^{I'}$ is a core of p, then the restriction $q^{I'}$ has Dynkin type Δ (by Proposition 3.20(a)) and $q^{I'} = (q^{I'})^{(i)}$. Hence we may assume from the beginning that p is a positive unit form with $\mathbf{Dyn}(p) = \Gamma$ such that $p = q^{(i)}$ for a nonnegative unit form q with $\mathbf{Dyn}(q) = \Delta$. Take for simplicity $i = n$. By Theorem 2.20 there is an iterated inflation T for p such that $pT = q_\Gamma$, and clearly $q_\Delta = (q(T \oplus [1]))^{(n)}$. Replacing q by $q(T \oplus [1])$, altogether we get a nonnegative unit form q with $\mathbf{Dyn}(q) = \Delta$ such that $q^{(n)} = q_\Gamma$. In particular by Lemma 3.17 we have $\mathbf{cork}(q) = 0$ or $\mathbf{cork}(q) = 1$.

By construction we have $q_{ij} \leq 0$ for all $i \neq j$ in $J - \{n\}$. If $q_{ni} > 0$ for some $i \neq n$ then the inflation T_{ni}^{+} does not modify the restriction $q^{(n)}$ and takes q into a root equivalent unit form $q' = qT_{ni}^{+}$. Iterating this process we consider two cases, when q is positive and when q has corank one. In the first case take a q-root w with $w_n > 0$ and observe that $w' = T_{ni}^{-}(w) = w + w_n e_i$ is a q'-root with $w'_n > 0$. Since the entries of all roots of a positive unit form are bounded in absolute value by 6 (Proposition 2.22), the process must stop after finitely many steps. Similarly, if q has corank one, take v to be the radical vector with $\mathbf{rad}(q) = \mathbb{Z}v$ and $v_n > 0$. Then $v' = T_{ni}^{-}v = v + v_n e_i$ is a generator of $\mathbf{rad}(q')$ with $v'_n > 0$. Now by Corollary 3.31, all entries of these generators are bounded in absolute value by 6, thus the process must be finite again.

Then we may assume that the associated bigraph G of q has no dotted edges, and by Propositions 2.2 and 3.1 we conclude that G is either Δ or the corresponding extended Dynkin diagram $\widetilde{\Delta}$, hence the result. □

Exercises 3.35.

1. Show that if Γ is an immediate predecessor of a Dynkin type Δ, then for any Dynkin type Σ we have that $\Sigma \sqcup \Gamma$ is an immediate predecessor of $\Sigma \sqcup \Delta$.
2. Let q be a nonnegative unit form with finite collections of roots $r = (r^i)_{i \in I}$ and $s = (s^i)_{i \in I}$. Prove that $q_r = q_s$ if and only if $r - s$ is a collection of radical vectors for q.
3. Show that if $\Gamma \sqcup \Sigma \leq \Delta \sqcup \Sigma$ then $\Gamma \leq \Delta$.
4. For a collection of q-roots $r = (r^1, \dots, r^m)$ where $q : \mathbb{Z}^n \to \mathbb{Z}$ is a nonnegative unit form, define $q[r]$ as $q_{e(r)}$ where $e(r) = (e_1, \dots, e_n, -r^1, \dots, -r^m)$, called the *multi-point extension* of q by the collection r. Show that the iterated one-point extension $q[r^1][r^2] \cdots [r^m]$ is a multi-point extension.
5. Prove or give a counterexample: A quadratic unit form q with $|q_{ij}| \leq 2$ for $i < j$ is nonnegative if and only if $q^{-1}(0)$ is an abelian group.

Chapter 4
Concealedness and Weyl Groups

Consider the quadratic form $q(x_1, x_2, x_3) = 5x_1^2+9x_1x_2+7x_2^2+5x_2x_3+x_3^2+4x_1x_3$, which is far from being a unit quadratic form. Nevertheless, the vectors $v^1 = (1, 0, -2)$, $v^2 = (0, 1, -3)$ and $v^3 = (0, 0, 1)$ form a basis in $\mathbb{Z}^3$, and the corresponding change of basis transformation T takes q into a unitary form,

$$qT(x_1, x_2, x_3) = x_1^2 - x_1x_2 + x_2^2 - x_2x_3 + x_3^2,$$

which is the positive unit form associated to Dynkin graph $\mathbb{A}_3$. The purpose of the first part of this chapter is to find criteria to 'uncover' unitary forms after a change of basis, with special interest in the positive case.

4.1 Completing Bases of $\mathbb{Z}^n$

In this section we consider conditions ensuring that a set of vectors $v^1, \ldots, v^r$ in $\mathbb{Z}^n$ may be completed to a basis $v^1, \ldots, v^r, v^{r+1}, \ldots, v^n$ of $\mathbb{Z}^n$. Recall that if S is a finitely generated free abelian group (written additively), then a subgroup S' of S is called a *pure subgroup* of S if the condition $ns \in S'$ for some $s \in S$ and nonzero $n \in \mathbb{Z}$ implies that s belongs to S'. We say that a vector v in $\mathbb{Z}^n$ is *irreducible* if its entries are relatively prime. Recall that in this case there are integers $\lambda_1, \ldots, \lambda_n$ such that $\sum_{i=1}^n \lambda_i v_i = 1$.

Proposition 4.1. *Let $v^1, \ldots, v^r$ be linearly independent vectors in $\mathbb{Z}^n$. Then the following are equivalent:*

a) There exists a $\mathbb{Z}$-basis $\{v^1, \ldots, v^r, v^{r+1}, \ldots, v^n\}$ of $\mathbb{Z}^n$.
b) If $\lambda_1, \ldots, \lambda_r$ are integers with $\gcd(\lambda_1, \ldots, \lambda_r) = 1$, then $\sum_{i=1}^r \lambda_i v^i$ is an irreducible vector in $\mathbb{Z}^n$.

M. Barot et al., *Quadratic Forms*, Algebra and Applications 25,
https://doi.org/10.1007/978-3-030-05627-8_4

c) *For each linear combination* $\sum_{i=1}^{r} \lambda_i v^i$ *in* $\mathbb{Z}^n$ *with* $\lambda_1, \ldots, \lambda_r \in \mathbb{Q}$, *we have* $\lambda_1, \ldots, \lambda_r \in \mathbb{Z}$.

d) *The subgroup* S *of* $\mathbb{Z}^n$ *generated by* $v^1, \ldots, v^r$ *is a pure subgroup of* $\mathbb{Z}^n$.

Proof. Assume that (a) holds and take integers $\lambda_1, \ldots, \lambda_r$ such that $x := \sum_{i=1}^{r} \lambda_i v^i$ is nonzero and not irreducible. Then there is a vector $y \in \mathbb{Z}^n$ and an integer $m \neq 1$ with $x = my$. Since $\{v^1, \ldots, v^n\}$ is a basis, there are integers $\mu_1, \ldots, \mu_n$ such that $\sum_{i=1}^{n} \mu_i v^i$, therefore

$$0 = x - my = \sum_{i=1}^{r} (\lambda_i - m\mu_i) + \sum_{i=r+1}^{n} \mu_i v^i.$$

Then $\mu_i = 0$ for $i = r+1, \ldots, n$ and $\lambda_i = m\mu_i$ for $i = 1, \ldots, r$ which shows that m divides $\gcd(\lambda_1, \ldots, \lambda_r)$ and (b) holds. That (b) implies (c) can be verified easily and is left as en exercise.

Assume (c) holds, and take a vector x in $\mathbb{Z}^n$ and a nonzero integer m with $mx \in S$. Then $mx = \sum_{i=1}^{r} \lambda_i v^i$ for some integers $\lambda_1, \ldots, \lambda_r$, that is, $x = \sum_{i=1}^{r} \frac{\lambda_i}{m} v^i$. By assumption in (c), the rational numbers $\frac{\lambda_i}{m}$ are actually integers, hence $x \in S$, that is, S is a pure subgroup of $\mathbb{Z}^n$.

Assume finally that S is a pure subgroup of $\mathbb{Z}^n$, and consider the canonical projection $\pi : \mathbb{Z}^n \to \mathbb{Z}^n/S$. Since S is pure, then $\mathbb{Z}^n/S$ is a (finitely generated) torsion free abelian group, therefore it is a free group with rank $n - r$. Take a basis $w^{r+1}, \ldots, w^n$ of $\mathbb{Z}^n/S$ and consider vectors v^i in $\mathbb{Z}^n$ with $\pi(v^i) = w^i$ for $i = r+1, \ldots, n$. We show that $v^1, \ldots, v^n$ is a $\mathbb{Z}$-basis of $\mathbb{Z}^n$. Indeed, if $0 = \sum_{i=1}^{n} \lambda_i v^i$ then

$$0 = \sum_{i=1}^{n} \lambda_i \pi(v^i) = \sum_{i=r+1}^{n} \lambda_i w^i,$$

hence $\lambda_i = 0$ for $i = r+1, \ldots, n$. This yields $0 = \sum_{i=1}^{r} \lambda_i v^i$, which shows that $\lambda_i = 0$ for $i = 1, \ldots, n$, that is, $v^1, \ldots, v^n$ are linearly independent vectors. Now, for $x \in \mathbb{Z}^n$ arbitrary, there are integers $a_{r+1}, \ldots, a_n$ with $\sum_{i=r+1}^{n} a_i w^i = \pi(x)$. Since $y := x - \sum_{i=r+1}^{n} a_i v^i$ is a vector in S (for $\pi(y) = 0$), there are integers $a_1, \ldots, a_r$ with $y = \sum_{i=1}^{r} a_i v^i$, that is, $x = \sum_{i=1}^{n} a_i v^i$, and $v^1, \ldots, v^n$ is a $\mathbb{Z}$-basis of $\mathbb{Z}^n$. □

The following observation is direct consequence of the definition of pure subgroup.

Remark 4.2. Let S' be a pure subgroup of $\mathbb{Z}^n$. Then for any subgroup S of $\mathbb{Z}^n$ containing S', the group S' is pure in S.

Lemma 4.3. *For a nonzero vector* v *in* $\mathbb{Z}^n$ *the following hold:*

a) *The vector* v *is irreducible if and only if* $\mathbb{Z}v$ *is a pure subgroup of* $\mathbb{Z}^n$.

b) If $q : \mathbb{Z}^n \to \mathbb{Z}$ is an integral quadratic form with $q(v) = 1$, then v is irreducible.

Proof. Consider the set $\{v\}$, which is linearly independent for $v \neq 0$. In this case, the equivalence of (b) and (d) in Proposition 4.1 is part (a) of the claim, and part (b) is immediate since $q(\alpha v') = \alpha^2 q(v')$ for any integer α and any v' in $\mathbb{Z}^n$. □

Proposition 4.1 is now used to provide a criterion for the completion to a basis of certain sets of vectors.

Proposition 4.4. *Let v be a vector in $\mathbb{Z}^n$ with $v_{r+1} \neq 0$ for some index $1 \leq r < n$. Then the set of vectors $\{v^1 = e_1, v^2 = e_2, \dots, v^r = e_r, v^{r+1} = v\}$ can be completed to a basis of $\mathbb{Z}^n$ if and only if $\gcd(v_{r+1}, \dots, v_n) = 1$.*

Proof. Consider the restriction $\pi : \mathbb{Z}^n \to \mathbb{Z}^{n-r}$ to the last $n - r$ entries of vectors in $\mathbb{Z}^n$ and the inclusion $\sigma : \mathbb{Z}^{n-r} \to \mathbb{Z}^n$ with $\pi \circ \sigma = \mathbf{Id}_{\mathbb{Z}^{n-r}}$. Then clearly we have a basis $\{v^1, \dots, v^r, v^{r+1}, \dots, v^n\}$ of $\mathbb{Z}^n$ if and only if we have a basis $\{v^1, \dots, v^r, \sigma(\pi(v^{r+1})), \dots, \sigma(\pi(v^n))\}$ of $\mathbb{Z}^n$, and since $\mathbb{Z}^n = \mathbb{Z}^r \oplus \mathbb{Z}^{n-r}$ and the set $\{v^1, \dots, v^r\}$ is basis of $\mathbb{Z}^r$, this happens if and only if $\sigma(\pi(v^{r+1})) = \sigma(\pi(v))$ can be extended to a basis of $\mathbb{Z}^{n-r}$.

Hence $\{v^1, \dots, v^{r+1}\}$ can be extended to a basis of $\mathbb{Z}^n$ if and only if $\sigma(\pi(v)) = (v_{r+1}, \dots, v_n)$ can be extended to a basis of $\mathbb{Z}^{n-r}$, which is equivalent by Proposition 4.1 to the vector $\sigma(\pi(v))$ being irreducible (that is, $\gcd(v_{r+1}, \dots, v_n) = 1$). □

Proposition 4.4 provides a theoretical criterion for completion of certain sets of vectors to $\mathbb{Z}$-bases of $\mathbb{Z}^n$, but the result does not provide a concrete constructive method. We will develop in the following result an explicit algorithm for the construction of a basis containing a given irreducible vector.

Algorithm 4.5 (Completing Bases). *Let $v = (v_1, \dots, v_n)$ be an irreducible vector. We will produce a $\mathbb{Z}$-invertible matrix M_n having the vector v as first column. For a vector $x \in \mathbb{Z}^i$ denote by $x|$ the vector in $\mathbb{Z}^{i-1}$ obtained from x by deleting the last entry x_i. Take also $\gcd(x) = \gcd(x_1, \dots, x_i)$.*

Step 1. Define recursively irreducible vectors $v^{(n)} = v, v^{(n-1)}, \dots, v^{(2)}, v^{(1)}$ with $v^{(i)} \in \mathbb{Z}^i$, and integers λ_i, μ_i and c_{i-1} for $i = n, n-1, \dots, 2$ satisfying the following equations,

$$\begin{aligned}
v^{(n)} &= v; & c_{n-1} &= \gcd(v^{(n)}|); & 1 &= \mu_n c_{n-1} + \lambda_n v_n; \\
c_{n-1} v^{(n-1)} &= v^{(n)}|; & c_{n-2} &= \gcd(v^{(n-1)}|); & 1 &= \mu_{n-1} c_{n-2} + \lambda_{n-1} v^{(n-1)}_{n-1}; \\
c_{n-2} v^{(n-2)} &= v^{(n-1)}|; & c_{n-3} &= \gcd(v^{(n-2)}|); & 1 &= \mu_{n-2} c_{n-3} + \lambda_{n-2} v^{(n-2)}_{n-2}; \\
&\dots & &\dots & &\dots \\
c_2 v^{(2)} &= v^{(3)}|; & c_1 &= \gcd(v^{(2)}|); & 1 &= \mu_2 c_1 + \lambda_2 v^{(2)}_2; \\
c_1 v^{(1)} &= v^{(2)}|.
\end{aligned}$$

Step 2. Define recursively (j) by $(j-1)$ matrices M'_j (for $j = n, n-1, \dots, 2, 1)$ as

$$M'_i = \begin{bmatrix} M'_{i-1} & -\lambda_i v^{(i-1)} \\ 0 \dots 0 & \mu_i \end{bmatrix}.$$

Consider also the j by j matrices $M_j = \begin{bmatrix} v^{(i)} & M'_j \end{bmatrix}$ for $j = 1, \dots, n$.

Then for $j = 1, \dots, n$ there exists a matrix Y_j such that $M_j Y_j = \mathbf{Id}_j$.

Proof. Define $c_{n-1} := \gcd(v|) = \gcd(v_1, \dots, v_{n-1})$. Since c_{n-1} and v_n are coprime, we find numbers $\mu_n, \lambda_n \in \mathbb{Z}$ such that

$$\mu_n c_{n-1} + \lambda_n v_n = 1.$$

Define a new vector $v^{(n-1)} \in \mathbb{Z}^{n-1}$ by $v_j^{(n-1)} c_{n-1} = v_j$ for $j = 1, \dots, n-1$. Then $\gcd(v_1^{(n-1)}, \dots, v_{n-1}^{(n-1)}) = 1$ and we introduce

$$c_{n-2} = \gcd(v_1^{(n-1)}, \dots, v_{n-2}^{(n-1)}), \quad \mu_{n-1} c_{n-2} + \lambda_{n-1} v_{n-1}^{(n-1)} = 1, \quad \text{and} \quad v^{(n-2)} \in \mathbb{Z}^{n-2},$$

with $v^{(n-2)} c_{n-2} = v^{(n-1)}$ for $j = 1, \dots, n-2$. By descending induction we suppose that $v^{(n)} := v$, $v^{(n-1)}, \dots, v^{(1)}$, $c_{n-1}, \dots, c_1$, $\mu_n, \dots, \mu_2$ and $\lambda_n, \dots, \lambda_2$ are well-defined with the corresponding properties.

We show inductively the existence of a $j \times j$ matrix Y_j such that $M_j Y_j = \mathbf{Id}_j$. Since $M_1 = [1]$, take $Y_1 = [1]$. Assume there exists a Y_{n-1} with $M_{n-1} Y_{n-1} = \mathbf{Id}_{n-1}$ and consider the matrix $B_n = \begin{bmatrix} \mu_n & 0 & \lambda_n \\ 0 & \mathbf{Id}_{n-2} & 0 \\ v_n & 0 & c_{n-1} \end{bmatrix}$. Then

$$\begin{aligned} M_n B_n &= \begin{bmatrix} v| & M'_{n-1} & -\lambda_n v^{(n-1)} \\ v_n & 0 \dots 0 & \mu_n \end{bmatrix} \begin{bmatrix} \mu_n & 0 & \lambda_n \\ 0 & \mathbf{Id}_{n-2} & 0 \\ v_n & 0 & c_{n-1} \end{bmatrix} \\ &= \begin{bmatrix} v^{(n)} & M'_{n-1} & 0 \\ 0 & 0 & 1 \end{bmatrix} = \begin{bmatrix} M_{n-1} & 0 \\ 0 & 1 \end{bmatrix}, \end{aligned}$$

since the first column is determined by

$$\mu_n v_i - \lambda_n c_{n-1} v_i^{(n-1)} = (\mu_n c_{n-1} + \lambda_n v_n) v_i^{(n-1)} = v_i^{(n-1)},$$

while the last column is determined by

$$\lambda_n v_i - \lambda_n c_{n-1} v_i^{(n-1)} = 0 \text{ (for } 1 \le i \le n-1) \quad \text{and} \quad \lambda_n v_n + \mu_n c_{n-1} = 1.$$

Consider the matrix $Y_n = B_n \begin{bmatrix} Y_{n-1} & 0 \\ 0 & 1 \end{bmatrix}$ and verify that

$$M_n Y_n = \begin{bmatrix} M_{n-1} & 0 \\ 0 & 1 \end{bmatrix} \begin{bmatrix} Y_{n-1} & 0 \\ 0 & 1 \end{bmatrix} = \begin{bmatrix} \mathbf{Id}_{n-1} & 0 \\ 0 & 1 \end{bmatrix} = \mathbf{Id}_n.$$

This completes the proof of the claim. □

4.2 Concealed Forms

An integral quadratic form $q : \mathbb{Z}^n \to \mathbb{Z}$ is called *concealed* if there exists a $\mathbb{Z}$-basis $\beta = (v^1, \dots, v^n)$ of $\mathbb{Z}^n$ such that $q_\beta := qT_\beta$ is a unit form, where $T_\beta = (v^1 \mid \cdots \mid v^n)$ is the matrix with columns given by vectors $v^1, \dots, v^n$.

Proposition 4.6. *Let q be an integral quadratic form, and take a $\mathbb{Z}$-basis $\beta = (v^1, \dots, v^n)$ of $\mathbb{Z}^n$. Then q_β is a unit form if and only if the basis β consists of roots of q (that is, q is a concealed form if and only if there exists a $\mathbb{Z}$-basis of $\mathbb{Z}^n$ consisting of q-roots).*

Proof. Take $q_\beta = qT_\beta$ where T_β has as columns the vectors in β. The claim follows from equation

$$(q_\beta)_{ii} = q_\beta(e_i) = q(T_\beta e_i) = q(v^i),$$

for $1 \leq i \leq n$. □

Keeping in mind Proposition 4.6 we will consider the following problems:

1. Given an integral quadratic form $q : \mathbb{Z}^n \to \mathbb{Z}$, when is it possible to find a $\mathbb{Z}$-basis of roots of q?
2. In special cases (the positive case, for instance), determine procedures to decide whether the form q is concealed.

We start with an inductive criterion to decide whether an integral quadratic form is concealed.

Lemma 4.7. *Let $q : \mathbb{Z}^n \to \mathbb{Z}$ be an integral quadratic form with a root v^1. Then v^1 may be completed to a $\mathbb{Z}$-basis $\beta = (v^1, \dots, v^n)$ of $\mathbb{Z}^n$. Moreover, if the restriction $q_\beta^{(1)} : \mathbb{Z}^{n-1} \to \mathbb{Z}$ is concealed, where $q_\beta := qT_\beta$ is the form in the basis β, then q itself is a concealed form.*

Proof. The first claim is consequence of Lemma 4.3 and Proposition 4.1.

Take $T_\beta = (v^1 | \cdots | v^n)$ and assume that $q_\beta^{(1)}$ is concealed. By Proposition 4.6 we may consider a $\mathbb{Z}$-basis $\{w^2, \dots, w^n\}$ of $\mathbb{Z}^{n-1}$ consisting of $q_\beta^{(1)}$-roots. Taking $\overline{w}^1 = e_1$ in $\mathbb{Z}^n$ and setting $\overline{w}^i$ as the vector in $\mathbb{Z}^n$ obtained from w^i by adding an entry zero in the first coordinate for $i = 2, \ldots, n$, then $\gamma = (\overline{w}^1, \ldots, \overline{w}^n)$ is a $\mathbb{Z}$-basis of $\mathbb{Z}^n$. Actually, γ is a basis of $\mathbb{Z}^n$ formed by q_β-roots, since

$$q_\beta(w^1) = q_\beta(e_1) = q(T_\beta(e_1)) = p(v^1) = 1,$$

and $q_\beta(\overline{w}^i) = q_\beta^{(1)}(w^i) = 1$ for $i = 2, \ldots, n$. Hence q_β is a concealed form by Proposition 4.6.

Now, the set $\{T_\beta \overline{w}^1, \ldots, T_\beta \overline{w}^n\}$ is a $\mathbb{Z}$-basis of $\mathbb{Z}^n$ (since T_β is $\mathbb{Z}$-invertible and γ is a basis), consisting of q-roots, therefore q is a concealed form. □

The converse of Lemma 4.7 is false. Consider the quadratic form $q(x_1, x_2) = x_1^2 + 2x_1x_2 + 2x_2^2$. The root $v^1 = e_1$ of q can be completed to a $\mathbb{Z}$-basis $\beta = (v^1 = e_1, v^2 = e_2)$ of $\mathbb{Z}^2$. Observe that $q_\beta^{(1)} = q^{(1)} = 2x_2^2$ is not concealed. However, q is a concealed form, since we have a $\mathbb{Z}$-basis $(w^1 = e_1, w^2 = (-1, 1))$ of $\mathbb{Z}^2$ consisting of q-roots.

Exercises 4.8.

1. In the proof of Algorithm 4.5 give an explicit construction of the inverse matrix Y_n as a function of the numbers $v_j^{(j)}, \mu_j, \lambda_j$ and c_{j-1} for $j = 2, \ldots, n$.
2. Find a $\mathbb{Z}$-basis of $\mathbb{Z}^4$ containing the vector $v = (3, 4, 6, 10)$. Do the same for the vector $w = (5, 2, 4, 6)$. Is it possible to find a basis containing both v and w?
3. Let $q : \mathbb{Z}^n \to \mathbb{Z}$ be an integral quadratic form with coefficients $q_{ii} \geq 1$ for each $1 \leq i \leq n$. Suppose that for indices $1 \leq i < j \leq n$ we have $q_{ii} < |q_{ij}| < q_{ii} + q_{jj}$. Show that we may apply a flation $T_{ij}^\epsilon = \mathbf{Id}_n + \epsilon E^{ij}$ (with $\epsilon = \pm 1$) in such a way that $q' = Tq$ has coefficient $q'_{ii} < q_{ii}$.
4. The reduction algorithm given in Exercise 3 is usually quite efficient, but stops when $|q_{ij}| \leq q_{ii}$, for all $1 \leq i, j \leq n$. Show that:

   i) The following form is concealed:

   $$q(x_1, x_2, x_3) = 3x_1^2 - 18x_1x_2 + 31x_2^2 + 9x_1x_3 - 33x_2x_3 + 9x_3^2.$$

   ii) The form

   $$q(x_1, x_2, x_3) = 2x_1^2 - 2x_1x_2 + 2x_2^2 - 2x_1x_3 - 2x_2x_3 + 3x_3^2$$

   cannot be further reduced using the procedure in Exercise 3.

4.3 Positive Concealed Forms

Let Δ be a graph with n vertices, no loop and no dotted edge. Let A_Δ be the symmetric matrix associated to q_Δ (that is, $x^t A_\Delta x = q_\Delta(x)$ for any vector x in $\mathbb{Z}^n$). Recall that the (symmetric) adjacency matrix M_Δ of Δ is defined by taking

M_{ij} as the number of edges between i and j (thus we have $A_\Delta = \mathbf{Id}_n - \frac{1}{2}M_\Delta$). We say that Δ is a *tree* if Δ is connected and $|\Delta_0| = |\Delta_1| + 1$.

Lemma 4.9. *Taking a tree graph Δ with n vertices and matrices A_Δ and M_Δ as above, the following are equivalent:*

a) The tree Δ is a Dynkin diagram.
b) $\mathbf{Spec}(M_\Delta) \subset (-2, 2)$.
c) $\mathbf{Spec}(A_\Delta) \subset (0, 2)$.

In particular, if Δ is a Dynkin diagram, then $0 < \det(A_\Delta) \leq 1$ with equality if and only if $n = 1$.

Sketch of Proof. Take for simplicity $A = A_\Delta$ and $M = M_\Delta$. Observe that there is a correspondence $\mathbf{Spec}(M) \to \mathbf{Spec}(A)$ given by $\lambda \mapsto 1 - \frac{1}{2}\lambda$. Recall that Δ is a Dynkin diagram if and only if q_Δ is a positive form (Proposition 2.2), that is, if and only if $\mathbf{Spec}(A)$ is contained in the set of positive real numbers (Proposition 1.32), or equivalently, if and only if $\lambda < 2$ for all λ in $\mathbf{Spec}(M)$.

It can be shown that if Δ is a tree, then the coefficients b_{2s+1} of the characteristic polynomial $p_M(t) = \det(t\mathbf{Id}_n - M) = \sum_{i=0}^{n} b_i t^i$ are all zero. In other words, $p_M(t) = p'(t^2)$ for a polynomial p' of degree $\frac{n}{2}$ if $n$is even, or $p_M(t) = tp'(t^2)$ for a polynomial p' of degree $\frac{n-1}{2}$ if n is odd (see Exercise 10 below). In this case, $\mathbf{Spec}(M)$ is symmetric with respect to 0 (i.e. $-\lambda \in \mathbf{Spec}(M)$ if and only if $\lambda \in \mathbf{Spec}(M)$).

This already shows that (a) implies (b), and that (b) implies (c). If Δ is not a Dynkin diagram, then Δ contains a full subgraph which is an extended Dynkin diagram Δ' (see Exercise 2.10.5). However, it is known that the adjacency matrix of any extended Dynkin graph Δ' has spectral radius $\rho_{\Delta'} = 2$ (cf. Theorem 3.11.1 in [20], see also Exercise 7 below). It is also known that $\rho_{\Delta'} < \rho_\Delta$ (see Propositions 1.3.9 and 1.3.10 in [20], or Exercises 6 and 9 below). We conclude that $\mathbf{Spec}(M) \not\subset (-2, 2)$, and by the above correspondence, that $\mathbf{Spec}(A) \not\subset (0, 2)$.

Finally, if Δ is a Dynkin graph and $\lambda_1, \ldots, \lambda_m$ are the *positive* eigenvalues of M (for $m \geq 0$), then $\mathbf{Spec}(A) \subset \{1, 1 - \frac{1}{2}\lambda_i, 1 + \frac{1}{2}\lambda_i \mid 1 \leq i \leq m\}$. Since $0 < \lambda_i < 2$ for $1 \leq i \leq m$, then by the above correspondence we have

$$0 < \det(A) = \prod_{i=1}^{m}(1 - \frac{1}{2}\lambda_i)(1 + \frac{1}{2}\lambda_i) = \prod_{i=1}^{m}(1 - \frac{1}{4}\lambda_i^2) \leq 1,$$

with equality only if $m = 0$ (that is, only if M has no nonzero eigenvalue). However, it is easy to show that a symmetric matrix M has no nonzero eigenvalue only when M is the zero matrix (see Exercise 4), which by connectedness happens only when Δ consists of a single isolated vertex. □

Using Proposition 4.9 we get a handy criterion to verify the concealedness of positive forms.

Proposition 4.10. *Let $q : \mathbb{Z}^n \to \mathbb{Z}$ be a concealed positive integral form. Consider the associated symmetric matrix A_q such that $x^t A_q x = q(x)$ for any vector x in $\mathbb{Z}^n$. Then $0 < \det(A_q) \le 1$.*

Proof. Since q is concealed, there exists an invertible $\mathbb{Z}$-matrix S such that $q' = qS$ is a unitary form. Reducing by inflations, as in Theorem 2.20, we get an invertible $\mathbb{Z}$-matrix T and a family of Dynkin graphs $\Delta_1, \dots, \Delta_s$ such that $q'T = q_{\Delta_1} \oplus \cdots \oplus q_{\Delta_s}$. Since $\det(A_q)$ is positive and $\det(T) = \pm 1 = \det(S)$ then

$$0 < \det(A_q) = \det(T)^2 \det(S)^2 \prod_{i=1}^{s} \det(A_{\Delta_i}) = \prod_{i=1}^{s} \det(A_{\Delta_i}),$$

where A_{Δ_i} is the symmetric matrix such that $x^t A_{\Delta_i} x = q_{\Delta_i}(x)$. Then the result follows from the last claim in Lemma 4.9. □

Remark 4.11. An alternative proof of Proposition 4.10 uses *Hadamard's Theorem* (see Exercises 2 and 3 below).

Recall that a positive form $q(x_1, \dots, x_n) = \sum_{i=1}^{n} q_{ii} x_i^2 + \sum_{i<j} q_{ij} x_i x_j$ can be written by Lagrange's Method 1.30 as

$$q(x_1, \dots, x_n) = \sum_{i=1}^{n} b_i X_i^2,$$

where $b_1, \dots, b_n$ are positive rational numbers and $X_i = \sum_{j=i}^{n} c_{ij} x_j$ for certain $c_{ij} \in \mathbb{Q}$. Moreover, these numbers can be inductively defined for $1 \le i < j \le n$ as

$$b_i = q_{ii} - \sum_{k<i} b_k c_{ki}^2,$$

$$c_{ii} = 1, \qquad \text{and} \qquad c_{ij} = \frac{1}{2} b_i^{-1} \left[q_{ij} - \sum_{k<i} 2 b_k c_{ki} c_{kj} \right].$$

As an application we get the following simple bound for the coordinates of roots of positive forms.

Proposition 4.12. *Let $q(x_1, \dots, x_n)$ be a positive integral quadratic form and define b_i (for $1 \le i \le n$) and c_{ij} (for $1 \le i \le j \le n$) as above. Let $v \in \mathbb{Z}^n$ be a root of q. Then the following holds for $1 \le i \le n$,*

$$|v_i| = \left\lfloor \sum_{i<j} |c_{ij} v_j| + \frac{1}{\sqrt{b_i}} \right\rfloor \le \left\lfloor \sum_{i \le j_1 < j_2 < \cdots < j_s \le n} \frac{|c_{ij_1} c_{j_1 j_2} \cdots c_{j_{s-1}, j_s}|}{\sqrt{b_{j_s}}} + \frac{1}{\sqrt{b_i}} \right\rfloor,$$

where $\lfloor \alpha \rfloor$ denotes the greatest integer m smaller than or equal to a real number α.

Table 4.1 Determinant of the symmetric matrix associated to the quadratic form of Dynkin diagrams

Δ	$\mathbb{A}_n$	$\mathbb{D}_m$ $(m \geq 4)$	$\mathbb{E}_6$	$\mathbb{E}_7$	$\mathbb{E}_8$
$\det(A_\Delta)$	$\frac{n+1}{2^n}$	$\frac{m-1}{2^{m-2}}$	$\frac{3}{16}$	$\frac{1}{16}$	$\frac{1}{64}$

Proof. For $i = n$ we have $b_n v_n^2 \leq 1$. By descending induction, assume that

$$|v_k| \leq \left\lfloor \sum_{j=k+1}^{n} |c_{kj} v_j| + \frac{1}{\sqrt{b_k}} \right\rfloor, \quad \text{for } k \geq i+1.$$

Since $b_i(v_i + \sum_{j>i} c_{ij} v_j)^2 \leq 1$, we have $-\frac{1}{\sqrt{b_i}} - \sum_{j=i+1}^{n} c_{ij} v_j \leq v_i \leq \frac{1}{\sqrt{b_i}} + \sum_{j=i+1}^{n} c_{ij} v_j$ and the claimed inequalities follow. □

Exercises 4.13.

1. Let A_Δ be the symmetric n by n matrix associated to q_Δ (that is, such that $x^t A_\Delta x = q_\Delta(x)$ for any vector x in $\mathbb{Z}^n$) where Δ is a Dynkin diagram. Determine the determinant of A_Δ. [Hint: see Table 4.1.]
2. Prove **Hadamard's Theorem**:
   Let S be a symmetric positive real n by n matrix. Show that $\det(S) \leq \prod_{i=1}^{n} S_{ii}$, with equality if and only if S is a diagonal matrix. [Hint: Define $d_i = S_{ii}^{-\frac{1}{2}}$ and let $D = \mathbf{diag}(d_1, d_2, \ldots, d_n)$. Therefore $\det(DSD) \leq 1$ if and only if $\det(S) \leq \prod_{i=1}^{n} S_{ii}$ and thus we may assume that all diagonal entries of S are equal to 1. If $0 \leq \lambda_1 \leq \lambda_2 \leq \cdots \leq \lambda_n$ are the eigenvalues of S, we have

$$\det(S) = \prod_{i=1}^{n} \lambda_i \leq \left(\frac{1}{n} \sum_{i=1}^{n} \lambda_i\right)^n = \left(\frac{1}{n}\,\mathbf{tr}(A)\right)^n = 1,$$

   applying the generalized arithmetic-geometric mean inequality for nonnegative real numbers.]
3. Prove Proposition 4.10 as a consequence of Hadamard's Theorem.
4. Let M be a real square matrix. Show that

   i) M is nilpotent if and only if M has no nonzero eigenvalue.
   ii) If M is symmetric, then M is nilpotent if and only if M is the zero matrix.

5. **Characteristic polynomial and spectral radius of a graph.** Consider a connected graph G_1 and a selected vertex x_2. Construct a new graph G by adding a vertex x_1 to G_1 and a single edge joining x_1 and x_2.

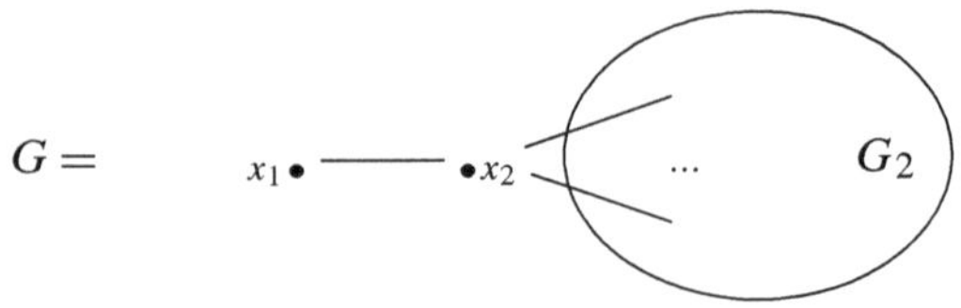

Show that the characteristic polynomial p_G of the graph G can be given as follows, $p_G(z) = zp_{G_1}(z) - p_{G_2}(z)$, where G_2 is obtained from G_1 by deleting the vertex x_2 (and all edges involving x_2).

6. Use Exercise 5 and the recursion formulas for Chebyshev polynomials given in Sect. 1.5 to complete and verify the contents of Table 4.2.
7. Let $\rho(\Delta)$ denote the *spectral radius* of the characteristic polynomial χ_Δ.

   a) Verify that for an extended Dynkin diagram $\widetilde{\Delta}$ we have $\rho(\widetilde{\Delta}) = 2$.
   b) Verify that for a Dynkin diagram Δ we have $\rho(\Delta) < 2$.

8. Show that $\rho(G) = \limsup \sqrt[k]{a_{ij}^{(k)}}$, where $A^k = (a_{ij}^{(k)})_{i,j=1}^n$ are the powers of $A = A_G$, the adjacency matrix of a connected graph G. [Hint: Take $\rho = \rho(G)$. There exists a $c > 0$ such that $a_{ij}^{(k)} \leq c\rho^k$. Use this to show that $\limsup a_{nn}^{(k)} = \rho$. Then by connectivity show that $\limsup a_{in}^{(k)} = \rho$.]
9. Use Exercise 8 to prove that $\rho(\Delta') \leq \rho(\Delta)$ for a full subgraph Δ' of Δ. [Hint: show that $a_{ij}^{(k)}$ is the number of walks of length k from i to j.]
10. Let G be a tree graph. Show that the characteristic polynomial p_G has the shape $p_G(t) = \widetilde{q}(t^2)$ if p_G has even degree, or $p_G(t) = t\widetilde{q}(t^2)$ if p_G has odd degree, for some polynomial $\widetilde{q}$. [Hint: observe that any tree G is a *bipartite graph* (that is, its vertex set G_0 can be partitioned into disjoint sets G_0' and G_0'' such that all edges in G have an end-point in G_0' and an end-point in G_0''). Those graphs admit an ordering of vertices such that their adjacency matrix has the shape $\begin{pmatrix} 0 & N \\ N^t & 0 \end{pmatrix}$ for some matrix N, see [16, 1.3.6].]
11. Prove or give a counter-example: for Δ a tree graph, $\det(A_\Delta) = 0$ if and only if Δ is an extended Dynkin diagram.

Table 4.2 Characteristic polynomials of Dynkin diagrams for $n \geq 1$, $m \geq 4$ and $p = 6, 7, 8$, where T_n and U_n are the Chebyshev polynomials of the first and second kind

Graph	Characteristic polynomial
$\mathbb{A}_n$	$U_n(t/2)$
$\mathbb{D}_m$	$U_m(t/2) - U_{m-4}(t/2)$
$\mathbb{E}_p$	?
$\widetilde{\mathbb{A}}_n$	$2[T_n(t/2) - 1]$
$\widetilde{\mathbb{D}}_m$	?
$\widetilde{\mathbb{E}}_p$	?

4.4 The Weyl Group of a Unit Form

For an integral quadratic form $q : \mathbb{Z}^n \to \mathbb{Z}$ we denote by $\mathcal{O}(q)$ the group of *isometries* of q, that is, the set of linear $\mathbb{Z}$-invertible transformations $T : \mathbb{Z}^n \to \mathbb{Z}^n$ (with product given by composition) such that $q(x) = q(Tx)$ for any vector x in $\mathbb{Z}^n$. For instance, if v is a reflection vector of q (see Sect. 1.2) then the reflection $\sigma_v : \mathbb{Z}^n \to \mathbb{Z}^n$ given by $\sigma_v(x) = x - \frac{q(x|v)}{q(v)}v$ is an isometry for q where $q(-|-) : \mathbb{Z}^n \times \mathbb{Z}^n \to \mathbb{Z}$ is the (symmetric) bilinear form associated to q (cf. Lemma 1.5(c)). Recall that for q a unit form the subgroup of $\mathcal{O}(q)$ generated by simple reflections, denoted $\mathbf{W}(q)$, is called the *Weyl group* of q.

In what follows we identify linear transformations from $\mathbb{Z}^n$ to $\mathbb{Z}^n$ with the $n \times n$ matrix corresponding to the canonical basis. If $T : \mathbb{Z}^n \to \mathbb{Z}^n$ is a $\mathbb{Z}$-invertible linear transformation and $q' = qT$, then the function $\Phi_T : \mathcal{O}(q') \to \mathcal{O}(q)$ given by $B \mapsto TBT^{-1}$ is an isomorphism of groups. Indeed, if A_q and $A_{q'}$ are the symmetric matrices associated to q and q' and $B \in \mathcal{O}(q')$, then $A_{q'} = T^t A_q T$ and $B^t A_{q'} B = A_{q'}$, thus we have

$$(TBT^{-1})^t A_q (TBT^{-1}) = T^{-t} B^t A_{q'} B T^{-1} = T^{-t} A_{q'} = T^{-1} = A_q,$$

that is, $\Phi_T(B)$ is an isometry of q. That Φ_T is a morphism of groups with inverse $\Phi_{T^{-1}}$ is clear. In particular, if q is a concealed form then the Weyl group $\mathbf{W}(q)$ is defined up to isomorphism (depending on the chosen basis of q-roots, see Proposition 4.6).

Proposition 4.14. *Assume that $q : \mathbb{Z}^n \to \mathbb{Z}$ is a positive integral quadratic form. The following hold:*

a) For $B \in \mathcal{O}(q)$ the set of eigenvalues of B is contained in the set $\{1, -1\}$.
b) The group of isometries $\mathcal{O}(q)$ is generated by the reflections σ_v with v a reflection vector.
c) If q is a concealed form then $\mathcal{O}(q)$ is a finite group.

Proof. (a) Let $A = A_q$ be the symmetric matrix associated to q. For $B \in \mathcal{O}(q)$ we have $B^t AB = A$. As in Sect. 1.6 there is a unitary matrix (that is, $U^{-1} = U^t$) such that

$$UAU^t = D := \mathbf{diag}(\lambda_1, \lambda_2, \ldots, \lambda_n),$$

where $0 < \lambda_1 \leq \lambda_2 \leq \cdots \leq \lambda_n$ are the eigenvalues of A (all of them positive by Proposition 1.32). Hence $(UBU^t)^t D(UBU^t) = D$, and for any eigenvector $0 \neq x \in \mathbb{C}^n$ of B with $Bx = \mu x$, we get an eigenvector $y = Ux$ of UBU^t with eigenvalue μ. Using complex conjugation $x \mapsto \overline{x}$ in $\mathbb{C}^n$ we get,

$$0 < \sum_{i=1}^{n} \lambda_i |y_i|^2 = \overline{y}^t Dy = \mu^2 \overline{y}^t (UBU^t)^t D(UBU^t)y = \mu^2 \overline{x}^t Ax,$$

and $\mu^2 \overline{x}^t Ax = \overline{x}^t B^t ABx = \overline{x}^t Ax$ implies that $\mu^2 = 1$. Hence $\mu = \pm 1$.

(*b*) Let B be an isometry of q, and take $0 \neq v \in \mathbb{Q}^n$ with $Bv = \varepsilon v$ and $\varepsilon \in \{1, -1\}$. Clearly we may assume that $v \in \mathbb{Z}^n$. Consider the lattice $v^\perp = \{w \in \mathbb{Z}^n \mid q(v|w) = 0\}$, which is a pure subgroup of $\mathbb{Z}^n$ isomorphic to $\mathbb{Z}^{n-1}$ since $v \notin v^\perp$ and $q(-|-)$ is an inner product in $\mathbb{Q}^n$ (for q is a positive form). Let $y \in \mathbb{Z}^n$ be such that $\mathbb{Z}^n = \mathbb{Z}y \oplus v^\perp$. Consider a basis $y^1 = y, y^2, \ldots, y^n$ of $\mathbb{Z}^n$ with $v^\perp = \bigoplus_{i=2}^n \mathbb{Z}y^i$.

Notice that $Bv^\perp \subset v^\perp$, since $q(v|Bw) = \varepsilon q(Bv|Bw) = \varepsilon q(v|w) = 0$ for $w \in v^\perp$. Therefore, in the basis $y^2, \ldots, y^n$, the restriction B' of the matrix B is an isometry of the restriction q' of q to $v^\perp$, that is, $B' \in \mathcal{O}(q')$. Since q' is positive, by induction $B' = \sigma'_{v_1} \ldots \sigma'_{v_m}$ for some reflection vectors $v_1, \ldots, v_m \in v^\perp$, where σ'_{v_i} denotes the corresponding reflection in $v^\perp$.

Observe that the matrix B and all reflections σ_w with $w \in v^\perp$ have the following shape,

$$B = \begin{bmatrix} \varepsilon & 0 \\ w & B' \end{bmatrix}, \qquad \sigma_w = \begin{bmatrix} 1 & 0 \\ w' & \sigma'_w \end{bmatrix},$$

since $\varepsilon \det B' = \det B = B_{11} \det B'$.

Assume $\varepsilon = 1$. Then B and the composition $\sigma_{v_1} \cdots \sigma_{v_m}$ are n by n matrices X satisfying

$$X = \begin{bmatrix} 1 & 0 \\ x & B' \end{bmatrix}, \qquad \text{and} \quad X^t A X = A = \begin{bmatrix} a_{11} & a^t \\ a & A' \end{bmatrix},$$

hence $x = (B'^t A')^{-1}(\mathbf{Id}_{\mathbb{Z}^{n-1}} - B'^t)a$. This implies that $B = \sigma_{v_1} \cdots \sigma_{v_m}$.

Assume $\varepsilon = -1$. Then $Y = B(\sigma_{v_1} \ldots \sigma_{v_m})^{-1}$ is an integral n by n matrix with shape

$$Y = \begin{bmatrix} -1 & 0 \\ y & \mathbf{Id}_{n-1} \end{bmatrix}, \qquad \text{and} \quad Y^t A Y = A.$$

Therefore $y = 2A'^{-1}a$ and we may consider the vector $v_0 = \begin{bmatrix} 1 \\ -A'^{-1}a \end{bmatrix}$ in $\mathbb{Z}^n$ satisfying

$$q(x|v_0) = \begin{bmatrix} x_1 & x' \end{bmatrix} \begin{bmatrix} a_{11} & a^t \\ a & A' \end{bmatrix} \begin{bmatrix} 1 \\ -A'^{-1}a \end{bmatrix} = x_1(a_{11} - a^t A'^{-1} a).$$

This shows that $\frac{q(x|v_0)}{q(v_0)} = 2x_1$ is always an integer, thus v_0 is a reflection vector.

Finally, $\sigma_{v_0} \begin{bmatrix} x_1 \\ x' \end{bmatrix} = \begin{bmatrix} x_1 \\ x' \end{bmatrix} - 2x_1 \begin{bmatrix} 1 \\ -A'^{-1}a \end{bmatrix} = \begin{bmatrix} -x_1 \\ x' + x_1 y \end{bmatrix} = Y \begin{bmatrix} x_1 \\ x' \end{bmatrix}$, which means that $B = \sigma_{v_0}\sigma_{v_1} \ldots \sigma_{v_m}$. This completes the proof of (*b*).

(c) Let $R(q)$ be the set of q-roots, and denoted by $\mathbf{Perm}(R(q))$ the group of permutations of $R(q)$. Then we have a function $\mathcal{O}(q) \to \mathbf{Perm}(R(q))$ since any element in $\mathcal{O}(q)$ preserves the roots of q. If two elements B, B' in $\mathbf{W}(q)$ satisfy $Bv^i = B'v^i$ for a $\mathbb{Z}$-basis $v^1, \ldots, v^n$ of $\mathbb{Z}^n$, then obviously $B = B'$. By Proposition 4.6 the set $R(q)$ contains a $\mathbb{Z}$-basis (for q is concealed), therefore $\mathcal{O}(q)$ may be seen as a subset of $\mathbf{Perm}(R(q))$ (actually a subgroup). Since q is positive, the set $R(q)$ of q-roots is finite (Proposition 2.3) and so is $\mathbf{Perm}(R(q))$, which completes the proof. □

Let $q : \mathbb{Z}^n \to \mathbb{Z}$ be a unit form. A root $v \in R(q)$ is called a *real root* of q if there exists an element w in the Weyl group $\mathbf{W}(q)$ of q such that $w(e_i) = v$ for some $1 \leq i \leq n$. Nonreal roots are also called *imaginary*.

Proposition 4.15. *Let $q : \mathbb{Z}^n \to \mathbb{Z}$ be a unit form and v be a root of q. The following hold:*

a) If q is positive then v is a real root of q and $\sigma_v \in \mathbf{W}(q)$.
b) If q is weakly positive and $v \in R^+(q)$ is a positive root of q, then there is a sequence of indices $i_1, \ldots, i_s$ and some $1 \leq j \leq n$ such that

$$e_j < \sigma_{i_1}(e_j) < \sigma_{i_2}\sigma_{i_1}(e_j) < \cdots < \sigma_{i_s} \cdots \sigma_{i_2}\sigma_{i_1}(e_j) = v.$$

In particular v is a real root of q and $\sigma_v \in \mathbf{W}(q)$.

Proof. (a) Assume that q is positive. By induction on the weight $|v| = \sum_{i=1}^n |v_i|$ of v we construct an isometry $w \in \mathbf{W}(q)$ such that $w(e_j) = v$ for some $1 \leq j \leq n$. If $|v| = 1$, then $v = \pm e_j$ for some $1 \leq j \leq n$ (notice that if $v = -e_j$ then $v = \sigma_j(e_j)$). Suppose $|v| > 1$. Then

$$2 = q(v|v) = \sum_{i=1}^n v_i q(e_i|v),$$

and there is some index $1 \leq k \leq n$ with $v_k q(e_k|v) > 0$. Assuming that $v_k > 0$ then $q(e_k|v) > 0$ and

$$0 < q(v - e_k) = 2 - q(v|e_k),$$

implies that $q(e_k|v) = 1$. Hence $\sigma_k(v) = v - e_k$ satisfies $|\sigma_k(v)| = |v| - 1$, and by induction hypothesis, $\sigma_k(v) = w(e_j)$ for some $w \in \mathbf{W}(q)$ and $1 \leq j \leq n$. Then $v = \sigma_k w(e_j)$ with $\sigma_k w \in \mathbf{W}(q)$, for σ_k is an involution (Lemma 1.5(b)). If $v_k < 0$ we get similarly $q(e_k|v) = -1$ and $\sigma_k(v) = v + e_k$ with $|\sigma_k(v)| = |v| - 1$, and the proof continues as in the first case.

To show that $\sigma_v \in \mathbf{W}(q)$ we proceed again by induction on the weight of v. If $v = \pm e_j$ for some $j \in \{1, \ldots, n\}$, then σ_v is the simple reflection σ_j. If $|v| > 1$ then as before there is an index $1 \leq k \leq n$ such that $|\sigma_k(v)| = |v| - 1$, and by induction hypothesis we have $w' := \sigma_{\sigma_k(v)} \in \mathbf{W}(q)$. Then $w' = \sigma_{\sigma_k(v)} = \sigma_k \sigma_v \sigma_k^{-1} = \sigma_k \sigma_v \sigma_k$

(see Lemma 1.5 and Exercise 1.7.4), which means that $\sigma_v = \sigma_k w' \sigma_k \in \mathbf{W}(q)$ as claimed.

(*b*) Both statements may be proved as in (*a*), following Proposition 2.5 and Remark 2.6. □

As direct consequence, if q is a positive concealed form then the Weyl group $\mathbf{W}(q)$ of q is uniquely determined, namely $\mathbf{W}(q) = \mathcal{O}(q)$.

Examples 4.16. The description of Weyl groups for some quadratic forms is known. Let us consider a few examples.

a) Consider $q = q_{\mathbb{A}_n}$ the form associated to the Dynkin diagram $\mathbb{A}_n$ for $n \geq 1$. Define a group homomorphism from the Weyl group $\mathbf{W}(q)$ to the group of permutations of $n+1$ elements S_{n+1}, sending σ_i to the transposition $(i, i+1)$. As an orthogonal transformation of $\mathbb{Z}^{n+1}$ the matrix associated to $(i, i+1)$ is the reflection at the vector $e_i - e_{i+1}$. Hence $\mathbf{W}(q)$ is isomorphic to the group generated by $(i, i+1)$, $1 \leq i \leq n$, that is, to the group of permutations S_{n+1}. In particular $\mathbf{W}(q)$ has $(n+1)!$ elements.
b) Let $q_m(x_1, x_2) = x_1^2 + x_2^2 - m x_1 x_2$ be the m-Kronecker form. Clearly,

$$\sigma_1 = \begin{bmatrix} -1 & 0 \\ m & 1 \end{bmatrix}, \quad \sigma_2 = \begin{bmatrix} 1 & m \\ 0 & -1 \end{bmatrix} \quad \text{and} \quad w = \sigma_1\sigma_2 = \begin{bmatrix} -1 & -m \\ m & m^2 - 1 \end{bmatrix},$$

(the matrix on the right is called *Coxeter matrix* corresponding to the Kronecker form q_m, see Theorem 4.31 below). For $|m| > 1$, the subgroup of $\mathbf{W}(q_m)$ generated by w is cyclic of infinite order, and since $\sigma_1 w \sigma_1 = \sigma_2 w \sigma_2 = \sigma_2\sigma_1 = (\sigma_1\sigma_2)^{-1}$, it is a normal subgroup. Moreover, $\sigma_1 w = \sigma_2$, which shows that $\mathbf{W}(q_m)$ is isomorphic to the semi-direct product of $\langle \sigma_2 \rangle \cong \mathbb{Z}/2\mathbb{Z}$ with $\langle w \rangle \cong \mathbb{Z}$ (written $\mathbf{W}(q_m) \cong (\mathbb{Z}/2\mathbb{Z}) \ltimes \mathbb{Z}$).
c) Let $\widetilde{\Delta}$ be an extended Dynkin graph with n vertices and take $\tilde{q} = q_{\widetilde{\Delta}} : \mathbb{Z}^n \to \mathbb{Z}$ the corresponding unit form. Consider the positive generator $\delta \in \mathbb{Z}^n$ of the radical of $\tilde{q}$ (see Table 2.2). The corresponding Dynkin graph Δ with $n-1$ vertices can be seen as a subgraph of $\widetilde{\Delta}$ such that, taking i the additional vertex, we have $\delta_i = 1$ and $w := \delta - e_i$, as a vector in $\mathbb{Z}^{n-1}$, is the maximal positive root of the restriction $\tilde{q}^{(i)} = q_\Delta$ (see Table 2.1). Hence

$$0 = \tilde{q}(\delta) = \tilde{q}(w) + \tilde{q}(e_i) + \tilde{q}(w|e_i), \quad \text{that is,} \quad \tilde{q}(w|e_i) = -2.$$

As indicated in Exercise 4, the element $\sigma_w\sigma_i$ has infinite order in $\mathcal{O}(\tilde{q})$. Moreover, since $\mathbf{W}(q_\Delta)$ is a finite group (Proposition 4.14), Proposition 4.15 yields $\sigma_w \in \mathbf{W}(q_\Delta) \subset \mathbf{W}(\tilde{q})$. Hence $\sigma_w\sigma_i \in \mathbf{W}(\tilde{q})$, which is an infinite group.

Lemma 4.17. *Let q and q' be non-negative unit forms such that there is an iterated flation $T : \mathbb{Z}^n \to \mathbb{Z}^n$ for q with $q' = qT$. Then the isomorphism $\Phi_T : \mathcal{O}(q') \to \mathcal{O}(q)$ given by $B \mapsto TBT^{-1}$ restricts to an isomorphism $\Phi_T : \mathbf{W}(q') \to \mathbf{W}(q)$.*

Proof. Assume first that T is a Gabrielov transformation, that is, there are vertices $i \neq j$ such that $T(x) = x - q_{ij}x_i e_j$ for any vector x in $\mathbb{Z}^n$. Then $T(e_k) = e_k$ if $k \neq i$ and $T(e_i) = \sigma_j(e_i)$, thus for any $1 \leq k \leq n$ there is an element w_k in $\mathbf{W}(q')$ with $T(e_k) = w_k(e_k)$. Using Lemma 1.5 and Exercise 1.7.4 we have

$$\Phi_T(\sigma'_k) = T\sigma'_k T^{-1} = \sigma_{T(e_k)} = \sigma_{w_k(e_k)} = w_k \sigma_k w_k^{-1},$$

which is an element in $\mathbf{W}(q)$, where σ' and σ denote reflections of q' and q respectively. Since T is invertible and its inverse is a Gabrielov transformation, by the above arguments $\Phi_{T^{-1}}$ is inverse of Φ_T.

Using now that q is nonnegative, then any iterated flation T for q such that qT is unitary is a composition of Gabrielov transformations (see Lemma 3.4), hence the result follows from the above. □

Theorem 4.18. *Let $q : \mathbb{Z}^n \to \mathbb{Z}$ be a unit form. Then $\mathbf{W}(q)$ is a finite group if and only if q is a positive form. In this case there is a finite collection of Dynkin diagrams $\Delta_1, \dots, \Delta_s$ such that $\mathbf{W}(q)$ is isomorphic to the direct product $\prod_{i=1}^s \mathbf{W}(q_{\Delta_i})$ (where the disjoint union $\Delta_1 \sqcup \dots \sqcup \Delta_s$ is the Dynkin type of q as defined in Sect. 2.4).*

Proof. In Proposition 4.14 we already showed that if q is positive then $\mathbf{W}(q) = \mathcal{O}(q)$ is a finite group. For the converse assume that $\mathbf{W}(q)$ is finite and that q is connected and nonpositive (see Exercise 2 below). Since for any restriction $q^{(i)}$ the Weyl group $\mathbf{W}(q^{(i)})$ of $q^{(i)}$ may be seen as a subgroup of $\mathbf{W}(q)$, we may assume by induction on n that $q^{(i)}$ is positive for $1 \leq i \leq n$.

Hence q is critical nonpositive, and by Theorem 2.12, since q is not the Kronecker form q_m for any integer m with $|m| \geq 3$ (by Example 4.16(b)), then q is nonnegative of corank one with a sincere radical vector. As shown in Theorem 3.5 (see also Step 1 in the proof of Theorem 3.8), there is an iterated flation T such that $qT = q_{\widetilde{\Delta}}$ for some extended Dynkin diagram $\widetilde{\Delta}$. Then by Lemma 4.17 we have $\mathbf{W}(q) \cong \mathbf{W}(q_{\widetilde{\Delta}})$, which is impossible since $\mathbf{W}(q_{\widetilde{\Delta}})$ is an infinite group as shown in Example 4.16(c).

For the second claim, if q is a positive unit form then by Theorem 2.20 there is an iterated flation T and a Dynkin type Δ (disjoint union of Dynkin diagrams $\Delta_1, \dots, \Delta_s$) such that $qT = q_\Delta$. Again by Lemma 4.17 and using Exercise 2 we conclude that

$$\mathbf{W}(q) \cong \mathbf{W}(q_\Delta) = \mathbf{W}(q_{\Delta_1} \oplus \cdots \oplus q_{\Delta_s}) = \prod_{i=1}^{s} \mathbf{W}(q_{\Delta_i}).$$

□

Exercises 4.19.

1. Let $q : \mathbb{Z}^n \to \mathbb{Z}$ be a unit form and take indices $1 \leq i < j \leq n$. Consider the simple reflections σ_j and σ_i in the Weyl group $\mathbf{W}(q)$ of q. Show that

a) $\sigma_i\sigma_j = \sigma_j\sigma_i$ (that is, $(\sigma_i\sigma_j)^2 = 1$) if and only if $q_{ij} = 0$.
b) $(\sigma_i\sigma_j)^3 = 1$ if and only if $q_{ij} \in \{1, -1\}$.

In general, let $m(i, j)$ be the minimal number m with $(\sigma_i\sigma_j)^m = 1$. Calculate $m(i, j)$ when $|q_{ij}| \geq 2$.

2. Show that if q is a unit form and $q = q^1 \oplus q^2$ then $\mathbf{W}(q) = \mathbf{W}(q^1) \times \mathbf{W}(q^2)$.
3. Let $q : \mathbb{Z}^n \to \mathbb{Z}$ be a unit form and take $i_0, i_1, \ldots, i_s$ a sequence of indices such that

$$e_{i_0} < \sigma_{i_1}(e_{i_0}) < \ldots < \sigma_{i_s} \cdots \sigma_{i_1}(e_{i_0}) =: v.$$

Show that $\sigma_v \in \mathbf{W}(q)$. Is the converse true?

4. Let v be a positive root of the unit form $q : \mathbb{Z}^n \to \mathbb{Z}$, and assume that $q(v|e_i) \leq -2$ for some index i. Show that $\sigma_v\sigma_i$ has infinite order as an element of $\mathcal{O}(q)$. [Hint: Let $\varphi = \sigma_v\sigma_i$ and consider $\varphi^k(e_i) = a_{11}^{(k)}e_i + a_{21}^{(k)}v$ and $\varphi^k(v) = a_{12}^{(k)}e_i + a_{22}^{(k)}v$. Show that integers $a_{11}^{(k)}, a_{12}^{(k)}, a_{21}^{(k)}$ and $a_{22}^{(k)}$ can be found as the coefficients of the k-th power of the following matrix,

$$B = \begin{bmatrix} -1 & -q(v|e_i) \\ q(v|e_i) & q(v|e_i)^2 - 1 \end{bmatrix}.$$

For $q(v|e_i) \leq -3$, the entries of B^k grow exponentially with k. For $q(v|e_i) = -2$, the matrix B is equivalent to a Jordan block $J_2(1)$.]

4.5 Cyclotomic Polynomials

We now prove a celebrated and useful result of Kronecker in number theory (1857), see for instance [45, Theorem 4.5.4(a)].

Theorem 4.20 (Kronecker). *Let f be a monic polynomial with integer coefficients. If every root λ of f lies in the unit disc $\{z \in \mathbb{C} \mid |z| \leq 1\}$, then any nonzero root of f is a root of unit.*

Sketch of Proof. Let n be the degree of f. The set of all monic polynomials of degree n with integer coefficients having all their roots in the unit disc is finite. To see this we write

$$f(z) = z^n + a_1z^{n-1} + a_2z^{n-2} + \ldots + a_n = \prod_{j=1}^{n}(z - z_j),$$

where $a_j \in \mathbb{Z}$ and z_j are the roots of the polynomial f for $j = 1, \ldots, n$. Using the hypothesis $|z_j| \leq 1$ we obtain

$$|a_{n-1}| = |z_1 + \ldots + z_n| \leq n = \binom{n}{1},$$

$$|a_{n-2}| = |\sum_{1 \leq i,j \leq n} z_i z_j| \leq \binom{n}{2},$$

$$\ldots$$

$$|a_0| = |z_1 z_2 \cdots z_n| \leq 1 = \binom{n}{n}.$$

Hence the claim follows since the coefficients a_j are integers.

Define for $\ell \geq 1$ the polynomial

$$f_\ell(z) = \prod_{j=1}^{n}(z - z_j^\ell) = \sum_{i=0}^{n}(-1)^{i+n} e_{n-i,\ell} z^i,$$

where $e_{n-i,\ell}$ is the elementary symmetric polynomial of degree $n - i$ in variables $z_1^\ell, \ldots, z_n^\ell$ (cf. Exercises 1 and 2 below for the definition of elementary symmetric polynomials and Newton's identities, see also [45, 3.1.1]). Using Newton's identities, the values $n! e_{i,\ell}$ are integers for $i = 0, \ldots, n$, thus as before the set of polynomials $\{f_1, f_2, f_3, \ldots\}$ is finite.

Therefore there exist $1 \leq k < m$ such that $f_k = f_m$. Since the roots of f_k are $z_1^k, \ldots, z_n^k$ and those of f_m are $z_1^m, \ldots, z_n^m$, then there is a permutation σ in S_n (the symmetric group in n letters) such that $z_r^k = z_{\sigma(r)}^m$ for $r = 1, \ldots, n$. Let s be the order of σ in S_n. Then we have for a nonzero root $\lambda = z_r$ of f,

$$\lambda^{k^s} = z_r^{kk^{s-1}} = z_{\sigma(r)}^{mk^{s-1}} = z_{\sigma^2(r)}^{m^2k^{s-2}} = \ldots = z_{\sigma^s(r)}^{m^s} = \lambda^{m^s}.$$

It follows that $\lambda^{k^s - m^s} = 1$, that is, λ is a root of unity. □

By the proof above, there are only finitely many monic polynomials of a fixed degree with integer coefficients and with roots lying in the unitary disc. For $n = 2$, for instance, $f(z) = z^2 - az + b$ implies $|a| \leq 2$ and $|b| \leq 1$, giving rise to 15 polynomials. Among them nine have their roots in the unitary disc:

$$z^2,\ z^2 - z,\ z^2 + z,\ z^2 - 1,\ z^2 + z + 1,\ z^2 + 1,\ z^2 - z + 1,\ z^2 - 2z + 1,\ z^2 + 2z + 1.$$

By *cyclotomic polynomial* we mean a monic polynomial with integer coefficients $f(z)$ that divides $(z^k - 1)^m$ for some $k, m \geq 1$ (that is, such that there is a polynomial $g(z)$ with integer coefficients and with $f(z)g(z) = (z^k - 1)^m$).

Corollary 4.21. *Let $f(z)$ be a monic polynomial with integer coefficients. Then f is a cyclotomic polynomial if and only if $|\lambda| = 1$ for any (complex) root λ of f.*

Proof. Let f be cyclotomic. Then there are $k, m \geq 1$ such that $(\lambda^k - 1)^m = 0$ for any root λ of f, thus in particular $|\lambda| = 1$. For the converse take $\lambda_1, \dots, \lambda_r$ the different complex roots of f, with multiplicities m_i for $i = 1, \dots, r$.

By Kronecker's Theorem 4.20 all λ_i are roots of unity, thus there exists $k \geq 1$ such that $\lambda_i^k = 1$ for $i = 1, \dots, r$. Taking $m = \max(m_1, \dots, m_r)$ we may factor the polynomial $(z^k - 1)^m$ as $f(z)g(z) = (z^k - 1)^m$ for some complex polynomial $g(z)$. It can be shown that $g(z)$ has integer coefficients (Exercise 5 below), that is, f is cyclotomic. □

We recall some facts on cyclotomic polynomials (see for instance [39] or [45] for more on this topic). The n-th cyclotomic polynomial Φ_n is inductively defined by the formula

$$z^n - 1 = \prod_{d|n} \Phi_d(z),$$

where the product runs over all divisors d of n (in particular $\Phi_1(z) = z - 1$ and $\Phi_2(z) = z+1$). Observe that all Φ_n are monic polynomials with integer coefficients (Exercise 5) and that the roots of Φ_n are primitive n-roots of unity, thus if $\phi(-)$ denotes *Euler's totient function*

$$\phi(n) = \sum_{\substack{1 \leq \ell \leq n \\ \gcd(\ell, n) = 1}} 1,$$

then the n-th cyclotomic polynomial Φ_n has degree $\phi(n)$. It is known that $\Phi_n(z)$ is irreducible in $\mathbb{Z}[z]$ (see [45, 3.3.3]), and that any cyclotomic polynomial $f(z)$ decomposes as product of irreducible cyclotomic polynomials $f(z) = \Phi_{n_1}(z) \cdots \Phi_{n_r}(z)$ for $1 \leq n_1 \leq \ldots \leq n_r$. For instance, the nine polynomials above factor as follows,

$$z^2,\ z\Phi_1(z),\ z\Phi_2(z),\ \Phi_1(z)\Phi_2(z),\ \Phi_3(z),\ \Phi_4(z),\ \Phi_6(z),\ \Phi_1(z)^2,\ \Phi_2(z)^2,$$

and in particular only the last six of them are cyclotomic.

Recall that a matrix X is called *weakly periodic* (resp. *periodic*) if the difference $\mathbf{Id} - X^k$ is a nilpotent matrix (resp. zero matrix) for some positive integer k.

Theorem 4.22. *Let X be an integral matrix and $p(z) = |\mathbf{Id}z - X|$ its characteristic polynomial. The following statements are equivalent:*

a) The matrix X is weakly periodic.
b) The characteristic polynomial $p(z)$ is cyclotomic.
c) For any eigenvalue λ of X we have $|\lambda| = 1$.

Proof. Let $q(z)$ be the minimal polynomial of X (with coefficients in $\mathbb{Q}$) and recall that there is an integer $\ell > 0$ such that p divides q^ℓ (for every divisor of p is a divisor of q). Then p is cyclotomic if and only if q is cyclotomic.

Take $k, m \geq 0$ and $f(z) = (z^k - 1)^m$. Then $f(X) = 0$ if and only if the minimal polynomial $q(z)$ divides $f(z)$, that is, X is weakly periodic if and only if q is cyclotomic. This and the above show the equivalence of (a) and (b), and their equivalence with (c) follows from Corollary 4.21. □

The following corollary describes a particular case of Theorem 4.22. Its proof is left as an exercise.

Corollary 4.23. *Let X be an integral matrix and $p(z) = |z\mathbf{Id} - X|$ its characteristic polynomial. The following statements are equivalent:*

a) The matrix X is periodic.
b) The matrix X is diagonalizable and $p(z)$ is a cyclotomic polynomial.
c) The matrix X is diagonalizable and $|\lambda| = 1$ for any eigenvalue λ of X.

We give now more explicit expressions for cyclotomic polynomials. Recall that the *Möbius function* is defined as follows (cf. [45, 3.3.2])

$$\mu(n) = \begin{cases} 1, & \text{if } n = 1, \\ 0, & \text{if } n > 1 \text{ is divisible by a square,} \\ (-1)^r, & \text{if } n = p_1 \cdots p_r \text{ is a factorization into distinct primes.} \end{cases}$$

For the rest of the chapter we set $v_n = 1 + z + z^2 + \ldots + z^{n-1}$ (notice that v_n has degree $n - 1$).

Lemma 4.24. *For each $n \geq 2$ we have*

$$\Phi_n(z) = \prod_{d|n} (z^{n/d} - 1)^{\mu(d)} \quad \textit{and} \quad \Phi_n(z) = \prod_{d|n} v_{n/d}(z)^{\mu(d)}.$$

Proof. The identity on the left follows directly from the definition of Φ_n and the Möbius inversion formula (see Exercise 4). For the identity on the right use that $\sum_{d|n} \mu(d) = 0$ (Exercise 3) to get

$$\prod_{d|n} v_{n/d}(z)^{\mu(d)} = \prod_{d|n} [v_{n/d}(z)(z-1)]^{\mu(d)} = \prod_{d|n} (z^{n/d} - 1)^{\mu(d)} = \Phi_n(z),$$

which shows our claim. □

Let $s(j)$ denote the number of irreducible cyclotomic polynomials of degree j, that is, for each $j \geq 1$ the number $s(j)$ is equal to the number of solutions x to the equation $\phi(x) = j$ (in other words, the cardinality of the set $\phi^{-1}(j)$). For example, $s(8) = 5$ since the equation $\phi(x) = 8$ has the following five solutions: 15, 16, 20, 24 and 30. We may have $s(j) = 0$, for instance if j is any odd integer greater than 1 (see Exercise 9), but also for some even integers such as 14, 26, 34, 38, 50... Carmichael conjectured in 1922 that $s(j) \neq 1$ for all j, that is, that equation $\phi(x) = j$ has either no solution or has at least two solutions. Schlafly and Wagon showed in [48] the validity of Carmichael's conjecture for all j below 10^{10^7}. An old conjecture of Sierpiński asserts that for each integer $k \geq 2$ there is an integer j for which $s(j) = k$. This conjecture was proved recently by Ford [25], while Erdös has shown that any value of the function s appears infinitely often [24].

A polynomial $p(z)$ of degree n is said to be *self-reciprocal* if $p(z) = z^n p(\frac{1}{z})$. In the following table $a(n)$ is the number of cyclotomic polynomials p of degree n for small n, the number of such polynomials which are additionally self-reciprocal is indicated by $b(n)$, and $c(n)$ is the number of those which are self-reciprocal and where $p(-1)$ is the square of an integer. Our interest in such polynomials will become clear below (see Lemma 4.28). There is an efficient algorithm to determine such polynomials of given degree n, based on a quadratic bound for $n \leq 4\phi(n)^2$ in terms of Euler's totient function, see [49, p. 248].

n	1	2	3	4	5	6	7	8	9	10	11	12
$a(n)$	2	6	10	24	38	78	118	224	330	584	838	1420
$b(n)$	1	5	5	19	19	59	59	165	165	419	419	1001
$c(n)$	1	3	5	12	19	34	59	99	165	244	419	598

Cyclotomic polynomials Φ_n and their products arc a natural source of self-reciprocal polynomials. Clearly, $\Phi_1(z) = z - 1$ is not self-reciprocal, but all remaining Φ_n (with $n \geq 2$) are. Hence, exactly the polynomials $(z-1)^{2k} \prod_{n \geq 2} \Phi_n^{e_n}$ with natural numbers k and e_n are self-reciprocal with spectral radio one and without eigenvalue zero.

n	15	20	25
$a(n)$	4514	30,532	152,170
$b(n)$	2257	20,399	76,085
$c(n)$	2257	12,526	76,085

It is not a coincidence that in the above tables we have $b(n) = c(n) = b(n-1)$ for $n > 1$ odd. Indeed, p is cyclotomic self-reciprocal of odd degree if and only if $p(z) = (z+1)q(z)$ for some cyclotomic self-reciprocal polynomial q of degree $n-1$, and in that case $p(-1) = 0$.

Exercises 4.25. One can see directly that for complex numbers $z_1, \ldots, z_n$, the following polynomial expansion holds,

$$\prod_{i=1}^{n}(z - z_i) = \sum_{i=0}^{n}(-1)^{n+i} e_{n-i}(z_1, \ldots, z_n) z^i,$$

where $e_i(z_1, \ldots, z_n)$ denotes the *elementary symmetric polynomial* in n variables

$$e_i(z_1, \ldots, z_n) = \begin{cases} 1, & \text{if } i = 0, \\ \sum\limits_{1 \le j_1 < \ldots < j_i \le n} z_{j_1} \cdots z_{j_i}, & \text{if } 1 \le i \le n, \\ 0, & \text{if } i > n. \end{cases}$$

For $\ell \ge 1$ define $e_{i,\ell} = e_i(z_1^\ell, \ldots, z_n^\ell)$ and consider the ℓ-th power sum:

$$p_\ell = \sum_{i=1}^{n} z_i^\ell.$$

The equations in Exercises 1 and 2 are known as **Newton's identities**.

1. Show that for any $k \ge 1$ we have $ke_{k,1} = \sum_{i=1}^k (-1)^{i-1} e_{k-i,1} p_i$, and conclude that for $\ell \ge 1$ we have

$$ke_{k,\ell} = \sum_{i=1}^{k} (-1)^{i-1} e_{k-i,\ell} p_{(i\ell)}.$$

   [Hint: Take the sum over $i = 1, \ldots, k$ of the substitutions $z = z_i$ in the polynomial $\prod_{i=1}^n (z - z_i)$, and consider its expansion as above.]
2. Show that for $\ell \ge 1$ we have

$$p_\ell = (-1)^{\ell-1} k e_\ell(z_1, \ldots, z_n) + \sum_{i=1}^{\ell-1} (-1)^{\ell+i-1} e_{\ell-i}(z_1, \ldots, z_n) p_i.$$

3. Prove that the Möbius function μ satisfies

$$\sum_{d|n} \mu(d) = \begin{cases} 1, \text{ if } n = 1, \\ 0, \text{ if } n \ge 2. \end{cases}$$

   [Hint: Take the product x of all primes dividing n and notice that

$$\sum_{d|n} \mu(d) = \sum_{d|x} \mu(d).$$

   Show the sum on the right must be zero for $n > 1$ (cf. [45, 3.3.2]).]
4. **Möbius inversion formula.** Let G be an abelian group (with multiplicative operation) and consider functions $f, g : \mathbb{N} \to G$ such that $f(n) = \prod_{d|n} g(d)$ for any $n \in \mathbb{N}$. Show that

$$g(n) = \prod_{d|n} f(n/d)^{\mu(d)},$$

where μ is the Möbius function. [Hint: Show that

$$\sum_{d|n} \mu(d) \sum_{t|(n/d)} g(t) = \sum_{t|n} g(t) \sum_{d|(n/t)} \mu(d),$$

since if d and t divide n then d divides n/t if and only if t divides n/d. Then use Exercise 3.]

5. Let $a(z) = \sum_{i\geq 0} a_i z^i$ and $c(z) = \sum_{i\geq 0} c_i z^i$ be polynomials with integer coefficients (thus $a_i = 0 = c_i$ for almost all i).

   i) Show that if $a \neq 0$ and $a(z)b(z) = c(z)$ for some complex polynomial $b(z)$, then $b(z)$ has rational coefficients.
   ii) Assume that $b(z) = \sum_{i\geq 0} b_i z^i$ is a polynomial with integer coefficients such that $a(z)b(z) = c(z)$. Show that if $\gcd(\{a_i\}_{i\geq 0}) = 1 = \gcd(\{b_i\}_{i\geq 0})$ then $\gcd(\{c_i\}_{i\geq 0}) = 1$.

   Conclude that if a is monic and $a(z)b(z)$ has integer coefficients for some complex polynomial $b(z)$, then b has integer coefficients.
6. Prove that there is no linear bound for n in terms of the totient function $\phi(n)$.
7. Show that the polynomials f_i in the proof of Theorem 4.20 actually have integer coefficients.
8. Give a proof of Corollary 4.23.
9. Show that if $n > 2$ then the n-th cyclotomic polynomial Φ_n has even degree.

4.6 Coxeter Matrices

Let $\langle -, - \rangle : \mathbb{Z}^n \times \mathbb{Z}^n \to \mathbb{Z}$ be a bilinear form. A linear transformation $X : \mathbb{Z}^n \to \mathbb{Z}^n$ satisfying

$$\langle v, w \rangle = -\langle w, Xv \rangle, \qquad \text{for all } v \text{ and } w \text{ in } \mathbb{Z}^n,$$

is called a *Coxeter transformation* for $\langle -, - \rangle$ (see for instance [34, 38] or [47]). If the bilinear form is nondegenerate then it has a unique Coxeter transformation. Indeed, if X and X' satisfy the above conditions, then for any v and w in $\mathbb{Z}^n$ we have

$$\langle w, X'v - Xv \rangle = \langle v, w \rangle - \langle v, w \rangle = 0,$$

therefore $X'w = Xw$. Fixing the canonical basis $e_1, \dots, e_n$ of $\mathbb{Z}^n$ consider the matrix $T = \mathbf{SM}_{\langle -,- \rangle}$ satisfying $\langle v, w \rangle = v^t T w$, with coefficients given by $\langle e_i, e_j \rangle$ for $1 \leq i, j \leq n$. In particular $\det(T) \neq 0$ if the bilinear form is nondegenerate, and taking $X = -T^{-1}T^t$ we have

$$-\langle w, Xv \rangle = w^t T(T^{-1}T^t)v = w^t T^t v = v^t T w = \langle v, w \rangle.$$

The matrix $X = -T^{-1}T^t$ is called *Coxeter matrix* of T, and corresponds to the Coxeter transformation of $\langle -,- \rangle$ with respect to the canonical basis. Observe that in the nondegenerate case, Coxeter transformations are $\mathbb{Z}$-invertible (in fact $\det(X) = (-1)^n$), and that $X = -\mathbf{Id}_n$ if and only if $\langle -,- \rangle$ is symmetric. Observe also that if $\langle -,- \rangle'$ is a bilinear form equivalent to $\langle -,- \rangle$ (that is, $T := \mathbf{SM}_{\langle -,- \rangle}$ and $T' := \mathbf{SM}_{\langle -,- \rangle'}$ are congruent matrices, say $T' = S^tTS$ for some $\mathbb{Z}$-invertible matrix S) then we have similar corresponding Coxeter matrices X and X',

$$X' = -(T')^{-1}(T')^t = -(S^{-1}T^{-1}S^{-t})(S^tT^tS) = S^{-1}(-T^{-1}T^t)S = S^{-1}XS,$$

where S^{-t} denotes the inverse of the transposed matrix S^t, which is just another way of expressing the uniqueness of the transformation defined by X. Conditions on the quadratic form $q(x) = \langle x, x \rangle$ impose severe restriction on the Coxeter matrix X, as we show next.

Theorem 4.26. *Let X be the Coxeter transformation of a nondegenerate bilinear form $\langle -,- \rangle : \mathbb{Z}^n \times \mathbb{Z}^n \to \mathbb{Z}$ and assume that the integral quadratic form $q(x) = \langle x, x \rangle$ is positive. Then X is periodic and $|\lambda| = 1$ with $\lambda \neq 1$ for any eigenvalue λ of X.*

Proof. Notice first that for $v \in \mathbb{Z}^n$ we have

$$q(Xv) = \langle Xv, Xv \rangle = \langle v, v \rangle = q(v).$$

By Remark 2.4 the set $\{X^k e_i\}_{k \geq 0}$ is finite for the canonical vectors $e_1, \ldots, e_n$, since $q(X^k e_i)$ is constant for all $k \geq 0$. Then there exists $k_i > 0$ with $X^{k_i} e_i = e_i$ for $i = 1, \ldots, n$. Take a common multiple k of $k_1, \ldots, k_n$ and observe that $X^k = \mathbf{Id}$.

Now, by Theorem 4.22 (see also Corollary 4.23) every eigenvalue λ of X satisfies $|\lambda| = 1$. That 1 is not an eigenvalue of X will be shown below in Lemma 4.28(c). □

The hypothesis that q is positive in Theorem 4.26 cannot be weakened, as shown by the following example given in [47]. Consider the matrix T with $\det(T) = 1$ and its associated Coxeter matrix $X = -T^{-1}T^t$ as given below,

$$T = \begin{pmatrix} 2 & 3 \\ 1 & 2 \end{pmatrix} \quad \text{and} \quad X = \begin{pmatrix} 5 & 4 \\ -4 & -3 \end{pmatrix}.$$

Then the quadratic form $q(v) = v^tTv = 2(v_1 + v_2)^2$ is nonnegative, and one can easily verify that X is not periodic. However $(X - \mathbf{Id})^2 = 0$, as expected from the following generalization of Theorem 4.26 for nonnegative quadratic forms, given by Sato in [47] (the assertion on the eigenvalues of X is shown in Theorem 4.22).

Theorem 4.27. *Let X be the Coxeter transformation of a nondegenerate bilinear form $\langle -,-\rangle : \mathbb{Z}^n \times \mathbb{Z}^n \to \mathbb{Z}$ and assume that the integral quadratic form $q(x) = \langle x, x\rangle$ is nonnegative. Then X is weakly periodic and $|\lambda| = 1$ for any eigenvalue λ of X.*

The converse of Theorems 4.26 and 4.27 does not hold in general (the Coxeter transformation X of any symmetric bilinear form satisfies $X^2 = \mathbf{Id}$). A similar example with T upper triangular and with ones on the main diagonal is given in [38], correcting an erroneous claim in [21].

$$T = \begin{pmatrix} 1 & 0 & -1 & -1 & -1 & 2 & 2 & 2 \\ 0 & 1 & -1 & -1 & -1 & 2 & 2 & 2 \\ 0 & 0 & 1 & 0 & 0 & -1 & -1 & -1 \\ 0 & 0 & 0 & 1 & 0 & -1 & -1 & -1 \\ 0 & 0 & 0 & 0 & 1 & -1 & -1 & -1 \\ 0 & 0 & 0 & 0 & 0 & 1 & 0 & 0 \\ 0 & 0 & 0 & 0 & 0 & 0 & 1 & 0 \\ 0 & 0 & 0 & 0 & 0 & 0 & 0 & 1 \end{pmatrix} \quad \text{and} \quad X = \begin{pmatrix} -4 & -3 & 2 & 2 & 2 & -1 & -1 & -1 \\ -3 & -4 & 2 & 2 & 2 & -1 & -1 & -1 \\ -5 & -5 & 2 & 3 & 3 & -1 & -1 & -1 \\ -5 & -5 & 3 & 2 & 3 & -1 & -1 & -1 \\ -5 & -5 & 3 & 3 & 2 & -1 & -1 & -1 \\ -2 & -2 & 1 & 1 & 1 & -1 & 0 & 0 \\ -2 & -2 & 1 & 1 & 1 & 0 & -1 & 0 \\ -2 & -2 & 1 & 1 & 1 & 0 & 0 & -1 \end{pmatrix}.$$

The upper triangular matrix T above has determinant one, and its Coxeter matrix $X = -T^{-1}T^t$ satisfies $X^3 = -\mathbf{Id}$, that is, the Coxeter transformation associated to the bilinear form determined by T is periodic. However, if v is the vector $(1, 1, 1, 1, 1, 0, 0, 0)$ then $q(v) = v^t T v = -1$.

The characteristic polynomial of X is usually known as *Coxeter polynomial* of the bilinear form, and by the above it is an invariant of the equivalence class to which $\langle -,-\rangle$ belongs. Some general properties of Coxeter polynomials are easy to determine.

Lemma 4.28. *Consider a Coxeter matrix $X = -T^{-1}T^t$ and its Coxeter polynomial $\chi(z) = \det(\mathbf{Id}z - X)$. The following assertions hold:*

a) The polynomial χ is self-reciprocal.
b) If $\det(T) = 1$ then $\chi(-1)$ is the square of an integer.
c) $\chi(1) = 0$ if and only if the integral quadratic form $q(x) = x^t T x$ has nontrivial corank $\mathbf{cork}(q) \neq 0$.

Proof. Point (a) is a direct calculation:

$$\begin{aligned} z^n \chi\left(\frac{1}{z}\right) &= \det(\mathbf{Id}z)\det(\mathbf{Id}\frac{1}{z} + T^{-t}T) \\ &= \det(\mathbf{Id} + T^{-t}Tz) = \det[T^{-t}(T^tT^{-1} + \mathbf{Id}z)T] \\ &= \det(\mathbf{Id}z + T^tT^{-1}) = \det(\mathbf{Id}z + T^{-t}T) \\ &= \chi(z). \end{aligned}$$

Assume now that $\det(T) = 1$. Then

$$\chi(z) = \det(\mathbf{Id}z + T^{-t}T) = \det(T^{-t})\det(zT^t + T) = \det(T + zT^t).$$

Hence $\chi(-1)$ is the determinant of a skew-symmetric matrix $S := T - T^t$. Using the skew-normal form of S (see for instance [41, Theorem IV.1]), we obtain $S' = U^t S U$ for some $\mathbb{Z}$-invertible matrix U, where S' is a block diagonal matrix whose first block is the zero matrix of certain size and the remaining blocks have the shape $\begin{bmatrix} 0 & m_i \\ -m_i & 0 \end{bmatrix}$ with integers m_i. Thus the claim (b) follows. For (c) observe simply that, since T is invertible, for any nontrivial v in $\mathbf{cork}(q)$ we have

$$0 = T^{-1}(T + T^t)v = (\mathbf{Id} + T^{-1}T^t)v = (\mathbf{Id} - X)v,$$

that is, v is an eigenvector of X with eigenvalue 1. Conversely, if 1 is an eigenvalue of X then $\det(\mathbf{Id} - X) = 0 = \det(T + T^t)$ and we may find a vector v in $\mathbb{Q}^n$ with $(T + T^t)v = 0$(using for instance the Gauss elimination process), and multiplying by an integer if necessary we may assume that v has integer entries. Therefore $v \in \mathbf{cork}(q)$. □

We associate a *Coxeter matrix* X_q to a given quadratic unit form q by means of the Gram matrix T_q and the bilinear form it determines $(x, y) \mapsto v^t T_q w$ (that is, we define $X_q := -T_q^{-1} T_q^t$), and take the *Coxeter polynomial* of q to be the characteristic polynomial $\chi_q(z)$ of X_q. This choice is somehow arbitrary, for equivalent unit forms q and q' might have different Coxeter polynomials χ_q and $\chi_{q'}$ (since the Gram matrices T_q and $T_{q'}$ need not be congruent even if q and q' are equivalent forms). Below we give an alternative construction of T_q^{-1} using the simple reflections $\sigma_1, \ldots, \sigma_n$ associated to a unit form q. Consider the coefficients

$$\gamma(i, j) := \begin{cases} \sum_{0 \le s} (-1)^{s+1} \gamma_s(i, j), & \text{if } i < j, \\ 1, & \text{if } i = j, \\ 0, & \text{if } i > j, \end{cases}$$

where for $i < j$ and $0 \le s$ we define

$$\gamma_s(i, j) := \begin{cases} \sum_{i < k_1 < \ldots < k_s < j} q_{ik_1} \cdots q_{k_r j}, & \text{if } 0 < s < j - i, \\ q_{ij}, & \text{if } s = 0, \\ 0, & \text{if } s \ge j - i. \end{cases}$$

and take the (upper triangular) matrix $C_q = (\gamma(i, j))_{i,j=1}^n$. First we verify that C_q is indeed the inverse of T_q.

Lemma 4.29. *Let T_q be the Gram matrix associated to a unit form $q : \mathbb{Z}^n \to \mathbb{Z}$ as defined in Sect. 1.1. Then $T_q C_q = \mathbf{Id}$.*

Proof. We proceed by induction on $n > 1$ (for $n = 1$ the claim is clear). Define the matrices T'_q and C'_q by deleting both the last column and the last row from T_q and C_q respectively. Take $q' = q^{(n)}$ and observe that T'_q is the Gram matrix $T_{q'}$ of q', and similarly C'_q is the matrix $C_{q'}$ constructed above corresponding to q'. Then by induction, and since $q_{nn}\gamma(n,n) = 1$, we only need to verify that for $1 \leq i < n$ we have

$$\sum_{r=i}^{n} q_{ir}\gamma(r,n) = 0.$$

Observe first that by definition, for $s \geq 0$ we have

$$\sum_{i<r<n} q_{ir}\gamma_s(r,n) = \sum_{i<r<n} q_{ir}\left[\sum_{r<k_1<\ldots<k_s<n} q_{rk_1}\cdots q_{k_r n}\right] = \gamma_{s+1}(i,n).$$

Then

$$\begin{aligned}\sum_{r=i}^{n} q_{ir}\gamma(r,n) &= q_{in} + \sum_{r=1}^{n-1}\sum_{s\geq 0}(-1)^{s+1} q_{ir}\gamma_s(r,n)\\ &= q_{in} + \sum_{s\geq 0}(-1)^{s+1}\left[q_{ii}\gamma_s(i,n) + \sum_{i<r<n} q_{ir}\gamma_s(r,n)\right]\\ &= q_{in} + \sum_{s\geq 0}(-1)^{s+1}\left[\gamma_s(i,n) + \gamma_{s+1}(i,n)\right]\\ &= q_{in} - \gamma_0(i,n) = q_{in} - q_{in} = 0,\end{aligned}$$

which completes the proof. □

Lemma 4.30. *Consider the simple reflections $\sigma_1, \ldots, \sigma_n$ associated to a unit form $q : \mathbb{Z}^n \to \mathbb{Z}$ for $1 \leq i \leq j \leq n$ and define the vectors $\alpha_{i,j} := \sigma_i \cdots \sigma_{j-1}(e_j)$ and $\beta_{i,j} := \sigma_j \cdots \sigma_{i+1}(e_i)$ in $\mathbb{Z}^n$. Then*

$$\alpha_{i,j} = \sum_{r=i}^{j} \gamma(r,j)e_r \quad \textit{and} \quad \beta_{i,j} = \sum_{r=i}^{j} \gamma(i,r)e_r,$$

where $\gamma(i,j)$ are the coefficients of the matrix C_q as defined above.

Proof. We show the claim for $\beta_{i,j}$, the proof for $\alpha_{i,j}$ goes similarly. Fix $1 \leq i \leq n$ and notice that $\beta_{i,i} = e_i = \gamma(i,i)e_i$. Assume the claim holds for $i \leq j < n$ and observe by induction that

$$\beta_{i,j+1} = \sigma_{j+1}(\beta_{i,j}) = \sigma_{j+1}\left(\sum_{r=i}^{j} \gamma(i,r)e_r\right)$$

$$= \sum_{r=i}^{j} \gamma(i,r)[e_r - q_{r,j+1}e_{i+1}]$$

$$= \sum_{r=i}^{j} \gamma(i,r)e_r + \left[-\sum_{r=i}^{j} \gamma(i,r)q_{r,j+1}\right] e_{j+1}.$$

Now, since $i \neq j+1$ and $C_q T_q = \mathbf{Id}$ by Lemma 4.29, the product of the i-th row of C_q with the $j+1$-th column of T_q is zero, that is,

$$0 = \sum_{r=i}^{j+1} \gamma(i,r)q_{r,j+1} = \sum_{r=i}^{j} \gamma(i,r)q_{r,j+1} + \gamma(i,j+1),$$

which shows that

$$\beta_{i,j+1} = \sum_{r=i}^{j} \gamma(i,r)e_r + \left[-\sum_{r=i}^{j} \gamma(i,r)q_{r,j+1}\right] e_{j+1} = \sum_{r=i}^{j+1} \gamma(i,r)e_r.$$

□

For $1 \leq i \leq n$ define the (column) vectors

$$\alpha_i = \sigma_1 \cdots \sigma_{i-1}(e_i), \quad \text{and} \quad \beta_i = \sigma_n \cdots \sigma_{i+1}(e_i),$$

or using the notation of Lemma 4.30, $\alpha_i = \alpha_{1,i}$ and $\beta_i = \beta_{i,n}$. When coming from certain algebraic settings (see for example [2] or [37]) α_i and β_i are called *projective* and *injective vectors*. The correspondence of Coxeter matrices with the product of simple reflections (in some order) was first observe by Howlett in [34] (see [38, Corollary 2.11]). In what follows we identify a transformation from $\mathbb{Z}^n$ to $\mathbb{Z}^n$ with its associated n by n matrix with respect to the canonical basis of $\mathbb{Z}^n$. Recall that the square matrix having as columns the vectors $x^1, \dots, x^n$ in $\mathbb{Z}^n$ is denoted by $(x^1|\cdots|x^n)$.

Theorem 4.31. *For a unit form $q : \mathbb{Z}^n \to \mathbb{Z}$ we have*

$$C_q = (\alpha_1|\dots|\alpha_n) \quad \textit{and} \quad C_q^t = (\beta_1|\dots|\beta_n).$$

In particular $X_q = -C_q C_q^{-t} = \sigma_1 \cdots \sigma_n$, *therefore the Coxeter transformation* X_q *is an isometry for* q *belonging to the Weyl group* $\mathbf{W}(q)$.

Proof. The first assertions follow directly from Lemma 4.30. For the second claim observe that both X_q and $\sigma_1 \cdots \sigma_n$ are linear transformations that send the vector β_i to the vector $-\alpha_i$ for $i = 1, \dots, n$. Thus the result follows since clearly both sets $\{\alpha_1, \dots, \alpha_n\}$ and $\{\beta_1, \dots, \beta_n\}$ are $\mathbb{Z}$-basis of $\mathbb{Z}^n$. □

Which self-reciprocal polynomials are Coxeter polynomials (of unitary forms)? Notice that the request for $p(-1)$ to be a square number discards many self-reciprocal polynomials (for instance Φ_4, Φ_6, Φ_8 and Φ_{10} in the cyclotomic case). What about the cyclotomic case? For example, the cyclotomic polynomial $p(z) = \Phi_2(z)\Phi_6(z) = z^3 + 1$ does not appear as a Coxeter polynomial X_q of any unitary form q, despite the fact that $p(-1) = 0$ is a square. Indeed, the matrix $T = \begin{bmatrix} 1 & a & b \\ 0 & 1 & c \\ 0 & 0 & 1 \end{bmatrix}$ yields the Coxeter matrix

$$X = \begin{bmatrix} a^2 + b^2 - abc - 1 & a - ac^2 + bc & b - ac \\ bc - a & c^2 - 1 & c \\ -b & -c & -1 \end{bmatrix},$$

with Coxeter polynomial $\chi(z) = z^3 + \alpha z^2 + \alpha z + 1$, where $\alpha = -a^2 - b^2 c^2 + abc + 3$. Therefore we look for integral solutions to the Markov–Hurwitz type equation

$$3 = a^2 + b^2 + c^2 - abc.$$

Any such solution, if exists, must satisfy $a, b, c \equiv 0 \bmod 3$. Thus the right side of the equation is divisible by 9, which is impossible (see Exercise 2).

We now consider the bilinear form $\langle -, - \rangle_{\vec{G}}$ determined by a *directed graph* (also known as *quiver*) $\vec{G}$, where every edge has a fixed ordering. In this case directed edges are called arrows and are depicted as such. The *adjacency matrix* $A_{\vec{G}}$ of a directed graph $\vec{G}$ has as coefficients the integers

$$(A_{\vec{G}})_{ij} = \text{Number of arrows from } i \text{ to } j.$$

Then for vectors v and w in $\mathbb{Z}^{\vec{G}_0}$ we take $\langle v, w \rangle_{\vec{G}} = v^t(\mathbf{Id} - A_{\vec{G}})w$. Observe that if G is the underlying graph of $\vec{G}$ (the graph obtained by ignoring the orientation of all edges), then $q_G(v) = \langle v, v \rangle_{\vec{G}}$ for any v in $\mathbb{Z}^{G_0}$. Recall that by a *path* in $\vec{G}$ we mean a walk such that all directed edges point in the same direction (trivial walks are also paths). An *oriented cycle* in $\vec{G}$ is a closed path, and a digraph without oriented cycles is called *acyclic*. In what follows we will assume that $\vec{G}$ is acyclic, thus in particular $\vec{G}$ has no loop. We will say that the ordering of vertices in $\vec{G}$

is admissible if the existence of an arrow $i \to j$ implies that $i < j$. Notice that the ordering of $\vec{G}_0$ is admissible if and only if the adjacency matrix $A_{\vec{G}}$ is upper triangular.

Lemma 4.32. *Let $\vec{G}$ be a directed graph with adjacency matrix $A_{\vec{G}}$. The number of paths in $\vec{G}$ of length $k \geq 0$ from vertex i to vertex j is given by the (i, j)-th entry in the matrix $A^k_{\vec{G}}$.*

Proof. We proceed by induction on $k \geq 0$. The claim is evident for $k = 0$ and $k = 1$, and assume it holds for $k > 1$. Every path w of length $k + 1$ from i to j is the concatenation of a path w' of length k from i to some vertex ℓ, and an arrow $\alpha : \ell \to j$. By induction the number of paths of length $k + 1$ from i to j is given by

$$\sum_{\ell \in \vec{G}_0} (A^k_{\vec{G}})_{i\ell}(A_{\vec{G}})_{\ell j} = (A^{k+1}_{\vec{G}})_{ij}.$$

□

In particular, if $\vec{G}$ is acyclic then the matrix $A_{\vec{G}}$ is nilpotent, therefore $(\mathbf{Id} - A_{\vec{G}})$ is an invertible matrix with inverse given by $\mathbf{Id} + A_{\vec{G}} + A^2_{\vec{G}} + A^3_{\vec{G}} + \ldots$ We also have

$$(\mathbf{Id} - A_{\vec{G}})^{-1}_{ij} = \text{Number of } \textit{paths} \text{ from } i \text{ to } j.$$

The Coxeter matrix $X_{\vec{G}}$ associated to an acyclic directed graph $\vec{G}$ is the Coxeter matrix of the nondegenerate bilinear form $\langle -, - \rangle_{\vec{G}}$. It has the following explicit shape

$$X_{\vec{G}} = -(\mathbf{Id} - A_{\vec{G}})^{-1}(\mathbf{Id} - A_{\vec{G}})^t = -\left(\sum_{\ell \geq 0} A^\ell_{\vec{G}}\right)(\mathbf{Id} - A^t_{\vec{G}}).$$

Remark 4.33. The following interpretation of the coefficients in the Coxeter matrix associated to an acyclic directed graph $\vec{G}$ can be verified directly (cf. A. Boldt [13]). By a *twisted path* from i to j we mean a path of length $\ell \geq 0$ from i to k for some vertex k, followed by a directed edge from k to j. Then we have

$$(X_{\vec{G}})_{ij} = \begin{pmatrix} \text{Number of twisted paths} \\ \text{from } i \text{ to } j \end{pmatrix} - \begin{pmatrix} \text{Number of (nontwisted) paths} \\ \text{from } i \text{ to } j \end{pmatrix},$$

for any pair of vertices i and j.

The *Coxeter polynomial* of the bilinear form associated to an acyclic directed graph $\vec{G}$ is denoted by $\chi_{\vec{G}}$. Recall that $\vec{G}$ is a forest if its underlying graph G is disjoint union of tree graphs. The following are well-known properties of Coxeter polynomials of directed graphs (see for instance [40]).

Proposition 4.34. *Let $\vec{G}$ be an acyclic directed graph.*

a) The Coxeter polynomial $\chi_{\vec{G}}$ does not depend on the ordering of vertices in $\vec{G}$.

b) If $\vec{G}$ is a forest, then the Coxeter polynomial $\chi_{\vec{G}}$ does not depend on the orientation of the arrows in $\vec{G}$.

Proof. Let $\{1, \dots, n\}$ be the set of vertices of $\vec{G}$, and take $\vec{G}'$ to be the directed graph obtained after a permutation π of the set of vertices. The adjacency matrix $A_{\vec{G}'}$ is given by the coefficients $(A_{\vec{G}'})_{ij} = (A_{\vec{G}})_{\pi(i)\pi(j)}$, that is, the permutation matrix $P = (e_{\pi(1)}| \cdots |e_{\pi(n)})$ yields a congruence

$$A_{\vec{G}'} = P^t A_{\vec{G}} P.$$

Since $P^t P = \mathbf{Id}$ we have equivalent bilinear forms $\langle -, - \rangle_{\vec{G}'}$ and $\langle -, - \rangle_{\vec{G}}$, hence similar Coxeter matrices $X_{\vec{G}'}$ and $X_{\vec{G}}$. Then $\chi_{\vec{G}'} = \chi_{\vec{G}}$.

The proof of (b) can be found in [40]. It uses a special type of transformation on directed graphs (sink-source reflections) as used by Bernstein, Gelfand and Ponomarev in the context of representations of algebras [10]. □

Lemma 4.35. *Consider an acyclic directed graph $\vec{G}^1$ with source x_2, and let $\vec{G}$ be the directed graph obtained from $\vec{G}^1$ by adding a vertex x_1 and an arrow from x_1 to x_2. The subgraph of $\vec{G}^1$ obtained by deleting vertex x_2 is denoted by $\vec{G}^2$.*

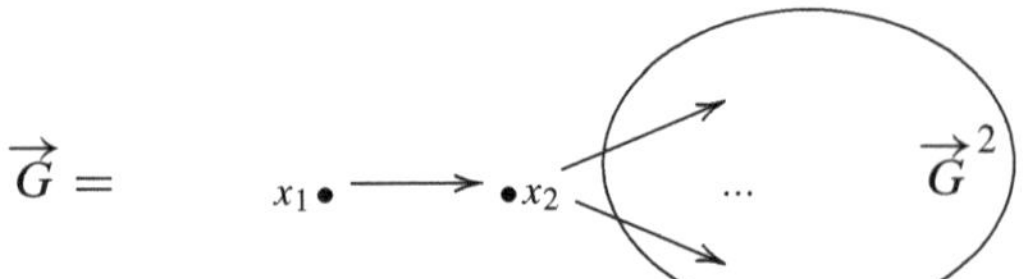

Then the Coxeter polynomial of $\vec{G}$ is given by

$$\chi_{\vec{G}}(z) = (z+1)\chi_{\vec{G}^1}(z) - z\chi_{\vec{G}^2}(z).$$

Proof. Let n be the number of vertices in $\vec{G}^1$. Enumerate the vertices of $\vec{G}$ taking $x_1 = n+1$ and $x_2 = n$. By the description of the coefficients of the Coxeter matrix $X_{\vec{G}}$ as a difference of twisted minus nontwisted paths (Remark 4.33), we have

$$X_{\vec{G}} = \begin{bmatrix} X_{\vec{G}^2} & b & 0 \\ a^t & \lambda & 1 \\ a^t & \lambda & 0 \end{bmatrix},$$

where a and b are column vectors with $n-1$ entries, $\lambda \in \mathbb{Z}$ and 0 is a zero matrix of appropriate size. Expanding $\det(z\mathbf{Id} - X_{\vec{G}})$ by minors along the last column we get

$$\chi_{\vec{G}}(z) = z\chi_{\vec{G}^1}(z) + \det(B),$$

where $B = \begin{bmatrix} \mathbf{Id}z - X_{\vec{G}^2} & -b \\ -a^t & -\lambda \end{bmatrix}$. Since $\chi_{\vec{G}^1}(z) = \det \begin{bmatrix} \mathbf{Id}z - X_{\vec{G}^2} & -b \\ -a^t & z-\lambda \end{bmatrix}$ (for x_2 is a source of $\vec{G}^1$), we conclude that

$$\chi_{\vec{G}^1}(z) = z\chi_{\vec{G}^2} + \det(B),$$

that is,

$$\chi_{\vec{G}}(z) = (z+1)\chi_{\vec{G}^1}(z) - z\chi_{\vec{G}^2}(z).$$

Hence the result. □

Recall that a directed graph $\vec{G}$ is called *bipartite* if there is a partition of the vertex set $\vec{G}_0 = P_1 \sqcup P_2$ such that every arrow in $\vec{G}$ starts at a vertex belonging to P_1 and ends in a vertex in P_2. Many important (spectral) properties of Coxeter matrices for bipartite directed graphs (a family of graphs that includes all trees) are consequences of A'Campo's Theorem [1], which describes Coxeter polynomials in terms of the characteristic polynomial associated to the underlying graph G of $\vec{G}$.

Theorem 4.36 (A'Campo). *Let $\vec{G}$ be a connected bipartite directed graph with n vertices. Then the Coxeter polynomial $\chi_{\vec{G}}$ of $\vec{G}$ satisfies*

$$\chi_{\vec{G}}(z^2) = z^n p_G(z + z^{-1}),$$

where G is the underlying graph of $\vec{G}$ and p_G denotes the characteristic polynomial of the (symmetric) adjacency matrix of G.

Proof. By Proposition 4.34(a) we may enumerate the vertices of $\vec{G}$ in such a way that vertices $1, \dots, m$ are the sources of all arrows in $\vec{G}$, while $m+1, \dots, n$ are the targets of all arrows, for some $1 < m < n$. Then the adjacency matrix $A_{\vec{G}}$ of the directed graph $\vec{G}$ is given as follows,

$$A_{\vec{G}} = \begin{pmatrix} 0_m & M \\ 0 & 0_{n-m} \end{pmatrix},$$

while the adjacency matrix A_G of its underlying graph G is $A_G = A_{\vec{G}} + A^t_{\vec{G}}$, where 0_m denotes the zero m by m matrix. Since $\vec{G}$ has no path of length 2, by Lemma 4.32 we have $A^2_{\vec{G}} = 0$. Since bipartite graphs have no oriented cycle, by Remark 4.33 the Coxeter matrix of $\vec{G}$ is given by

$$X_{\vec{G}} = -(\mathbf{Id} + A_{\vec{G}})(\mathbf{Id} - A^t_{\vec{G}}).$$

Taking $A = A_{\vec{G}}$ and considering again that $A^2 = 0$, we have

$$\begin{aligned}
\chi_{\vec{G}}(z^2) &= \det(z^2\mathbf{Id} - X_{\vec{G}}) = \det(z^2\mathbf{Id} + (\mathbf{Id} + A)(\mathbf{Id} - A^t)) \\
&= \det(z^2\mathbf{Id} + (\mathbf{Id} + A)(\mathbf{Id} - A^t)) \det(\mathbf{Id} - A) \\
&= \det(z^2(\mathbf{Id} - A) + (\mathbf{Id} - A^2)(\mathbf{Id} - A^t)) \\
&= z^n \det((z + z^{-1})\mathbf{Id} - z^{-1}A^t - zA) \\
&= z^n \det((z + z^{-1})\mathbf{Id} - (A^t + A)),
\end{aligned}$$

where the last equality follows from the specific shape of the matrix $zA + z^{-1}A^t$,

$$zA + z^{-1}A^t = \begin{pmatrix} 0 & zM \\ z^{-1}M^t & 0 \end{pmatrix}.$$

This completes the proof. □

As consequence of Theorem 4.36 it can be shown that the eigenvalues of the Coxeter matrix $X_{\vec{G}}$, for $\vec{G}$ a bipartite directed graph, are either positive real numbers, or complex numbers with modulus one (cf. [1] and Exercise 7 below).

The formula in Lemma 4.35 can be used to find the Coxeter polynomials of a wide variety of oriented graphs. Take for instance the *star graph* $[p_1, \ldots, p_t]$ with t branches of length $p_1, \ldots, p_t$ and the following orientation of arrows

$[2, 2, 3, 3, 4] =$ 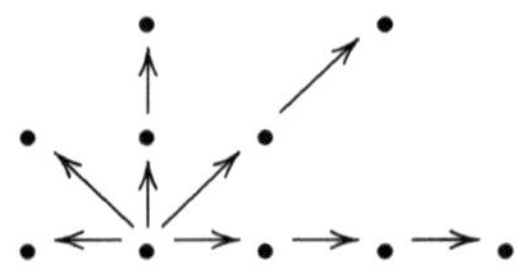

The sequence of numbers $(p_1, \ldots, p_t)$ is referred to as the *star type* of the star. By Lemma 4.35 (see Exercise 3) we find that the Coxeter polynomial of a star $[p_1, \ldots, p_t]$ has the form

$$\chi_{[p_1,\ldots,p_t]}(z) = \left[(z+1) - z \sum_{i=1}^{t} \frac{v_{p_i-1}(z)}{v_{p_i}(z)} \right] \prod_{i=1}^{t} v_{p_i}(z),$$

which gives an explicit formula for the sum of coefficients of $\chi_{[p_1,\ldots,p_t]}$ as follows,

$$\chi_{[p_1,\ldots,p_t]}(1) = \left[2 - t + \sum_{i=1}^{t} \frac{1}{p_i}\right] \prod_{i=1}^{t} p_i.$$

Remark 4.37. The special value $\chi(1) = \chi_{[p_1,\ldots,p_t]}(1)$ has a specific combinatorial meaning, which can be directly verified:

1) $\chi(1) > 0$ if and only if the star $[p_1, \ldots, p_t]$ is of Dynkin type.
2) $\chi(1) = 0$ if and only if the star $[p_1, \ldots, p_t]$ is of extended Dynkin type.
3) $\chi(1) < 0$ if and only if the star $[p_1, \ldots, p_t]$ is neither of Dynkin nor extended Dynkin type.

In particular, Tables 4.3 and 4.4 show factorizations of the Coxeter polynomial associated to Dynkin and extended Dynkin diagrams without oriented cycles. Observe that, for the extended Dynkin graph $\widetilde{\mathbb{A}}$, the Coxeter polynomial depends on the orientation of arrows: If a (resp. b) denotes the number of arrows in clockwise

Table 4.3 Coxeter polynomials of Dynkin diagrams, expressed with a v-factorization and a cyclotomic factorization

Dynkin diagram	Star symbol	v-Factorization	Cyclotomic factorization	Coxeter number
$\mathbb{A}_n$	$[n]$	v_{n+1}	$\prod_{d \mid n+1, \, d>1} \Phi_d$	$n+1$
$\mathbb{D}_m$	$[2, 2, m-2]$	$v_2 \frac{v_{2(m-1)}}{v_{m-1}}$	$\Phi_2 \prod_{d \mid 2(m-1), \, d>m-1} \Phi_d$	$2(m-1)$
$\mathbb{E}_6$	$[2, 3, 3]$	$\frac{v_2 v_3}{v_4 v_6} v_{12}$	$\Phi_3 \Phi_{12}$	12
$\mathbb{E}_7$	$[2, 3, 4]$	$\frac{v_2 v_3}{v_6 v_9} v_{18}$	$\Phi_2 \Phi_{18}$	18
$\mathbb{E}_8$	$[2, 3, 5]$	$\frac{v_2 v_3 v_5}{v_6 v_{10} v_{15}} v_{30}$	Φ_{30}	30

Table 4.4 Coxeter polynomials of extended Dynkin diagrams

Extended Dynkin diagram	Star type	Coxeter polynomial	Cyclotomic factorization
$\widetilde{\mathbb{A}}_{p,q}$	–	$(z-1)^2 v_p v_q$	$\left(\prod_{d \mid p} \Phi_d\right)\left(\prod_{d \mid q} \Phi_d\right)$
$\widetilde{\mathbb{D}}_m$	–	$(z-1)^2 v_2^2 v_{m-2}$	$\Phi_1 \Phi_2^2 \prod_{d \mid m-2} \Phi_d$
$\widetilde{\mathbb{E}}_6$	$[3, 3, 3]$	$(z-1)^2 v_2 v_3^2$	$\Phi_1^2 \Phi_2 \Phi_3^2$
$\widetilde{\mathbb{E}}_7$	$[2, 4, 4]$	$(z-1)^2 v_2 v_3 v_4$	$\Phi_1^2 \Phi_2^2 \Phi_3 \Phi_4$
$\widetilde{\mathbb{E}}_8$	$[2, 3, 6]$	$(z-1)^2 v_2 v_3 v_5$	$\Phi_1^2 \Phi_2 \Phi_3 \Phi_5$

(resp. anti-clockwise) orientation (thus $a, b \geq 1$ and $a + b = n + 1$), that is, the quiver has type $\widetilde{\mathbb{A}}_{a,b}$, the Coxeter polynomial is $\chi_{a,b}(z) = (z-1)^2 v_a v_b$.

In a slightly different context, in 1982 Howlett [34] gave further properties on the spectrum of Coxeter matrices. Recall that an *M-matrix* C is a square matrix with nonpositive off diagonal coefficients such that each of its principal minors is positive. For the proof of the following result we refer interested readers to [34].

Theorem 4.38 (Howlett). *Let T be an M-matrix and assume that the quadratic form $q(v) = v^t T v$ is not positive. Then the Coxeter matrix $X = -T^{-1}T^t$ has a real eigenvalue λ with $\lambda \geq 1$. Moreover,*

a) If q is not nonnegative then $\lambda > 1$.
b) If q is nonnegative then all eigenvalues of X have modulus one, the real number 1 is a repeated eigenvalue and X is not diagonalizable.

Exercises 4.39.

1. Find the Coxeter matrix and Coxeter polynomial of the Kronecker form q_m and the Pell form $q_{[d]}$.
2. Show that the equation $a^2 + b^2 + c^2 \equiv abc \bmod 3$ has only trivial solutions in the finite field $\mathbb{F}_3$.
3. Let $\vec{G}$ be a directed graph without oriented cycle. Show that the Coxeter polynomial of the directed graph $\vec{G}^{op}$ (obtained by changing the orientation of all directed edges) is equal to the Coxeter polynomial of $\vec{G}$.
4. Let q' be a unit form obtained from a quadratic unit form q after a point inversion. Show that $\chi_q = \chi_{q'}$.
5. Find the Coxeter polynomial of the maximal star $\mathbb{S}_m$ (of type $[2, 2, \dots, 2]$ with m entries).
6. Let C be the companion matrix of a monic polynomial p. Find the inverse of C if p is self-reciprocal.
7. Use A'Campo's Theorem 4.36 to show that for any bipartite directed graph $\vec{G}$, all eigenvalues of Coxeter matrix $X_{\vec{G}}$ are either positive real numbers or complex numbers with modulus one.
8. Give an example of a tree graph T that is not an extended Dynkin graph and such that 1 is a root of the Coxeter polynomial χ_{q_T} associated to the quadratic form q_T.
9. Prove the statements in Remark 4.37.
10. Show that the matrix $X = \begin{pmatrix} 5 & 4 \\ -4 & -3 \end{pmatrix}$ is not periodic.
11. Let $\vec{G}$ be an acyclic directed graph, with adjacency matrix $A_{\vec{G}}$. Show that if the order of vertices in $\vec{G}$ is admissible, then $\mathbf{Id} - A_{\vec{G}}$ is an M-matrix. Is it true in general?
12. Let X be the Coxeter transformation of a nondegenerate bilinear form $\langle -, - \rangle$, and let v and w be eigenvectors of X with eigenvalues λ and μ such that $\lambda\mu \neq 1$. Show that $\langle v, w \rangle = 0$.

Chapter 5
Weakly Positive Quadratic Forms

Consider the quadratic form q associated to the bigraph G below (left). On one hand we observe that q is not a positive form, since $T = T_{12}^{-}T_{13}^{-}T_{31}^{+}$ is an iterated flation for q such that qT is the form associated to extended Dynkin diagram $\widetilde{\mathbb{D}}_4$ (alternatively calculate $q(-1, 0, 1, 1, 1) = 0$). In particular q has infinitely many roots (Theorem 2.16).

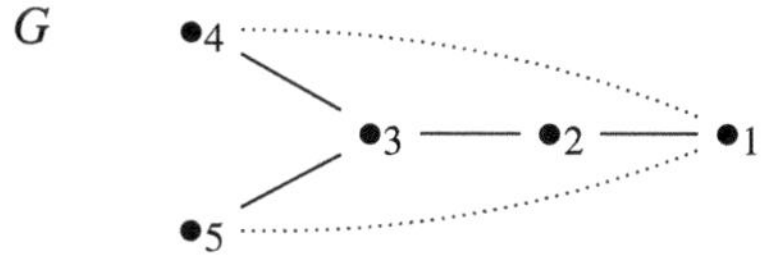

On the other hand, the positive roots $R^+(q)$ of q are contained in the set of positive q_Δ-roots $R^+(q_\Delta)$, where Δ is the Dynkin diagram $\mathbb{D}_5$. Indeed, for a vector $x \in \mathbb{Z}^5$ we have $q(x) = q_\Delta(x) + x_1(x_4 + x_5)$, and if x is a positive root of q, then $x_1(x_4 + x_5) = 0$ and x is a positive root of q_Δ. Hence the set of positive roots $R^+(q)$ of q is finite. The equality $q(x) = q_\Delta(x) + x_1(x_4 + x_5)$ also shows that if $x \in \mathbb{Z}^5$ is a positive vector, then $q(x) > 0$.

A semi-unit form $q : \mathbb{Z}^n \to \mathbb{Z}$ is said to be *weakly positive* if $q(x) > 0$ for every positive vector $x \in \mathbb{Z}^n$ (recall that a vector $x \in \mathbb{Z}^n$ is positive given $x \neq 0$ and $x_i \geq 0$ for $i = 1, \dots, n$).

Examples 5.1. The following are examples of weakly positive unit forms.

a) A positive unit form is weakly positive.
b) Let B be a bigraph with only dotted edges, and take q_B to be its associated quadratic form (that is, q_B is a unit form with $(q_B)_{ij} \geq 0$ for $i \neq j$). Then q_B is weakly positive.

M. Barot et al., *Quadratic Forms*, Algebra and Applications 25,
https://doi.org/10.1007/978-3-030-05627-8_5

c) Consider the quadratic form $q^{a,b}$ associated to the following bigraph, with integers $a, b \geq 1$

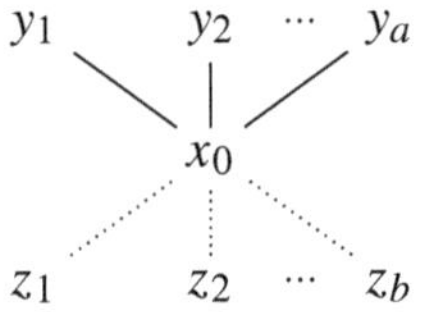

Then $q^{a,b}$ is a unit form in $a + b + 1$ variables which is weakly positive exactly when $a \leq 3$. Indeed, we may write

$$q(x_0, y_1, \ldots, y_a, z_1, \ldots, z_b) = \sum_{i=1}^{a} \left(y_i - \frac{1}{2} x_0 \right)^2 + \frac{4-a}{4} x_0^2 + \sum_{j=1}^{b} (z_j^2 + x_0 z_j),$$

and verify the claim.

5.1 Critical Unit Forms

A unit form q is called *critical nonweakly positive*, or for short just *critical*, if every proper restriction of q is weakly positive, but the form q itself is not weakly positive (compare with critical nonpositive forms defined in Sect. 2.3). The following characterization of critical forms was shown by Ovsienko in [43] (see also [52]). For the proof we follow Ringel in [46].

Theorem 5.2. *Let q be a unit form. Then q is critical if and only if q is the Kronecker form $q_m(x_1, x_2) = x_1^2 - m x_1 x_2 + x_2^2$ for some $m \geq 3$, or q is nonnegative of corank one with radical generated by a sincere positive vector (referred to as a* critical vector *of q).*

Proof. Clearly the stated conditions are sufficient (see proof of Theorem 2.12). For the converse let $q : \mathbb{Z}^n \to \mathbb{Z}$ be a critical form and $v > 0$ a positive vector with minimal weight $|v| := \sum_{i=1}^{n} |v_i|$ such that $q(v) \leq 0$.

Since any proper restriction of q is weakly positive, the vector v is sincere. Then for each vertex $i \in \{1, \ldots, n\}$ we have by minimality

$$0 < q(v - e^i) = q(v) + 1 - q(v|e^i),$$

and therefore $q(v|e^i) \leq q(v)$ for all i.

If $q(v) = 0$ then it follows from $q(v) = \frac{1}{2} \sum_{i=1}^{n} v_i q(v|e^i)$ that $q(v|e^i) = 0$ for all i (since v is sincere and positive), that is, v is a radical vector. For any other positive w with $q(w) \leq 0$ we choose an index a such that $\frac{w_a}{v_a} \leq \frac{w_i}{v_i}$ for all $1 \leq i \leq n$.

Take $z := v_a w - w_a v$ and notice that z is a positive vector in $\mathbb{Z}^n$ with $z_a = 0$. Then

$$0 \leq q^{(a)}(z) = q(v_a w - w_a v) = v_a^2 q(w) \leq 0,$$

and since $q^{(a)}$ is weakly positive, $q^{(a)}(z) = 0$ implies $v_a w = w_a v$. Again by minimality of v all its entries are mutually coprime, therefore v_a divides w_a, that is, w is an integral multiple of v. This shows that if $q(v) = 0$ then q is nonnegative with radical generated by v.

If $q(v) < 0$ then we proceed as in the proof of Theorem 2.12 to obtain $n = 2$ and $q(x_1, x_2) = x_1^2 + q_{12}x_1x_2 + x_2^2$ with $q_{12} \leq -3$. □

In particular notice that all critical forms q in $n \geq 3$ variables are nonnegative with radical generated by a sincere positive vector. Using Theorem 3.5, if q is connected there exists an iterated inflation T for q and an extended Dynkin graph $\widetilde{\Delta}$ such that $qT = q_{\widetilde{\Delta}}$.

Corollary 5.3. *A critical unit form is always critical nonpositive, that is, any proper restriction of a critical unit form is positive.*

Proof. The claim is clear for critical forms $q : \mathbb{Z}^n \to \mathbb{Z}$ with $n = 2$ (Kronecker forms q_m with $m \geq 2$). If $n > 2$, it follows from Theorem 5.2 that a critical unit form is nonnegative with radical generated by a sincere vector. Therefore any proper restriction of q is positive, that is, q is critical nonpositive. □

Using Corollary 5.3 we are ready now to correct the picture drawn in Sect. 2.3.

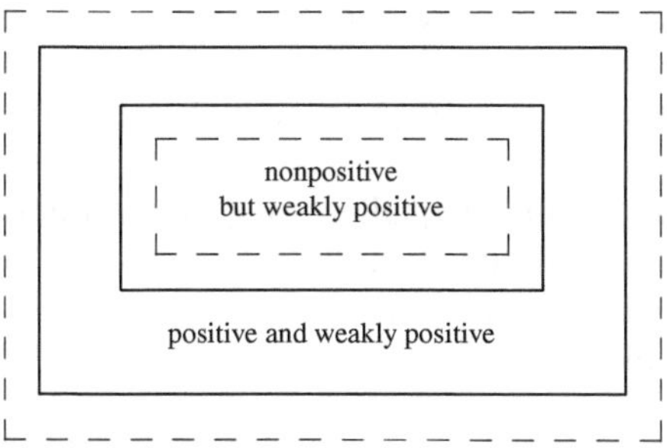

Recall that the *one-point extension* $q[v] : \mathbb{Z}^{n+1} \to \mathbb{Z}$ of a unit form $q : \mathbb{Z}^n \to \mathbb{Z}$ with respect to a q-root v is defined as

$$q[v](x_1, \ldots, x_n, x_{n+1}) = q\left(\sum_{i=1}^{n} x_i e_i - x_{n+1}v\right),$$

which is again unitary, see Lemma 3.26.

Proposition 5.4. *Let q be a unit form in more than two variables.*

a) The form q is critical nonpositive if and only if $q = p[v]$, where p is a positive unit form and v is a sincere root of p.
b) The form q is critical if and only if $q = p[v]$, where p is a positive unit form and v is a positive sincere root of p.

In both cases the radical $\mathbf{rad}(q)$ *of q is generated by a vector z having a vertex i with $z_i = 1$, while for all vertices j we have $|z_j| \leq 6$.*

Proof. This is a direct consequence of Theorems 2.12 and 5.2, since $q[v]$ is a nonnegative unit form with $\mathbf{rad}(q[v])$ generated by the vector $v + e_n$ (cf. Lemma 3.26).

For the last statement, see Corollary 3.31. □

The following technical lemma will be widely used throughout this chapter. Recall that for $v \in \mathbb{Z}^n$, the support of v is given by $\mathbf{supp}(v) = \{i \in \{1, \dots, n\} \mid v_i \neq 0\}$.

Lemma 5.5. *For a weakly positive semi-unit form $q : \mathbb{Z}^n \to \mathbb{Z}$ the following statements hold:*

a) The form q is a unitary.
b) For every pair of indices $i \neq j$ with $q_{ij} < 0$ we have $q_{ij} = -1$.
c) If $v \in \mathbb{Z}^n$ is a positive q-root then $-1 \leq q(v|e_i)$. Moreover, if i and j are different indices in the support of v, then $q(v|e_i) \leq 1$ and $q_{ij} \leq 2$.

Proof. Point (a) is clear. For (b) we evaluate q at the vector $e_i + e_j$ to get

$$0 < q(e_i + e_j) = q(e_i) + q(e_j) + q(e_i|e_j) = 2 + q(e_i|e_j) = 2 + q_{ij}.$$

To show (c) notice that the inequality $-1 \leq q(v|e_i)$ holds in general (evaluate q at $v + e_i$). Now, if $i, j \in \mathbf{supp}(v)$, the nonzero vector $v - e_i$ has no negative coordinates, therefore $0 < q(v - e_i) = 2 - q(v|e_i)$. For the second inequality assume that $q_{ij} \geq 3$, and notice that

$$q(e_i - e_j) = q(e_i) + q(e_j) - q(e_i|e_j) = 2 - q_{ij} < 0.$$

Since we may assume that $q(v|e_i - e_j) \leq 0$ (change the roles of i and j otherwise), for $y = v + e_i - e_j$ we have

$$q(y) = q(v) + q(e_i - e_j) + q(v|e_i - e_j) \leq 0,$$

a contradiction since y is a positive vector. □

We say that a weakly positive unit form $q : \mathbb{Z}^n \to \mathbb{Z}$ is *sincere* if there exists a positive sincere root v of q.

Corollary 5.6. *For $n \geq 1$ there are finitely many sincere weakly positive unit forms in n variables.*

Proof. If $q : \mathbb{Z}^n \to \mathbb{Z}$ is a sincere weakly positive unit form, then by Lemma 5.5 we have $-1 \leq q_{ij} \leq 2$ for all $1 \leq i \neq j \leq n$. Thus the result follows. □

In Sect. 1.2 we have defined, for a quadratic unit form q, the i-th simple reflection $\sigma_i : \mathbb{Z}^n \to \mathbb{Z}^n$ given as $\sigma_i(x) = x - q(x|e_i)e_i$ for x in $\mathbb{Z}^n$. In the following Proposition we resume some basic facts related to reflections when applied to weakly positive unit forms. We need some preliminary observations.

Lemma 5.7. *Let q be a unit form.*

a) If v is a q-root, then $\sum_{i=1}^{n} v_i q(v|e_i) = 2q(v) = 2$.

If moreover q is a weakly positive form and v is a nonsimple positive root, then:

*b) For all $i \in$ **supp**(v) we have $|q(v|e_i)| \leq 1$.*
*c) There exists an $i \in$ **supp**(v) with $q(v|e_i) = 1$.*

Proof. Part (a) is a direct calculation. For (b), by hypothesis we have $v \pm e_i > 0$. Therefore $0 < q(v \pm e_i) = 2 \pm q(v|e_i)$, which implies that $|q(v|e_i)| \leq 1$. Part (c) follows directly from (b) and (c). □

Let q be a unit form. Recall that a positive q-root v is called *maximal* if for any q-root w with $w \geq v$ (that is, such that $w - v$ is a nonnegative vector) we have $w = v$. Maximal roots play a key role in understanding weakly positive roots. Furthermore, since the restriction of a weakly positive unit form is again weakly positive, we may want to first understand those forms which are sincere.

Proposition 5.8. *The following are equivalent for a positive root v of a weakly positive unit form $q : \mathbb{Z}^n \to \mathbb{Z}$.*

a) The q-root v is maximal.
b) We have $\sigma_i(v) \leq v$ for all $i = 1, \ldots, n$.
c) We have $q(v|e_i) \geq 0$ for all $i = 1, \ldots, n$.

Proof. Assume (a) holds. By definition $\sigma_i(v) = v - q(v|e_i)e_i$, thus we have either $\sigma_i(v) \leq v$ or $\sigma_i(v) > v$. Since $\sigma_i(v)$ is also a root of q (Lemma 1.5(c)), by maximality of v we have $\sigma_i(v) \leq v$, therefore (b) holds.

That (b) implies (c) is obvious. We show that (c) implies (a). Let w be a q-root with $w \geq v$. Then $w_i \geq v_i$ and $q(v|e_i) \geq 0$ for any index i, therefore

$$0 \leq q(w-v) = q(w)+q(v)-q(w|v) = 2-\sum_{i=1}^{n} w_i q(v|e_i) \leq 2-\sum_{i=1}^{n} v_i q(v|e_i) = 0,$$

showing that $q(w - v) = 0$, that is, $w = v$ since q is weakly positive. □

The hypothesis that q is weakly positive is essential to show that (c) implies (a) in Proposition 5.8, as the following example shows. Let $q = q_B$ be the form associated to the bigraph B below and take v and w to be the vectors as indicated by the integers at the vertices.

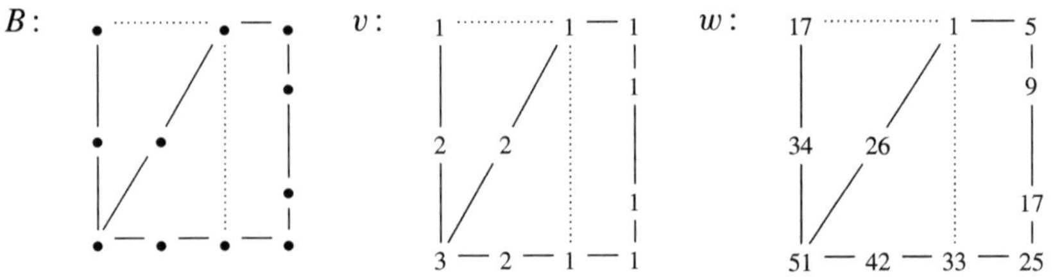

Then it is easy to show that in fact $q(v) = q(w) = 1$ and $q(v|e_i) \geq 0$ for any vertex i, but clearly $v < w$. It also clear that q is not weakly positive since $q(w-v) = -16$.

In view of the preceding result, for a maximal q-root v it is natural to distinguish between vertices i for which $q(v|e_i) > 0$ and those vertices j where $q(v|e_j) = 0$. A vertex i is called *exceptional* for the maximal q-root v if $q(v|e_i) > 0$. The following result was observed by Ringel [46] in the context of sincere representation finite algebras.

Lemma 5.9. *Let v be a sincere maximal positive root of a weakly positive unit form. If $v \neq e_i$ for $1 \leq i \leq n$ then either there exist exactly two exceptional vertices $i \neq j$ with $v_i = v_j = 1$, or there is exactly one exceptional vertex i with $v_i = 2$.*

Proof. By Proposition 5.8(c) and Lemma 5.7(b) we have $q(v|e_i) = 0, 1$ for any vertex i. Hence the result follows from $\sum_{i=1}^{n} v_i q(v|e_i) = 2$, see Lemma 5.7($a$). □

Notice that a vertex is exceptional with respect to a maximal root. Since there might exist several maximal roots, exceptional vertices are in general not inherent to unit forms (but to specific maximal roots), as the following example shows.

Example 5.10. Consider the quadratic form q_B associated to the bigraph depicted below.

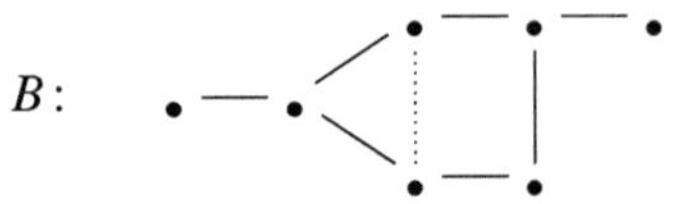

Then there are two maximal roots (indicated by the numbers at the vertices)

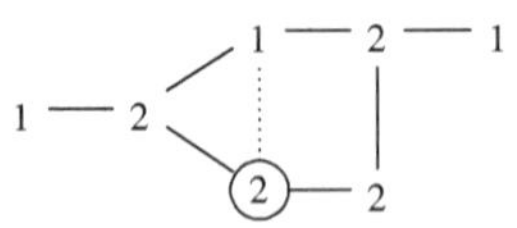

and

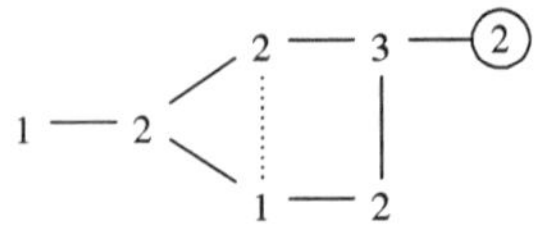

where encircled numbers indicate the exceptional vertex in each case.

The following important result will be used below in Theorem 5.13.

Corollary 5.11. *For any critical unit form $q : \mathbb{Z}^n \to \mathbb{Z}$ and any positive q-root v there is a vertex $i \in \{1, \dots, n\}$ with $q(v|e_i) < 0$. In particular q has infinitely many positive roots.*

Proof. By Theorem 5.2, if q is critical then either q is the Kronecker form $q_m(x_1, x_2) = x_1^2 + x_2^2 - mx_1x_2$ with $m \geq 3$, or q is nonnegative with radical generated by a positive vector z.

Consider first the former case, and take $v = (v_1, v_2)$ a positive root of q_m, thus in particular $v_1^2 + v_2^2 = 1 + mv_1v_2$. Then

$$\begin{aligned} q(v|e_1)q(v|e_2) &= (2v_1 - mv_2)(2v_2 - mv_1) = (4 + m^2)v_1v_2 - 2m(v_1^2 + v_2^2) \\ &= (4 + m^2)v_1v_2 - 2m(1 + mv_1v_2) \\ &= (2 + m)(2 - m)v_1v_2 - 2m, \end{aligned}$$

and since v is positive and $m > 2$ we have $q(v|e_1)q(v|e_2) < 0$.

Now, for the second case consider a positive root v with $q(v|e_i) \geq 0$ for all i and take $w := v + z$. Since z is positive and sincere we have $w_i > v_i > 0$ for $i = 1, \dots, n$. Then

$$q(z) = q(w-v) = q(w)+q(v)-q(w|v) = 2-\sum_{i=1}^{n} w_i q(v|e_i) < 2-\sum_{i=1}^{n} v_i q(v|e_i) = 0,$$

which is a contradiction. We conclude in any case that there is an index i with $q(v|e_i) < 0$. For the last claim, for any positive root v with $q(v|e_i) < 0$ we have that $\sigma(v)$ is a positive root larger than v, thus the assertion follows. □

If q is a critical unit form in more than two variables, then q is connected and nonnegative by Theorem 5.2. As defined in Sect. 3.2 the Dynkin type **Dyn**(q) of q is a Dynkin graph. For instance, in Table 5.1 we exhibit all critical forms of Dynkin type $\mathbb{E}_6$.

5.2 Checking for Weak Positivity

As a first (rather obvious) criterion to verify weak positivity notice that a unit form q is weakly positive if and only if it does not contain as a restriction any critical form. The following nontrivial characterization is due to Drozd and Happel (cf. [30]). We need a preliminary observation.

Lemma 5.12. *Let $v^1, v^2, v^3, \dots$ be an infinite sequence of different positive vectors in $\mathbb{Z}^n$. Then there exist $0 < s < t$ such that $v_s < v_t$ (in other words, the poset of positive vectors in $\mathbb{Z}^n$ has* finite width*).*

Table 5.1 Critical forms of type $\mathbb{E}_6$

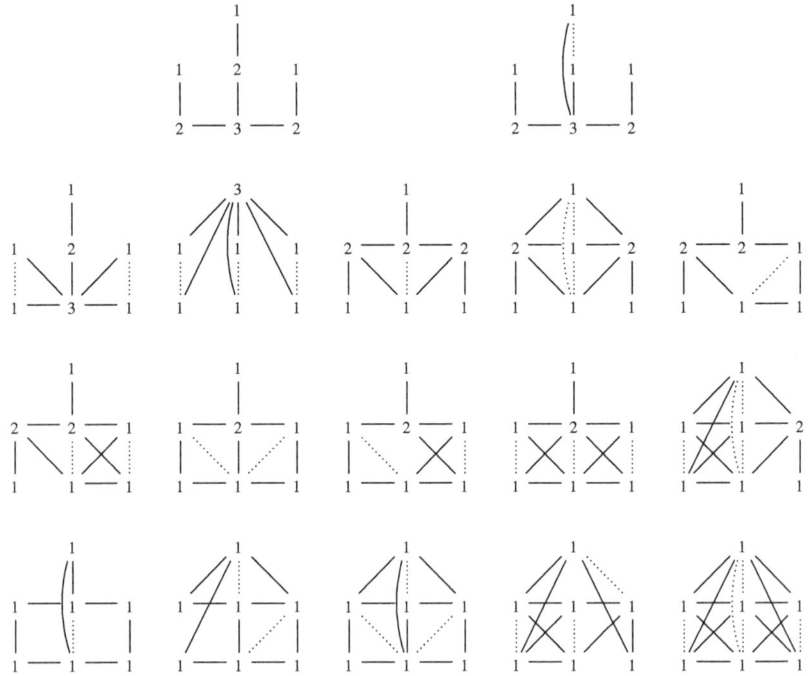

Proof. We proceed by induction on $n > 0$ (the case $n = 1$ is evident). Consider the nonnegative integer $m^k := \min(v_1^k, \ldots, v_n^k)$ for $k \geq 1$. If the sequence $\{m^k\}_{k \geq 1}$ is unbounded the claim is clear (taking t such that $m^t > \max(v_1^1, \ldots, v_n^1)$ then we guarantee that $v^1 < v^t$), hence we may assume that $\{m^k\}_{k \geq 1}$ is bounded.

Taking subsequences if necessary, we may further assume that there is an index $1 \leq i \leq n$ such that the sequence $\{v_i^k\}$ is bounded, thus we may actually assume that the integer v_i^k is fixed for all $k \geq 1$. Consider the vector $\widehat{v}^k = (v_1^k, \ldots, v_{i-1}^k, v_{i+1}^k, \ldots, v_n^k)$ in $\mathbb{Z}^{n-1}$ for $k \geq 1$, and observe that $\{\widehat{v}^k\}_{k \geq 1}$ is a sequence of different positive vectors in $\mathbb{Z}^{n-1}$. Hence the result follows by induction. □

Theorem 5.13 (Drozd–Happel). *Let $q : \mathbb{Z}^n \to \mathbb{Z}$ be a unit form. Then q is weakly positive if and only if q accepts only finitely many positive roots. Moreover, in this case there is an iterated deflation T such that $qT = q_G$ where G is a bigraph with only dotted edges and no loop.*

Proof. We start by proving the last statement. Suppose that $R^+(q)$ is a finite set and that q is weakly positive. If $q_{ij} < 0$ for some $i \neq j$ then $q_{ij} = -1$ by Lemma 5.5(*b*). Consider the deflation $T = T_{ij}^-$ for q and take $q_1 = qT$ (which is a unit form by Proposition 2.17). Then $T : R^+(q_1) \to R^+(q)$ is a proper embedding (Lemma 2.19) thus $R^+(q_1)$ is a finite set. To continue we will look for indices $k \neq \ell$

such that $(q_1)_{k\ell} < 0$. Since this procedure may be iterated only a finite number of times, we get a composition of deflations T taking q to q_G where G has only dotted edges.

Assume first that q is not weakly positive. Then there exists a critical restriction q^I of q, and by Corollary 5.11 the forms q^I and q have infinitely many positive roots.

Assume now that $q : \mathbb{Z}^n \to \mathbb{Z}$ is a weakly positive unit form such that $R^+(q)$ is an infinite set. Let n be minimal with this property, so that for each index i the weakly positive unit form $q^{(i)}$ has finitely many positive roots. In particular, q has infinitely many *sincere* positive roots, and by Lemma 5.5(*c*), for any such root v we have $q(v|e_i) \in \{-1, 0, 1\}$. Therefore there should be an infinite sequence $\{v^k\}_{k\geq 1}$ of positive q-roots with $(q(v^k|e^i))_{i=1}^n$ a fixed vector in $\mathbb{Z}^n$. By Lemma 5.12 we can find two comparable roots $v^s < v^t$, and we have

$$\begin{aligned} 0 < q(v^t - v^s) &= \frac{1}{2} q(v^t - v^s | v^t - v^s) \\ &= \frac{1}{2} \sum_{i=1}^{n} (v_i^t - v_i^s)[q(v^t|e_i) - q(v^s|e_i)] = 0, \end{aligned}$$

which is impossible. Therefore $R^+(q)$ is a finite set. □

The iterated simple reflections of a unit form q may also be used to check for weak positivity of q (cf. Proposition 2.5 and Remark 2.6).

Proposition 5.14. *A unit form $q : \mathbb{Z}^n \to \mathbb{Z}$ is weakly positive if and only if there exists $N > 0$ such that for every sequence of q-roots with the shape*

$$e_i < \sigma_{\ell_1}(e_i) < \sigma_{\ell_2}\sigma_{\ell_1}(e_i) < \ldots < \sigma_{\ell_r} \cdots \sigma_{\ell_1}(e_i),$$

we have $r < N$.

Proof. For q weakly positive the condition is necessary since $R^+(q)$ is a finite set (Theorem 5.13).

If q is not weakly positive then there is a critical restriction q' of q. By Corollary 5.11, for any positive q'-root v there is a vertex ℓ such that $q'(v|e_\ell) = q(v|e_\ell) < 0$. In particular, if $v = \sigma_{\ell_r} \cdots \sigma_{\ell_1}(e_i)$ is a positive root, then $v < \sigma_\ell(v)$ which completes the result. □

Ovsienko's Theorem (see Theorem 5.25 below) claims that if $v \in \mathbb{Z}^n$ is a positive root of a weakly positive unit form $q : \mathbb{Z}^n \to \mathbb{Z}$, then $v_i \leq 6$ for $i = 1, \ldots, n$. This establishes a priori the bound $N = 6^n$ in the algorithm of Proposition 5.14.

Recall from Proposition 2.5 and Remark 2.6 that we may construct all positive roots (inductively using reflections) for a unit form q known to be weakly positive. However, we usually do not know beforehand that q is weakly positive. Still, we could start to construct q-roots inductively using reflections, and find a way to stop the process using the following simple criterion.

Proposition 5.15. *If $q : \mathbb{Z}^n \to \mathbb{Z}$ is a nonweakly positive unit form with $q_{ij} \geq -2$ for $1 \leq i, j \leq n$, then there exists a positive q-root w and a vertex i such that $q(w|e_i) \leq -2$.*

Proof. Since q is not weakly positive there exists a critical restriction q^I of q which, by the hypothesis $q_{ij} \geq -2$ and Theorem 5.2, has a positive sincere radical vector z. For an index i in I and identifying $z \in \mathbb{Z}^I$ with its inclusion in $\mathbb{Z}^n$, we have $q(z|e_i) = q^I(z|e_i) = 0$. Take $w = z - e_i \in \mathbb{Z}^n$ which is a positive root of q, and calculate

$$q(w|e_i) = q(z|e_i) - q(e_i|e_i) = 0 - 2. \qquad \square$$

Algorithm 5.16. *By iteratively calculating positive q-roots using reflections one of the following two situations appear after a finite number of steps: either one finds a positive root w and a vertex i such that $q(w|e_i) \leq -2$ and conclude that the form was not weakly positive, or we end up with a finite number of positive roots unable to produce any new positive roots using reflections, hence concluding that the form is weakly positive (and we have reached all positive roots).*

The last result of this section will be heavily used in the rest of this chapter.

Lemma 5.17. *Let q be a sincere weakly positive unit form and consider its associated bigraph B_q. Then the subgraph of B_q determined by all solid edges is connected.*

Proof. Suppose that the opposite holds, namely, that the set of vertices may be divided into two disjoint subsets I and J such that $q_{ij} \geq 0$ whenever $i \in I$ and $j \in J$. Consider a positive sincere root v and write $v = v^I + v^J$ where $\mathbf{supp}(v^I) = I$ and $\mathbf{supp}(v^J) = J$. Then each summand on the right side of the following equation is nonnegative,

$$1 = q(v^I + v^J) = q(v^I) + q(v^J) + \sum_{i \in I,\ j \in J} v_i^I v_j^J q_{ij},$$

hence we must have $q(v^I) = 0$ or $q(v^J) = 0$, that is, either $I = \emptyset$ or $J = \emptyset$. $\square$

Exercises 5.18.

1. Find the exceptional vertices of the maximal positive root of the quadratic form associated to each Dynkin diagram.
2. Calculate the root-picture for q_B (that is, the Hasse diagram of the poset of positive q_B-roots) where B is the following bigraph,

B: 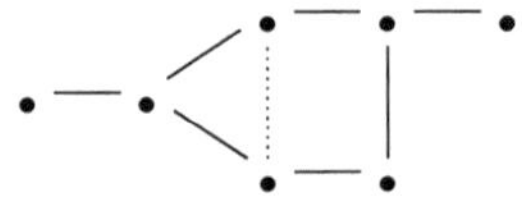

3. Find all the maximal roots of q_B and their exceptional vertices, where

B:
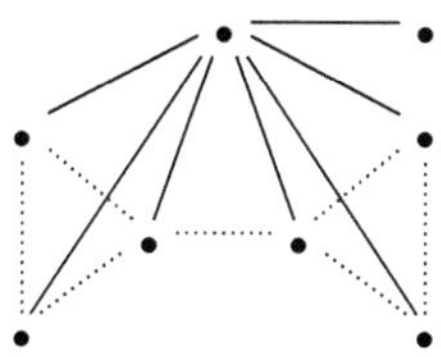

4. Determine which of the following bigraphs correspond to a weakly positive unit form.

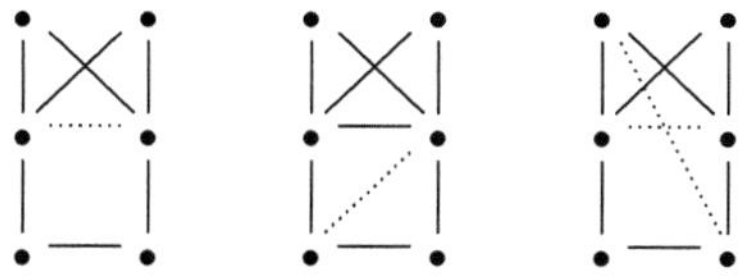

Which of these forms is sincere?

5. Give an iterated deflation T such that the bigraph associated to the form qT has no dotted edges, where q is the following weakly positive unit form,

$$q(x) = x_1^2 + \ldots + x_7^2 - x_2(x_1 + x_3 + x_4) - x_3(x_5 + x_6) + x_4(x_1 + x_6 - x_7) - x_6x_7.$$

5.3 Edge Reduction

Let $q : \mathbb{Z}^n \to \mathbb{Z}$ be a semi-unit form and take different indices i and j such that $q_{ij} < 0$. Define a new unit form $q' : \mathbb{Z}^{n+1} \to \mathbb{Z}$ by the formula

$$q'(x) = q(\rho(x)) + x_i x_j, \quad \text{where} \quad \rho(e_k) = \begin{cases} e_k, & \text{if } 1 \leq k \leq n; \\ e_i + e_j, & \text{if } k = n + 1. \end{cases}$$

We say that q' is obtained from q by *edge reduction* with respect to indices i and j (see [53]). The quadratic form q can be recovered from q' using the nonlinear map $\pi : \mathbb{Z}^n \to \mathbb{Z}^{n+1}$ defined as

$$\pi(x)_k = x_k, \quad \text{for } k \notin \{i, j, n + 1\} \text{ and}$$

$$(\pi(x)_i, \pi(x)_j, \pi(x)_{n+1}) = \begin{cases} (0, x_j - x_i, x_i), & \text{if } x_i \leq x_j; \\ (x_i - x_j, 0, x_j), & \text{if } x_i > x_j. \end{cases}$$

Indeed, we have $\rho \circ \pi = \mathbf{Id}_{\mathbb{Z}^n}$ and $q(x) = q(\rho(\pi(x))) = q'(\pi(x)) - \pi(x)_i \pi(x)_j = q'(\pi(x))$ for every $x \in \mathbb{Z}^n$.

Lemma 5.19. *If q is a unit form and q' is an edge reduction of q with respect to i and j, then q' is again a unit form if and only if $q_{ij} = -1$.*

Proof. The claim follows from the observations $q'(e_k) = q(\rho(e_k)) = q(e_k) = 1$ for $1 \le k \le n$, and

$$q'(e_{n+1}) = q(\rho(e_{n+1})) = q(e_i + e_j) = 2 - q_{ij}. \qquad \square$$

Proposition 5.20. *Let $q : \mathbb{Z}^n \to \mathbb{Z}$ and $q' : \mathbb{Z}^{n+1} \to \mathbb{Z}$ be unit forms such that q' is obtained from q by edge reduction with respect to vertices i and j. The following hold:*

a) The function $\pi : \mathbb{Z}^n \to \mathbb{Z}^{n+1}$ induces an injection $\pi : R^+(q) \to R^+(q')$.
b) The form q is weakly positive if and only if q' is weakly positive. In this case $\pi : R^+(q) \to R^+(q')$ is a bijection.

Proof.

(a) If x is a positive q-root then

$$q'(\pi(x)) = q(\rho(\pi(x))) + \pi(x)_i \pi(x)_j = q(x) = 1,$$

since by definition either $\pi(x)_i = 0$ or $\pi(x)_j = 0$. Clearly π is an injection.

(b) Assume q is weakly positive and take y to be a positive vector in $\mathbb{Z}^{n+1}$. Then clearly $\rho(y)$ is a positive vector in $\mathbb{Z}^n$ and $q'(y) = q(\rho(y)) + y_i y_j$, where the first summand is strictly positive and the second nonnegative. Hence $q'(y) > 0$.

For the converse, assume that q' is weakly positive and take x a positive vector in $\mathbb{Z}^n$. By construction $\pi(x)$ is a positive vector in $\mathbb{Z}^n$ and $q(x) = q'(\pi(x)) > 0$.

Finally suppose q is weakly positive and take a positive root $y \in R^+(q')$. Then $1 = q'(y) = q(\rho(y)) + y_i y_j > y_i y_j \ge 0$, which means that $y_i y_j = 0$ and $\rho(y) \in R^+(q)$ with $\pi(\rho(y)) = y$, that is, $\pi : R^+(q) \to R^+(q')$ is a bijection $\square$

Examples 5.21. Next we illustrate graphically the edge reduction procedure.

a) Consider $q^1 = q_{B^1}$, where B^1 is the bigraph

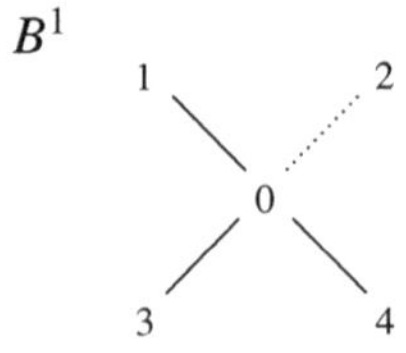

Reducing q^1 with respect to 0 and 1 yields $q^2 = q_{B^2}$, where B^2 is the bigraph below (and the added vertex is labeled 5).

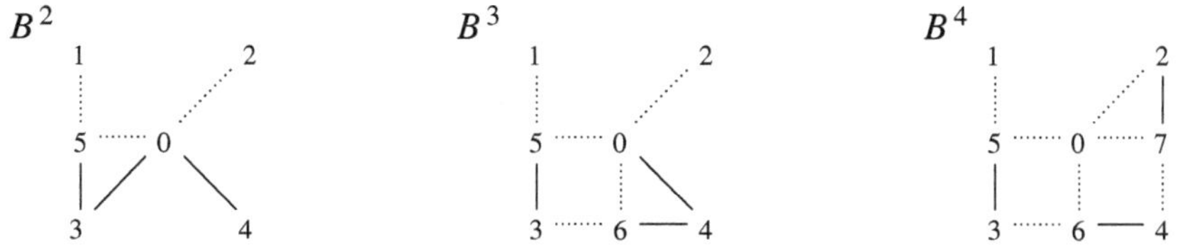

Reduce q^2 with respect to 0 and 3 to get $q^3 = q_{B^3}$, after reducing bigraph B^3 to avoid both types of edges between two vertices (regularization). Continue with edge 0 and 4 to get B^4, and similarly as indicated below:

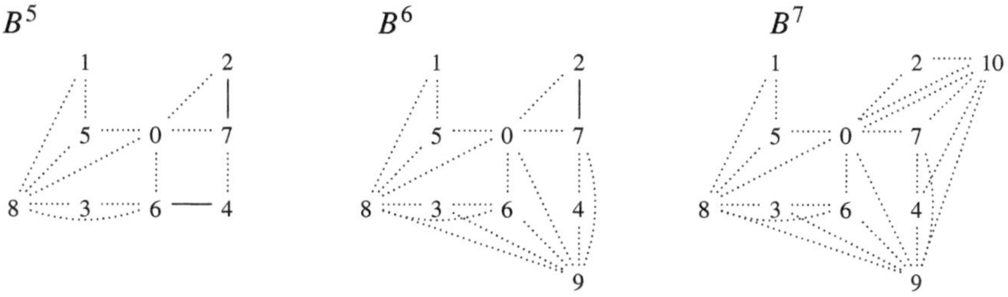

At the end we get a bigraph B^7 containing only dotted edges, hence we cannot continue to perform reductions. According to Proposition 5.20, all quadratic forms $q^7, q^6, \ldots, q^1$ are weakly positive. Observe also that B^7 has a double dotted edge between a pair of vertices (0 and 10).

b) As an illustration of Lemma 5.19, consider the unit form $q = q_{C^1}$ where C^1 is the bigraph

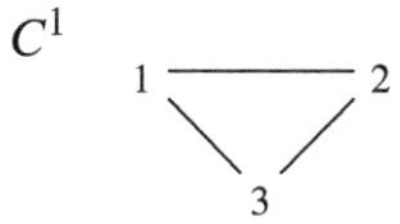

After applying edge reduction with respect to 1 and 2 we get $q^2 = q_{C^2}$, where C^2 is the bigraph below

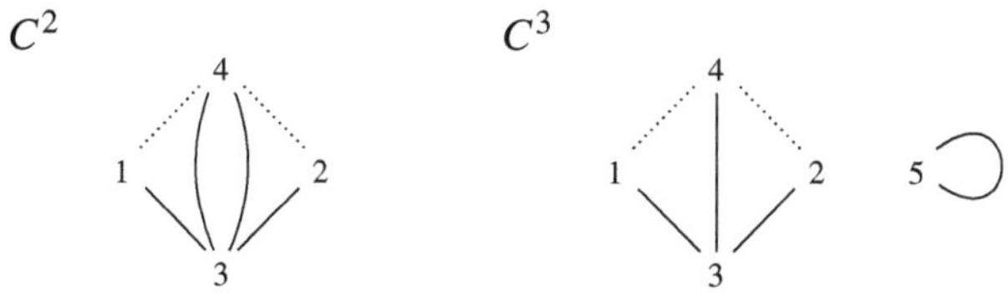

Now we get the following interesting situation: Reduction with respect to vertices 3 and 4, which are joined by a couple of solid edges. This reduction yields quadratic form $q^3 = q_{C^3}$, where C^3 is (the regularization of) the bigraph above, which has an isolated loop. Hence q^3 is not a unit form, and therefore none of the forms q, q^2 and q^3 is weakly positive.

The following *reduction procedure* for weakly positive unit forms is presented by von Höhne in [53], and it forms the basis of the algorithm described below.

Theorem 5.22. *Let $q : \mathbb{Z}^n \to \mathbb{Z}$ be a weakly positive unit form and consider a sequence of forms $q = q^n, q^{n+1}, \dots, q^s$ such that q^i is obtained from q^{i-1} by edge reduction (hence $q^i : \mathbb{Z}^i \to \mathbb{Z}$) for $i > n$. The following hold:*

a) Each q^i is a weakly positive unit form for $i \geq n$.
b) We have $s \leq |R^+(q)|$ and if $s = |R^+(q)|$ then q^s has coefficients $q^s_{ij} \geq 0$ for every pair of indices $1 \leq i, j \leq s$.

Proof. By Proposition 5.20 each q^i is a weakly positive unit form, and the mapping $\pi^i : \mathbb{Z}^{n+1} \to \mathbb{Z}^{i+1}$ induces a bijection $\pi^i : R^+(q^i - 1) \to R^+(q^i)$ for every $i > n$. Since each canonical vector e^j is a positive q^i-root, we have

$$i \leq |R^+(q^i)| = |R^+(q)|.$$

If $s = |R^+(q)|$ and if for some pair of indices $i < j$ we have $q^s_{ij} < 0$, then $q^s_{ij} = -1$ by Lemma 5.5, hence $q^s(e_i + e_j) = 1$, which is impossible since $R^+(q^s) = \{e^i \mid 1 \leq i \leq n + s\}$. □

As consequence of Theorem 5.22, if q is a weakly positive unit form there is a bound for the length of any possible iterated edge reduction for q, namely $|R^+(q)| - n$ where n is the number of variables of q. The converse is false. For instance, the (classical) Kronecker unit form q_2 admits iterated edge reductions of length at most two, although q_2 is not weakly positive. Now we describe an algorithm to verify weak positivity for a unit form $q : \mathbb{Z}^n \to \mathbb{Z}$, constructing on the way all positive q-roots.

Algorithm 5.23. *Let $q : \mathbb{Z}^n \to \mathbb{Z}$ be a unit form.*

Step 1. *Construct a sequence of quadratic forms $q^n, q^{n+1}, \dots, q^N$, where $q^n = q$ and q^{k+1} is obtained from q^k by edge reduction with respect to vertices i_k and j_k (in particular, $q^{k+1} : \mathbb{Z}^{k+1} \to \mathbb{Z}$ is a quadratic form), for $n \leq k < N$.*
Step 2. *Define the sequence of vectors $z^1, \dots, z^N$ in $\mathbb{Z}^n$ as follows: For $k = 1, \dots, n$ take $z^k = e_k$ the canonical vector, and for $k \geq n$ define*

$$z^{k+1} = z^{i_k} + z^{j_k}.$$

Step 3. *For each $N \geq n$ verify the following stopping conditions:*

a) $q_{ij}^N \geq 0$ *for all* $1 \leq i, j \leq N$.
b) $q_{ij}^N \leq -2$ *for some* $i \neq j$.
c) $N > 6^n$.

Then the algorithm must stop after finitely many steps, and the unit form q is weakly positive if condition (a) in Step 3 is satisfied at some point.

Proof. First, if case (a) arises for some $N \geq n$, then q^N is weakly positive with $R^+(q^N) = \{e_1, \ldots, e_N\}$. By Theorem 5.22 we conclude that q itself is a weakly positive unit form and $|R^+(q)| = N$. Moreover, it can be shown that

$$R^+(q) = \{z^1, \ldots, z^N\}.$$

If one of the cases (b) or (c) holds, the form q^N is not weakly positive (respectively by Lemma 5.5(b) and Ovsienko's Theorem 5.25 below). In any case, q is not weakly positive. □

In practice it is never necessary to go so far as the bound 6^n in the algorithm above, and in the next chapter we will review this algorithm and see how to improve it to make it one of the fastest of all.

Theorem 5.24. *A unit form q is weakly positive if and only if any iterated edge reduction q' of q is unitary.*

Proof. The necessity is a consequence of Theorem 5.22(a). Let us assume that the quadratic form $q : \mathbb{Z}^n \to \mathbb{Z}$ is unitary, but not weakly positive.

Assume first that there are vertices $i \neq j$ such that $q_{ij} \leq -3$. If q' is edge reduction of q with respect to i and j, then clearly $q'(e_{n+1}) = 2 - m \leq -1$, that is, q is a nonunitary form.

Assume now that $q_{ij} \geq -2$ for all indices i and j, and take a critical restriction q^I of q. By Theorem 5.2 the restriction q^I is nonnegative and has a critical vector z in $\mathbb{Z}^I$ (which will be identified with its inclusion in $\mathbb{Z}^n$). Since q is unitary the weight $|z| = \sum_{i=1}^n z_i$ of z is larger than 1. Consider the following evident facts:

i) If $|v| > 1$ for a positive vector with $q(v) = 0$, then there are vertices $i \neq j$ in the support $\mathbf{supp}(v)$ of v such that $q_{ij} < 0$.
ii) If moreover q' is the edge reduction of q with respect to vertices i and j, and $v' = \pi(v) \in \mathbb{Z}^{n+1}$, then v' is a positive vector with $q'(v') = 0$ and $|v'| < |v|$.

Starting with the critical vector z, the result follows by induction using points (i) and (ii) above. □

It follows from the proof of Theorem 5.24 that in the reduction process we may find quadratic forms q with $q_{ii} \leq 1$ for some vertex i. These are called *pre-unit* forms, and will be considered again in next chapter when addressing the weakly nonnegative setting.

5.4 Ovsienko's Theorem

As shown in Proposition 2.22, the absolute values of the entries of any root v of a positive unit form are bounded by 6. This is now extended to positive roots of weakly positive unit forms, the celebrated Ovsienko's Theorem. The proof given closely follows Ringel in [46] (see also Gabriel and Roiter [26]).

Theorem 5.25 (Ovsienko). *For any vertex $i \in \{1, \ldots, n\}$ and any positive root $v \in \mathbb{Z}^n$ of a weakly positive unit form $q : Z^n \to \mathbb{Z}$ we have $v_i \leq 6$.*

Proof. The proof is combinatorial and done in several steps. We have already seen a positive (hence weakly positive) unit form with a root having an entry 6, namely $q_{\mathbb{E}_8}$ (see Table 2.2).

Let $s \geq 6$ be an integer. Among those weakly positive unit forms q with

$$s = M(q) := \max\{v_i \mid i \text{ is an index of } q \text{ and } v \text{ is a positive root of } q\},$$

we choose one, say q, having minimal number of positive roots. Fix a maximal positive q-root v such that $v_k = s$ for some index k. By minimality, v is a sincere root.

Step 1. *We show that $q_{ij} \geq 0$ for all $i, j \neq k$.*
Suppose $v_i \leq v_j$. We see from $0 < q(e_i + e_j) = 2 + q_{ij}$ that $q_{ij} \geq -1$. If $q_{ij} = -1$ then we know from Lemma 2.19 that $q^- = qT_{ij}^-$ has fewer positive roots than q. Take $v' = (T_{ij}^-)^{-1}v = v - v_ie_j$, which is a positive root of q^- satisfying $v'_k = v_k = s$. This contradicts the assumed minimality of q, since $M(q^-) = M(q)$.

Step 2. *We show that $q_{ij} \leq 1$ for all $i, j \neq k$.*
It follows from Lemma 5.5(c) that $q_{ij} \leq 2$. If $q_{ij} = 2$, assuming that $q(v|e_i) \leq q(v|e_j)$ and taking the positive vector $w = v - v_je_j + v_je_i$, we obtain

$$1 \leq q(w) = q(v) + 2v_j^2 - v_j[q(v|e_j) - q(v|e_i)] - v_j^2 q_{ij} \leq 1.$$

Hence w is a positive root of the restriction $q^{(j)}$ with $w_k = s$. But $q^{(j)}$ certainly has fewer positive roots than q, again in contradiction to minimality.

Step 3. *We have $q_{ki} = -1$ for every vertex $i \neq k$.* This is a direct consequence of Steps 1 and 2, and Lemma 5.17.

Step 4. *The root v has exactly one exceptional vertex ℓ and $v_\ell = 2$.*
Otherwise Lemma 5.9 implies that there are precisely two exceptional vertices ℓ and ℓ' with $q(v|e_\ell) = q(v|e_{\ell'}) = 1$ and $v_\ell = v_{\ell'} = 1$. But in that case, $\sigma_\ell(v) = v - e_\ell$ is a sincere positive root of $q^{(\ell)}$ with $\sigma_\ell(v)_k = s$, in contradiction to the assumed minimality of q.

Step 5. *Define the sets of vertices $I = \{i \neq \ell \mid q_{i\ell} = 1\}$ and $J = \{i \neq \ell \mid q_{i\ell} = 0\}$, where ℓ is the exceptional vertex for v. Then we have $q_{ij} = 1$ for all i, j in I.*

Indeed, consider the positive vector $w = v + e_k - e_\ell + e_i + e_j$. Since ℓ is the unique exceptional vertex for v we have

$$q(v|e_k) = q(v|e_i) = q(v|e_j) = 0, \quad \text{and} \quad q(v|e_\ell) = 1,$$

thus we deduce from $w_k = v_k + 1 > s$ that

$$2 \le q(w) = 5 - q(v|e_\ell) - q_{k\ell} + q_{ki} + q_{kj} - q_{\ell i} - q_{\ell j} + q_{ij} = 1 + q_{ij}.$$

Hence $q_{ij} = 1$ by Step 2.

Step 6. *We have $v_k = 3 + \sum_{i\in I} v_i = -1 + \sum_{i\in J} v_i$.*

Indeed, from $1 = q(v|e_\ell) = 2v_\ell - v_k + \sum_{i\in I} v_i$ we get $v_k = s = 3 + \sum_{i\in I} v_i$, while from

$$\begin{aligned} 0 = q(v|e_k) &= 2s - \sum_{i\neq k} v_i \\ &= s + \left(v_k - \sum_{i\in I} v_i\right) - \sum_{i\in J} v_i - v_\ell = s + 3 - \sum_{i\in J} v_j - 2, \end{aligned}$$

we obtain $s = -1 + \sum_{i\in J} v_i$.

Step 7. *For all $i \in I$ and $j \in J$ we have $v_i = 1$ and $q_{ij} = 0$.*

Indeed, we calculate

$$\begin{aligned} 0 &= q(v|e_i) \\ &= 2v_i + \sum_{m\neq i} q_{im} v_m \\ &= v_i + \sum_{m\in I} v_m + \sum_{j\in J} q_{ij} v_j + v_\ell - v_k \\ &= v_i + (v_k - 3) + \sum_{j\in J} q_{ij} v_j + 2 - v_k \\ &= -1 + v_i + \sum_{j\in J} q_{ij} v_j. \end{aligned}$$

Since $q_{ij} \ge 0$ for all $j \in J$ we must have $v_i = 1$ and $q_{ij} = 0$.

Step 8. *Let $z \in J$ be such that $v_z \ge v_j$ for all $j \in J$. Then there exist two vertices $j_1 \neq j_2$ in J with $q_{zj_1} = q_{zj_2} = 0$.*

By Step 7 we have

$$0 = q(v|e_z) = 2v_z + \sum_{j\in J,\, j\neq z} q_{zj} v_j - v_k = v_z + \sum_{j\in J} q_{zj} v_j - \sum_{j\in J} v_j + 1.$$

Thus we infer that

$$v_z < \sum_{j \in J}(1 - q_{zj})v_j \leq \sum_{j \in J}(1 - q_{zj})v_z,$$

hence $2 \leq \sum_{j \in J}(1 - q_{zj})$, which implies the claim.

Step 9. *For vertices j_1 and j_2 as in Step 8 we have $q_{j_1 j_2} = 1$.*
Otherwise the restriction of q to vertices $\{k, \ell, z, j_1, j_2\}$ equals $q_{\mathbb{D}_4}$ in contradiction to the weak positivity of q.

We have now collected enough information to conclude the proof. Assume $s \geq 7$. Then by Steps 6 and 7 the set I has at least four vertices i_1, i_2, i_3, i_4. Hence the restriction of q to the set $\{k, \ell, i_1, i_2, i_3, i_4, z, j_1, j_2\}$ has exactly the following associated bigraph (left)

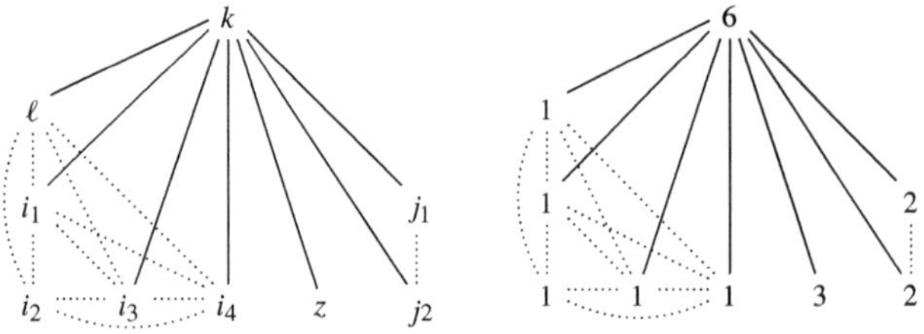

But q evaluates to zero at the positive vector indicated by the number on the vertices in the figure above (right), a contradiction (the bigraph above corresponds to a critical unit form, see figure $\mathscr{C}(6)$ in Table 5.3). □

We now present a suitable generalization of Proposition 1.32 to the weakly positive case due to Zeldych [55] and based on unpublished notes by S. Brenner, where the assumption of q being unitary is dropped. Recall that the *adjugate* $\mathbf{ad}(B)$ of a square matrix B is the transpose of the matrix of cofactors of B.

Theorem 5.26 (Zeldych). *Let A be the associated symmetric matrix of an integral quadratic form $q : \mathbb{Z}^n \to \mathbb{Z}$ (that is, $q(x) = x^t Ax$ for any vector x in $\mathbb{Z}^n$). Then the following conditions are equivalent:*

a) The form q is weakly positive.
b) For each principal submatrix B of A we have either $\det(B) > 0$, *or* $\mathbf{ad}(B)$ *is not positive (that is,* $\mathbf{ad}(B)$ *has a nonpositive entry).*

Proof. Assume (a) holds, let B be a principal submatrix of A and suppose that $\mathbf{ad}(B)$ is a positive matrix. By the Perron–Frobenius Theorem 1.36 there exists a positive eigenvector $v \in \mathbb{R}^n$ of $\mathbf{ad}(B)$ with eigenvalue $\rho > 0$. Considering q as a real function $q_{\mathbb{R}} : \mathbb{R}^n \to \mathbb{R}$ it is clear that $q_{\mathbb{R}}(x) \geq 0$ for any positive vector x in $\mathbb{R}^n$. That actually $q_{\mathbb{R}}(x) > 0$ can be argued as in the proof of Proposition 2.3. Then the inequality $\det(B) > 0$ is deduced from

$$0 < q_{\mathbb{R}}(v) = v^t Bv = \frac{1}{\rho} v^t B(\mathbf{ad}(B)v) = \frac{1}{\rho} \det(B)\|v\|^2,$$

since we have $(B)\mathbf{ad}(B) = \det(B)\mathbf{Id}$.

For the converse we assume that q satisfies (b) but is not weakly positive. Take such a form minimal in the number of variables. Since taking principal submatrices corresponds to restrictions, we infer from minimality that q is critical. Hence each proper restriction of q is positive (see Corollary 5.3), and thus by Proposition 1.32 we have $\det(B) > 0$ for each proper principal submatrix B of A.

Thus $\det(A) \leq 0$ since otherwise q would be positive (again by Proposition 1.32). Take $\mathbf{ad}(A) = (v_{ij})_{i,j=1}^n$, thus by hypothesis there must exist i, j with $v_{ij} \leq 0$. Let v be the j-th column of $\mathbf{ad}(A)$, so that $Av = \det(A)e_j$ and $q(v) = \det(A)v_{jj}$. Further, let $w > 0$ be a sincere positive vector with $q(w) \leq 0$. For $\lambda = -\frac{v_{ij}}{w_i} \geq 0$ we have $(v+\lambda w)_i = 0$ and (since the restriction $q^{(i)}$ is a positive form)

$$\begin{aligned} 0 &< q(v + \lambda w) \\ &= q(v) + 2\lambda w^t Av + \lambda^2 q(w) \\ &\leq \det(A)[v_{jj} + 2\lambda w_j] \\ &= \frac{\det(A)}{w_i}[v_{jj}w_i - 2v_{ij}w_j]. \end{aligned}$$

If $v_{jj} < 0$ (thus we may take $i = j$) then

$$0 < q(v + \lambda w) \leq \det(A)(-v_{jj}) \leq 0,$$

a contradiction. If $v_{jj} \geq 0$ then $v_{jj}w_i - 2v_{ij}w_j \geq 0$ and the following equation yields another contradiction

$$0 < q(v + \lambda w) \leq \frac{\det(A)}{w_i}[v_{jj}w_i - 2v_{ij}w_j] \leq 0,$$

which completes the proof. □

For convenience in what follows we collect the different **Criteria for Weak Positivity** shown in this chapter.

Theorem 5.27. *For a quadratic unit form $q : \mathbb{Z}^n \to \mathbb{Z}$ the following claims are equivalent:*

a) The form q is weakly positive.
b) The form q admits only finitely many positive roots.
c) For any positive root v and any vertex i we have $v_i \leq 6$.

d) *For any positive nonsimple root* v *and any vertex* i *we have* $q(v|e_i) \geq -1$.
e) *For each principal submatrix* B *of* A_q *we have* $\det(B) > 0$ *or* $\mathbf{ad}(B)$ *is positive.*
f) *For all vertices* $i \neq j$ *we have* $q_{ij} \geq -2$ *and* $q^{-1}(0) \cap \mathbb{N}_0^n = \{0\}$.
g) *For all vertices* $i \neq j$ *we have* $q_{ij} \geq -2$ *and for all subset of vertices* I *we have* $\mathbf{rad}(q^I) \cap \mathbb{N}_0^n = \{0\}$.

Proof. The equivalence of (a) and (b) was shown in Theorem 5.13, that of (a) and (e) in Proposition 5.26, that (a) implies (c) is Ovsienko's Theorem 5.25 and that (c) implies (b) is obvious. That (a) implies (d) is shown in Lemma 5.5(c) and that (d) implies (a) is Proposition 5.15. This already show the equivalence of $(a - e)$.

Now, (f) and (g) are reformulations of the fact that no critical form can be contained in a weakly positive unit form: Suppose q is not weakly positive. Then there exists a restriction $p = q^I$ which is critical, that is, p is either an m-Kronecker form $p(x_i, x_j) = x_i^2 - mx_ix_j + x_j^2$ for some $-m = p_{ij} < -2$, or p is nonnegative with a positive sincere radical vector. Therefore (f) and (g) imply (a). Conversely, if (f) or (g) do not hold then q admits a critical restriction, which completes the proof. □

Exercises 5.28.

1. Consider a sequence of quadratic forms $q^n, q^{n+1}, \dots, q^N$, where $q^n = q$ and q^{k+1} is obtained from q^k by edge reduction with respect to vertices i_k and j_k, as in Algorithm 5.23. Also take vectors $z^{(k)} = e_k$ for $k = 1, \dots, n$ and $z^{(k+1)} = z^{(i_k)} + z^{(j_k)}$ for $k \geq n$. For $k > n$ define recursively transformations $\rho^{k-n} : \mathbb{Z}^k \to \mathbb{Z}^n$ as

$$\rho^1 = \rho_{i_n, j_n} \qquad \text{and} \qquad \rho^{k+1-n} = \rho^{k-n} \circ \rho_{i_{n+k}, j_{n+k}},$$

where ρ_{ij} is the transformation associated to the edge reduction with respect to vertices i and j.

a) Show that $z^{(k)} = \rho^{N-n}(e_k)$ for $k = 1, \dots, N$.
b) Conclude that if $q_{ij}^N \geq 0$ for all $1 \leq i, j \leq N$, then

$$R^+(q) = \{z^{(1)}, \dots, z^{(N)}\}.$$

2. Give an example of a weakly positive unit form with corank two.
3. Find an iterated edge reduction σ for the following forms q such that the bigraph associated to $q\sigma$ has no solid edge.

i) $q(x) = x_1^2 + \ldots + x_4^2 - x_1(x_2 + x_3 + x_4)$.
ii) $q(x) = x_1^2 + \ldots + x_5^2 - x_1(x_2 + x_3 + x_4 + x_5) + x_2x_3$.

4. Give a weakly positive unit form q such that $M(q) = 5$ (see proof of Theorem 5.25).
5. Provide an example of a weakly positive unit form that fails to be nonnegative.

6. Consider the quadratic form q associated to the following bigraph, and show that q is weakly positive.

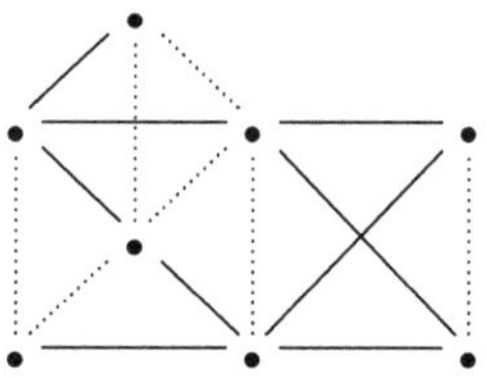

5.5 Explosions and Centered Forms

Let $q : \mathbb{Z}^n \to \mathbb{Z}$ be a unit form. We say that a unitary form $\overline{q}$ is a *(radical) explosion* of q if q is a particular type of restriction of $\overline{q}$, namely: There is a vector $s = (s_1, \dots, s_n)$ in $\mathbb{N}^n$ such that the set

$$I_s = \{(i, k) \mid 1 \leq i \leq n \text{ and } 1 \leq k \leq s_i\},$$

is an index set for $\overline{q}$ satisfying $e_{i,k} - e_{i,1} \in \mathbf{rad}(\overline{q})$ for $1 < k \leq s_i$ (where $\{e_{i,k}\}_{(i,k)\in I_s}$ denotes the canonical basis of $\mathbb{Z}^{I_s}$) and q is the restriction of $\overline{q}$ to the indices $(1, 1), \dots, (n, 1)$. If $s_i > 1$ for some index i we will say that the vertex i is *exploded* $s_i - 1$ *times*. If $s_\omega = 1$ we say that $\overline{q}$ is an *explosion of* q *with respect to* ω, for $\omega \in \{1, \dots, n\}$. If $s_j = 1$ for $j \neq i$ and $s_i = 2$, then we say that $\overline{q}$ is obtained from q by *doubling* vertex i (cf. Exercise 3.32.4). Below we show a small example, doubling vertex 2 in the Dynkin graph $\mathbb{A}_3$.

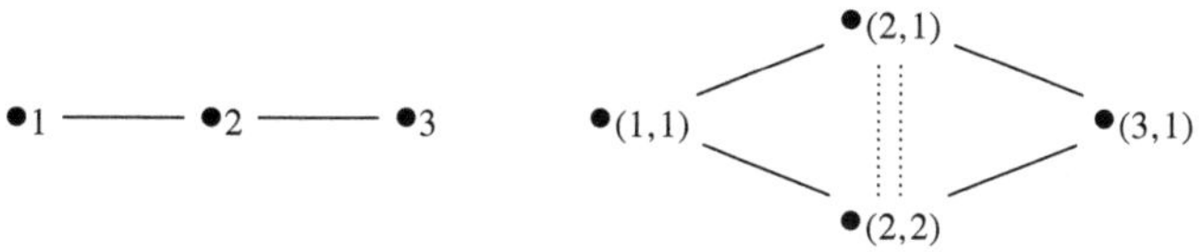

The following result collects some elementary properties of explosions of weakly positive unit forms. For instance, it shows that the new quadratic form in the example above has no sincere root.

Proposition 5.29. *Let $q : \mathbb{Z}^n \to \mathbb{Z}$ be a weakly positive unit form and $\overline{q}$ an explosion of q with index set I_s for $s = (s_1, \dots, s_n)$. The following hold:*

a) The form $\overline{q}$ is weakly positive.

b) If q has a maximal sincere root, then $\overline{q}$ has a maximal sincere root $\overline{z}$ if and only if $s \leq v$ for a sincere maximal positive root v of q. Moreover:

i) If $s = v$ then $\overline{z} = \sum_{(i,k)\in I_s} e_{i,k}$.

ii) If $s_\omega = 1$ *for some* ω *in* $\{1, \ldots, n\}$ *and* $s_i = v_i$ *for* $i \neq \omega$, *then*

$$\overline{z} = v_\omega e_{\omega,1} + \sum_{\substack{(i,k)\in I_s \\ (i,k)\neq(\omega,1)}} e_{i,k}.$$

In both cases $\overline{z}$ *is uniquely determined.*

In situation (i) *we say that* $\overline{q}$ *is a* full explosion of q *(with respect to the maximal root* v*). In situation* (ii) *we say that* $\overline{q}$ *is a* full explosion of q with respect to vertex ω *(and the maximal root* v*).*

Proof. Consider $r_{i,k} = e_{i,k} - e_{i,1} \in \mathbf{rad}(\overline{q})$ for $k = 1, \ldots, s_i$ (notice that $r_{i,1} = 0$ for $i = 1, \ldots, n$), and the function $\Phi : \mathbb{Z}^{I_s} \to \mathbb{Z}^n$ given by

$$\overline{z} \mapsto \Phi(\overline{z}) = \overline{z} - \sum_{(i,k)\in I_s} \overline{z}_{i,k} r_{r,k}.$$

Considering $\mathbb{Z}^n$ as a subgroup of $\mathbb{Z}^{I_s}$ by means of the inclusion $e_i \mapsto e_{i,1}$, we observe that Φ is a projection of $\mathbb{Z}^{I_s}$ onto $\mathbb{Z}^n$ satisfying $\overline{q}(\overline{z}) = q(\Phi(\overline{z}))$, and that $\overline{z} > 0$ implies $\Phi(\overline{z}) > 0$. Therefore $\overline{q}$ is weakly positive if so is q.

Assume now that $z \in \mathbb{Z}^n$ is a sincere positive vector. Clearly there is a sincere positive vector $\overline{z} \in \mathbb{Z}^{I_s}$ such that $\Phi(\overline{z}) = z$ if and only if $s \leq v$. In this case $\overline{z}$ is a maximal sincere root of $\overline{q}$ if and only if z is a maximal sincere root of q, which shows (b). The description of $\overline{z}$ can be easily verified. □

For instance, the full explosion of $q_{\mathbb{E}_6}$ with respect to the star center is given by

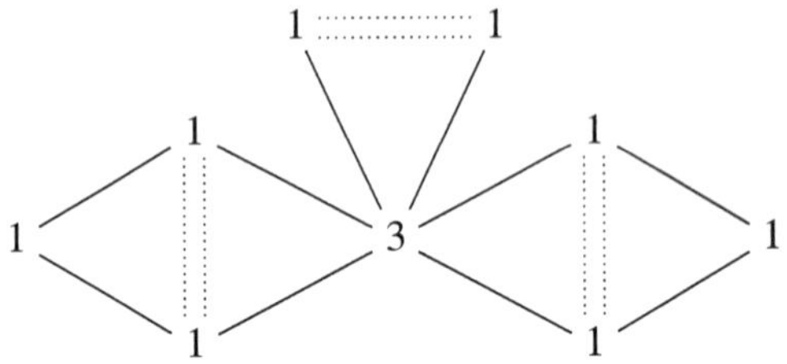

where the numbers at the vertices indicate the maximal positive root.

A unit form q is said to be *centered at vertex* c if $q_{ci} = -1$ for all $i \neq c$ and $q_{ij} \geq 0$ for all $i, j \neq c$. The importance of centered forms (already used in the proof of Theorem 5.25) relies on the following result. Recall that

$$M(q) := \max\{v_i \mid i \text{ is an index of } q \text{ and } v \text{ is a positive root of } q\}.$$

Proposition 5.30. *For each $S \in \{2, \dots, 6\}$ let q_S be a weakly positive unit form with $M(q_S) = S$ such that*

$$|R^+(q_S)| = \min\{|R^+(q)| \text{ such that } q \text{ is weakly positive with } M(q) = s\}.$$

Then q_S is a centered form, with a maximal sincere positive root having a unique exceptional vertex.

Proof. Arguing as in Steps 1 and 2 of Ovsienko's Theorem 5.25, we see that there exists a vertex c such that $0 \leq q_{ij} \leq 1$ for all $i, j \neq c$. Let v be a root with $v_c = M(q)$.

Since for each $i \notin \mathbf{supp}(v)$ the restriction $q^{(i)}$ has fewer roots than q, but still $M(q^{(i)}) = M(q)$, we deduce from the minimality in the number of positive roots of q that v is sincere. As a consequence of Lemma 5.17,we obtain $q_{ci} = -1$ for all $i \neq c$.

If v has two exceptional vertices $i \neq j$ then $v_i = 1$ and $q(v|e_i) = 1$. Hence $\sigma_i(v) = v - e^i$ is a sincere root of $q^{(i)}$ and again $q^{(i)}$ has fewer roots than q, but still $M(q^{(i)}) = M(q)$, contradicting minimality. Thus the result. □

It is important to observe that the maximal value $M(q)$ may not be attained at a sincere root of q. To see this, define

$$M_{sin}(q) := \max\{v_i \mid i \text{ is an index of } q \text{ and } v \text{ is a positive sincere root of } q\},$$

and observe that $M_{sin}(q) \leq M(q)$. Let us consider some examples where the inequality is strict. For each bigraph B in Table 5.2 observe that there is a unique sincere root v of q_B, the one displayed by the integers at the vertices. However, there exists another positive root w satisfying

$$\max\{w_i \mid i \in \mathbf{supp}(w)\} > \max\{v_i \mid i \in \mathbf{supp}(v)\}.$$

Indeed, the bigraph on top fully contains the Dynkin graph $\mathbb{D}_4$, those in the middle fully contain Dynkin graphs $\mathbb{E}_6$ and $\mathbb{E}_7$, and both in the bottom fully contain $\mathbb{E}_8$.

The unit forms q in Table 5.2 are examples of the situation $M_{sin}(q) < M(q)$ for $M_{sin}(q) = 1, \dots, 5$. By Ovsienko's Theorem we cannot expect to find a similar example for $M_{sin}(q) = 6$.

We will now determine those centered forms which are critical (nonweakly positive). Since critical Kronecker forms are not centered, by Theorem 5.2 any critical centered form q is nonnegative of corank one with a sincere positive radical vector. We can say even more:

Proposition 5.31. *If q is a critical centered unit form then $q = p[w]$ where p is a positive centered unit form and w is a sincere positive root of p.*

Proof. Denote by c the center of q, and let v be a sincere positive radical vector of q with mutually coprime entries. Then there exists an index i with $v_i = 1$ (an omissible vertex, see Proposition 3.20).

Table 5.2 Some examples of weakly positive unit forms q with $M_{sin}(q) \in \{1, \ldots, 5\}$ and satisfying $M_{sin}(q) < M(q)$

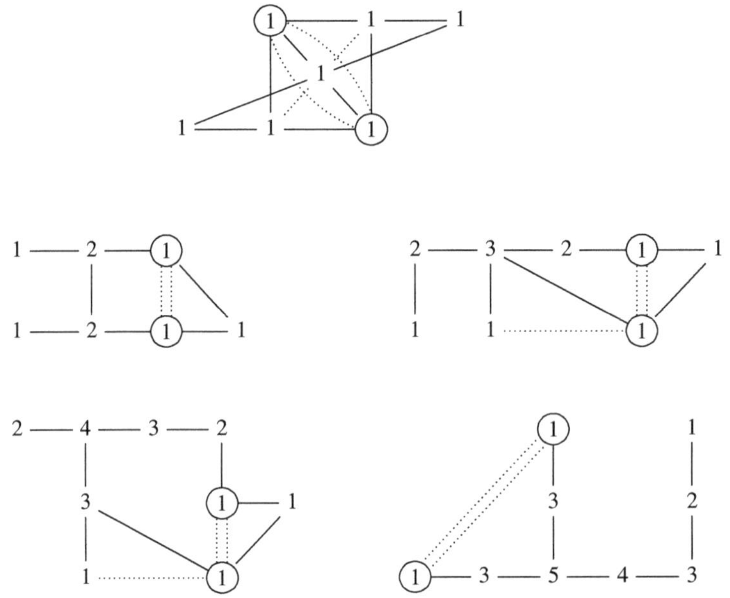

Encircled numbers correspond to exceptional vertices of the displayed maximal root

For $j \neq c$ we have $0 = q(v|e_j) = 2v_j + \sum_{\ell \neq j,c} q_{j\ell}v_\ell - v_c$, that is, $v_c = 2v_j + \sum_{\ell \neq j,c} q_{j\ell}v_\ell > 1$, therefore $i \neq c$. Hence $q^{(i)}$ is a positive connected centered unit form with **Dyn**$(q^{(i)}) =$ **Dyn**(q) (again by Proposition 3.20) and $v' = v - e^i$ may be seen as a positive sincere $q^{(i)}$-root. From Lemma 3.26 we have $q = q^{(i)}[v']$, thus the result. □

Since any root of a positive connected unit form of Dynkin type $\mathbb{A}_n$ has as support a line (see Proposition 2.39), there are only two positive centered unit forms p of Dynkin type $\mathbb{A}_n$ which admit a sincere positive root v, namely $q_{\mathbb{A}_2}$ and $q_{\mathbb{A}_3}$,

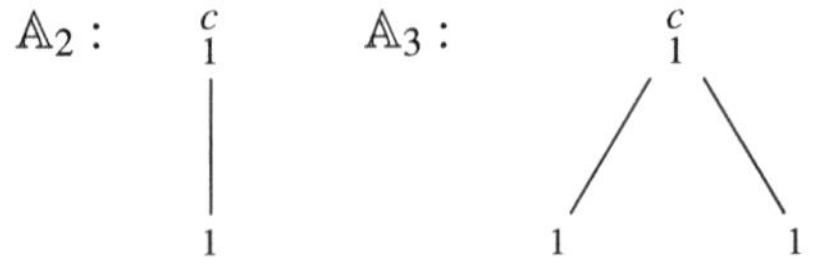

In any case, however, the form $p[v]$ is not centered.

In order to ensure that $p[v]$ is centered again we need the condition $p(v|e_c) = 1$ and $p(v|e_i) \leq 0$ for all $i \neq c$. From Lemma 5.9, the only possibility for a centered positive form of type $\mathbb{D}_m$ is $\mathbb{D}_4$, with centered critical extension $\widetilde{\mathbb{D}}_4$,

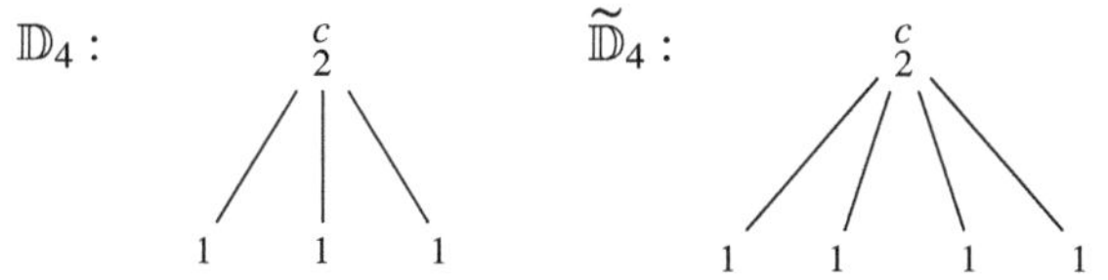

In a similar way, we calculate all cases for $\mathbb{E}_p$ and obtain the list in Table 5.3.

Since the approach in this book is based on algorithms, we do not present a 'paper proof' of the fact that Table 5.4 contains all weakly positive centered forms q admitting a sincere positive root and satisfying $q_{ij} \leq 1$ for all vertices $i \neq j$ (*graphical forms*). By induction any sincere weakly positive centered form admits a restriction to a sincere weakly positive centered form in one less variable. Hence a paper proof could show that no form q in the list admits an extension to a centered form $\overline{q}$ by a vertex k with $\overline{q}(w|e_k) = -1$ for any sincere q-root w not containing any of the critical centered forms above.

Our list is not entirely complete, since we removed from it all forms which can be obtained by explosions of noncentered vertices. For a weakly positive unit form q with associated bigraph belonging to Table 5.3, and vector v with entries as indicated in the vertices, the maximal number of times a noncentered point may be exploded is $v_i - 1$. This is due to the fact that this number is the corresponding entry of the (unique) maximal sincere positive root of (any) restricted centered form $q^{(k)}$ with $v_k = 1$, cf. Proposition 5.29(*b*).

Table 5.3 Critical centered forms

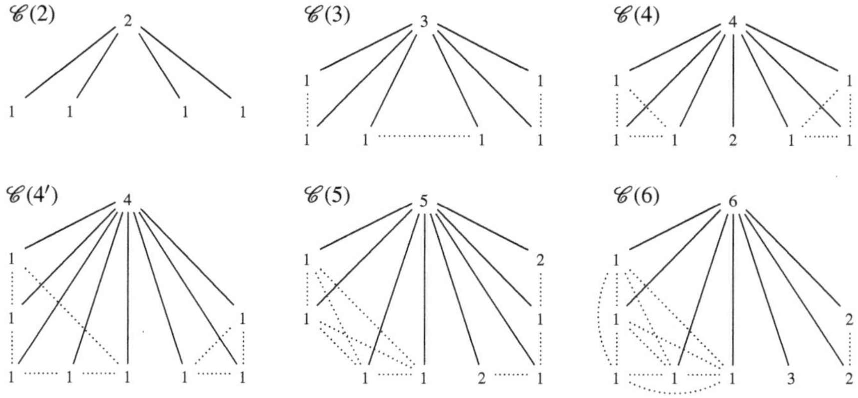

The minimal positive radical vector is indicated by the values at the vertices

Table 5.4 Sincere weakly positive centered forms without multiple edges (graphical forms)

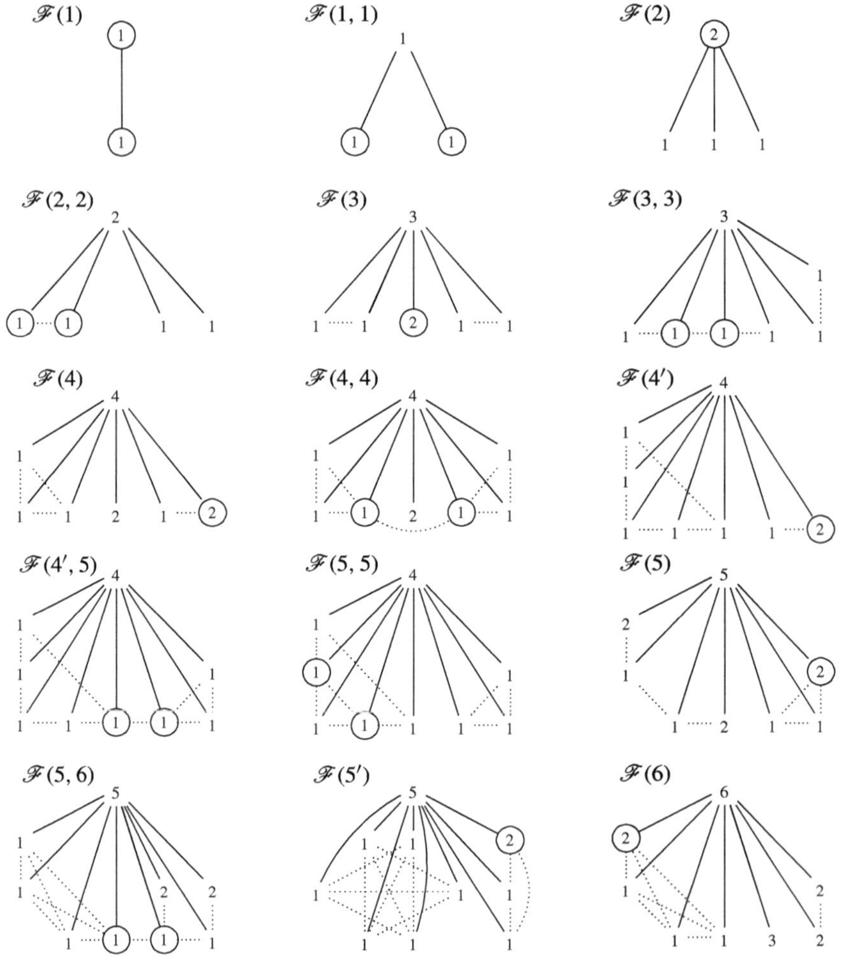

5.6 Roots with an Entry 6

By direct inspection of the list of sincere centered weakly positive unit forms (Table 5.4), we observe that some of these forms are indefinite. However, there need not exist an indefinite weakly positive form q with $M(q) = s$ for all possible values $s = 1, \dots, 6$. In fact, in the following we will prove that if q is a weakly positive unit form having a sincere positive root v with $v_\omega = 6$ for some vertex ω, then q is a nonnegative unit form (Theorem 5.38 due to Ostermann and Pott [42]).

A brief description of the proof is in order. The starting point is Ringel's Lemma 5.32 below, where centered weakly positive unit forms having a positive root v with an entry $v_i = 6$ for some vertex i (plus certain additional properties) are described. One of these properties, indicating that all other entries v_j for $j \neq i$ are equal either to 1 or 0, is the main technical condition of so-called *regular pairs*. This definition is meant to keep track of forms having positive roots with this particular shape. In Lemmas 5.33, 5.34 and Proposition 5.35 it is shown how iterated deflations can be used to reduce our problem to *centered forms*. With the help of Lemma 5.36 we prove the main technical result in [42] (Theorem 5.37 below), ensuring the existence of radical vectors that somehow control vertices outside the support of the maximal root in a centered regular pair. This result is used to sketch the proof of Ostermann and Pott's Theorem 5.38.

Let q be a weakly positive unit form and v a maximal sincere positive q-root with $v_\omega = 6$ for some vertex ω. Denote by $\tilde{q}$ the unit form obtained from q by exploding each vertex $i \neq \omega$ exactly v_i times (that is, a full explosion with respect to vertex ω as in Proposition 5.29) and let $\widetilde{v}$ be the maximal root of $\widetilde{q}$ given in Proposition 5.29$(b)(ii)$. Notice that $\tilde{v}_\omega = 6$ and $\tilde{v}_x = 1$ for any other vertex x. Since q is nonnegative if and only if so is $\tilde{q}$, we can restrict our attention to the case where $v_i = 1$ for any $i \neq \omega$. Explosion was our first reduction step. Our second step will be reduction to centered forms by means of deflations for full edges $i - j$ with $i \neq \omega \neq j$. After each such deflation T_{ij}^-, the corresponding vector $(T_{ij}^-)^{-1}v = T_{ij}^+ v$ will have smaller support than v, so we have to keep track of the points running out of the support of v. This motivated the definition of regular pairs as given in [42]. For simplicity, for the rest of this chapter we consider pairs (q, v) where q is a unit form and v is a root of q, and referred to them simply as *(unit) pairs*. The following terminology will be useful for the technical results below.

a) A pair (q, v) is *weakly positive* if q is a weakly positive form and v is a positive root.
b) A pair (q, v) is *sincere* if v is a sincere root.
c) A weakly positive pair (q, v) is *centered* if q is a centered form.
d) A weakly positive pair (q, v) is *regular* if

    i) v is a maximal q-root.
    ii) $v_\omega = 6$ and $0 \leq v_i \leq 1$ for all $i \neq \omega$.
    iii) $q_{ij} \leq 2$ for all $i \neq j$.
    iv) $q_{\omega i} = -1$ and $q(v|e_i) = 0$ for all $i \notin \mathbf{supp}(v)$.

Notice that a pair (q, v) is regular and sincere if and only if v is a maximal root of q with $v_\omega = 6$ and $v_i = 1$ for $i \neq \omega$ (cf. Lemma 5.5(c)). In view of Lemma 5.7 and Proposition 5.8, for a positive q-root v condition (i) is equivalent to having $0 \leq q(v|e_i) \leq 1$ for all $i \in \mathbf{supp}(v)$. By an *exceptional vertex* of a regular pair (q, v) we mean an exceptional vertex of the maximal q-root v, that is, a vertex $i \in \mathbf{supp}(v)$ such that $q(v|e_i) = 1$ (cf. Lemma 5.9).

Proposition 5.29(b) may be reinterpreted as follows: To any sincere positive maximal q-root z with $z_\omega = 6$, where q is a weakly positive unit form q, we can assign a regular sincere pair $(\overline{q}, \overline{z})$ where $\overline{q}$ is a full explosion of q with respect to vertex ω.

Table 5.5 Weakly positive centered forms $g_{(8)} = q_{\mathscr{G}(8)}$ and $g_{(13)} = q_{\mathscr{G}(13)}$ having a maximal sincere positive root $z_{(8)}$ and $z_{(13)}$ with an entry 6

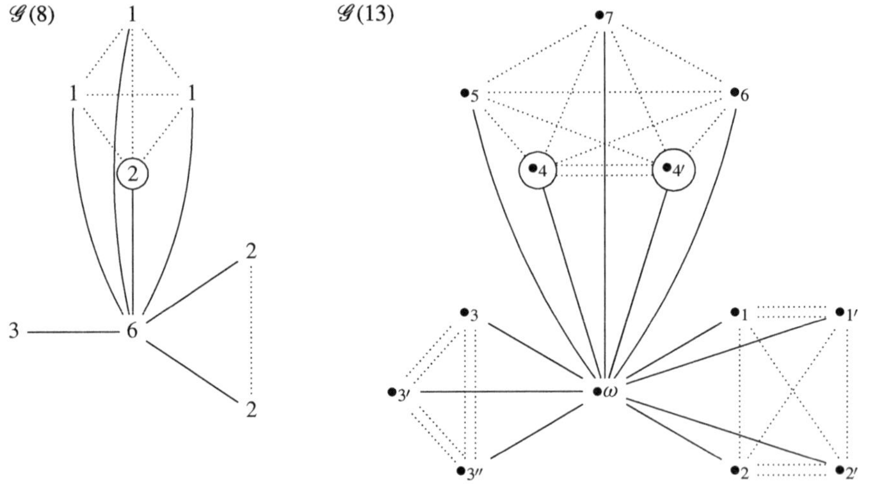

On the left we have $|(g_{(8)})_{ij}| \leq 1$ for all $1 \leq i, j \leq 8$ (numbers on the vertices indicate vector $z_{(8)}$). On the right the pair $(g_{(13)}, z_{(13)})$ is regular. Encircled points indicate exceptional vertices

The following lemma, whose proof we skip (Part (a) is shown by Ovsienko in [43] whereas Part (b) is Lemma 4.2 in [42]), is a fundamental part of (and perhaps the inspiration behind) Ostermann and Pott's results concerning weakly positive unit forms having a positive root with entry 6.

Lemma 5.32. *Let (q, v) be a sincere maximal centered pair with $v_\omega = 6$ for ω the center vertex of q.*

a) If $|q_{ij}| \leq 1$ for all i, j then (q, v) is, up to a permutation of vertices, the pair $(g_{(8)}, z_{(8)})$ given in Table 5.5.
b) If (q, v) is a regular pair then (q, v) is, up to a permutation of vertices, the pair $(g_{(13)}, z_{(13)})$ given in Table 5.5.

Next we prove the basic results for our second reduction step. Notice that Lemma 5.32 plays a key role in the proof of Lemma 5.34. If (q, v) is a unit form and T is a flation for q such that qT is a unit form, then we denote by $(q, v)T$ the unit pair $(qT, T^{-1}v)$.

Lemma 5.33. *Let (q, v) be a regular pair and $i, j \in$ **supp**$(v) - \{\omega\}$ two different vertices with $q_{\omega j} = q_{ij} = -1$.*

*a) Then the restriction of qT_{ij}^- and $T_{ij}^+ v$ to **supp**$(T_{ij}^+ v)$ is a sincere regular pair.*
b) If moreover $q(v|e_j) = 0$, then $(q, v)T_{ij}^-$ is a regular pair.

Proof. Let $q' = qT_{ij}^-$ and $v' = T_{ij}^+ v = v - v_i e_j$. If v' is not maximal then there exists a root $w > v'$ and hence $T_{ij}^- w = w + w_i e_j > v' + v_i e_j = v$, in contradiction to the maximality of v. This shows point (i) in the definition of a regular pair for both (a) and (b), whereas (ii) is obvious, since $v'_\ell = v_\ell$ for all $\ell \neq j$ and $0 = v'_j = v_j - v_i$. Hence (a) holds by the discussion after the definition of a sincere pair.

Let us assume now that $q(v|e_j) = 0$ to show (b). For (iii), observe that $q'_{\ell k} = q_{\ell k}$ for all $\ell, k \neq i$. Now, for $\ell \notin \mathbf{supp}(v')$ we have

$$\begin{aligned}1 &\leq q'(v' + e_\ell - e_i) = 3 - q'(v'|e_i) - q'_{i\ell}\\ &\leq 3 - q'_{i\ell},\end{aligned}$$

where the last inequality is due to the maximality of v'. Therefore $q'_{i\ell} \leq 2$, and for $\ell \in \mathbf{supp}(v')$ the same inequality holds by Lemma 5.5(c).

Finally, for (iv) observe that $\mathbf{supp}(v') = \mathbf{supp}(v) - \{j\}$. So, if $\ell \notin \mathbf{supp}(v')$ then $\ell \neq i$ and we have $q'_{\omega\ell} = q_{\omega\ell}$ and $q'(v'|e_\ell) = q(v|e_\ell)$. For $\ell \neq j$, we use that (q, v) is regular whereas for $\ell = j$, it follows directly from the hypothesis that $q'_{\omega\ell} = -1$ and $q'(v'|e_\ell) = 0$. □

The previous result gives an inductive tool as long as we can find different vertices $i, j \in \mathbf{supp}(v) - \{\omega\}$ with $q_{\omega j} = q_{ij} = -1$ and $q(v|e_j) = 0$. Now, if q is not centered, then it follows from Lemma 5.17 that there exist different vertices $i, j \in \mathbf{supp}(v) - \{\omega\}$ with $q_{\omega j} = q_{ij} = -1$. So the question is whether we can always find such vertices for which, in addition, $q(v|e_j) = 0$. This is affirmatively shown in the following lemma.

Lemma 5.34. *Let (q, v) be a regular, noncentered pair. Then there exist $i, j \in \mathbf{supp}(v) - \{\omega\}$ with $q_{\omega j} = -1 = q_{ij}$ such that $q(v|e_j) = 0$.*

Proof. Assume that v is a sincere q-root. Since v is a maximal positive root, recall from Lemma 5.9 that v has exactly two exceptional vertices, say k and k'. Assume on the contrary that (q, v) satisfies the following:

[$*$] The pair (q, v) is a sincere regular noncentered pair such that for any $i, j \neq \omega$ with $q_{\omega j} = -1 = q_{ij}$ we have $q(v|e_j) = 1$.

Consider the set $\mathscr{A}_{(q,v)} = \{\ell \in \mathbf{supp}(v) - \{\omega\} \mid q_{\ell\omega} \geq -1\}$, which by hypothesis is nonempty. Since the bigraph of q is connected by solid walks (cf. Lemma 5.17), there are $\ell \in \mathscr{A}_{(q,v)}$ and $k'' \in \mathbf{supp}(v) - \{\omega\}$ with $q_{k''\ell} = -1 = q_{\omega k''}$. By hypothesis $q(v|e_{k''}) = 1$, therefore $k'' \in \{k, k'\}$. Let us say that $\ell, \ell' \in \mathscr{A}_{(q,v)}$ are such that $q_{k\ell} = -1$ and $q_{k'\ell'} = -1$ (possibly $\ell = \ell'$).

Take $\widetilde{q} = (qT^-_{\ell k})|_{\mathbf{supp}(T^+_{\ell k}v)}$ and $\widetilde{v} = (T^+_{\ell k}v)|_{\mathbf{supp}(T^+_{\ell k}v)}$, and notice by Lemma 5.33(a) that $(\widetilde{q}, \widetilde{v})$ is a sincere regular pair.

Step 1. *The sincere regular pair $(\widetilde{q}, \widetilde{v})$ satisfies condition [$*$] above.*

Take $p, r \in \mathbf{supp}(\widetilde{v}) - \{\omega\}$ with $\widetilde{q}_{\omega p} = -1 = \widetilde{q}_{pr}$.

If $p = \ell$ then

$$\widetilde{q}(\widetilde{v}|e_\ell) = q(v|T^-_{\ell k}e_\ell) = q(v|e_\ell) + q(v|e_k) = q(v|e_k) = 1.$$

If $p \neq \ell$, assume first that $r \neq \ell$. Then $q_{\omega p} = \widetilde{q}_{\omega p} = -1 = \widetilde{q}_{pr} = q_{pr}$, and by hypothesis [*] we have $p = k'$. Calculate

$$\widetilde{q}(\widetilde{v}|e_p) = q(v|T_{\ell k}^{-} e_p) = q(v|_{k'}) = 1.$$

Assume finally that $p \neq \ell$ and $r = \ell$. Since $-1 = \widetilde{q}_{p\ell} = q_{p\ell} + q_{pk}$, and $q_{\omega p} = q_{\omega k} = -1$, we must have $q_{p\ell} = -1$. Hence $p = k'$ and

$$\widetilde{q}(\widetilde{v}|e_p) = 1.$$

Step 2. *The vertex ℓ is exceptional for $(\widetilde{q}, \widetilde{v})$. In particular $\widetilde{q}_{\omega\ell} = 1$, thus $q_{\omega\ell} = 0$.* We calculate

$$\widetilde{q}(\widetilde{v}|e_\ell) = (qT_{\ell k}^{-})(T_{\ell k}^{+} v|e_\ell) = q(v|T_{\ell k}^{-} e_\ell) = q(v|e_\ell) + q(v|e_k) = 1.$$

Consider now k'', the second exceptional vertex of $\widetilde{v}$, and take $w = \sigma_{k''}(\widetilde{v}) = \widetilde{v} - e_{k''}$. By connectedness with solid walks (Lemma 5.17), and the fact that [*] holds for $(\widetilde{q}, \widetilde{v})$, we notice that if $q_{\omega\ell} \geq 0$ then there is a solid walk from ℓ to ω that does not pass through the exceptional vertex k''. Hence [*] implies that there must be a third exceptional vertex, a contradiction. Then $\widetilde{q}_{\omega\ell} = -1$, and therefore $q_{\omega\ell} = 0$.

Step 3. *We have $|\mathscr{A}_{(\widetilde{q},\widetilde{v})}| = |\mathscr{A}_{(q,v)}| - 1$.*
This follows from Step 2 considering that after applying a flation T_{ij}^{ϵ} to a quadratic form q, all modified edges in the bigraph $B_{qT_{ij}^{\epsilon}}$ have as end-point vertex j.

Using Steps 1–3 as many times as necessary we may assume that (q, v) is a sincere regular and centered pair satisfying [*] with $\mathscr{A}_{(q,v)} = \{\ell\}$. We next observe that $q_{kk'} = 2$, and deduce from $q_{\omega k} = q_{\omega k'} = -1 = q_{k\ell} = q_{k'\ell}$ and $q_{\omega\ell} = 0$ (by Step 2) that $0 < q_{kk'} \leq 2$. Assume that $q_{kk'} = 1$, and notice that $\sigma_\omega(\sigma_k(v)) = v - e_k - e_\omega$ (for $q(\sigma_k(v)|e_\omega) = q(v - e_k|e_\omega) = -q_{\omega k} = 1$). Moreover, we have

$$q(\sigma_\omega\sigma_k(v)|e_{k'}) = q(v - e_k - e_\omega|e_{k'}) = q(v|e_{k'}) - q_{kk'} - q_{\omega k'} = 1,$$

and therefore $w := \sigma_{k'}\sigma_\omega\sigma_k(v) = v - e_k - e_\omega - e_{k'}$. Since $k, k' \notin \mathbf{supp}(w)$, there must exist a vertex $k'' \in \mathbf{supp}(w) - \{\omega\}$ connecting ℓ with ω, that is, $q_{\omega k''} = -1$. However, by [*] the vertex k'' is exceptional for (q, v), a contradiction.

So far we have shown that we may assume that the restriction of q to the set $\{\omega, k, k', \ell\}$ has the following associated bigraph (left):

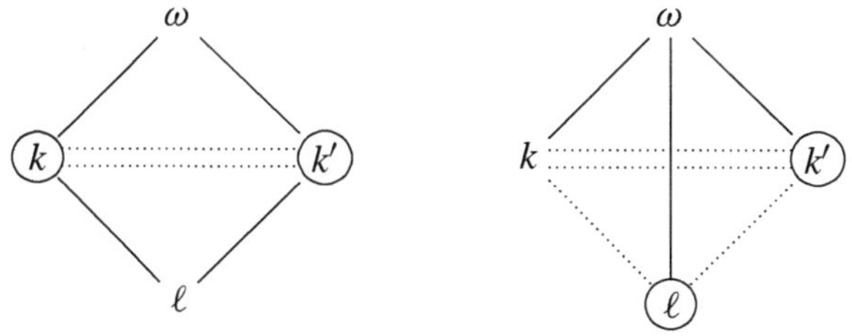

Apply once more deflation $T^-_{\ell k}$ to the pair (q, v) and restrict to the support of $T^+_{\ell k} v$ to obtain a sincere regular pair $(\widetilde{q}, \widetilde{v})$ as before (bigraph on the right above), which is centered by Step 3. The same step shows that k' and ℓ are the exceptional vertices of $(\widetilde{q}, \widetilde{v})$. But notice that in this case we have $\widetilde{q}_{k\ell} = 1$ (since $q_{k'k} = 2$ and $q_{k'\ell} = -1$).

On the other hand, by Lemma 5.32(b) the pair $(\widetilde{q}, \widetilde{v})$ coincides with the pair $(g_{(13)}, z_{(13)})$, where $(g_{(13)})_{k'\ell} = 2$ (the exceptional vertices of the maximal $g_{(13)}$-root $z_{(13)}$ are joined by a double dotted edge, see Table 5.5). This is a contradiction, which completes the proof. □

Proposition 5.35. *Let q be a weakly positive unit form and v a maximal positive q-root such that $v_\omega = 6$ and $v_i = 1$ for $i \neq \omega$. Then there is an iterated deflation T for q such that $(q, v)T := (qT, T^{-1}v)$ is a regular centered pair.*

Proof. Since v is a sincere vector, by assumption (q, v) is a regular pair. If (q, v) is a noncentered pair, use Lemmas 5.33(b) and 5.34 to find a deflation T^-_{ij} such that $(qT^-_{ij}, T^+_{ij} v)$ is a regular pair. This process has to stop, since

$$|v| = \sum_i v_i > \sum_i v_i - 1 = |T^+_{ij} v|.$$

Hence the result. □

We need a final preliminary result.

Lemma 5.36. *Let (q, v) be a regular centered pair, and take $j \in$ **supp**(v) and $k \notin$ **supp**(v) such that $q_{jk} = 2$. Then*

a) Vertex j is nonexceptional for v.
b) Vector $e_j - e_k$ is radical for the form $q|_{\mathbf{supp}(v) \cup \{k\}}$.
*c) For $\ell \notin$ **supp**$(v) \cup \{k\}$ we have $q_{j\ell} \leq q_{k\ell}$.*

Proof. For (a) we have

$$0 < q(v - e_j + e_k) = 3 - q(v|e_j) + q(v|e_k) - q_{jk} = 1 - q(v|e_j),$$

therefore $q(v|e_j) = 0$.

Notice now that $y = v - e_j + e_k$ is a positive q-root. Let a and a' be the exceptional vertices of v, and observe that they are also the exceptional vertices for y (indeed, by Lemma 5.32(b) the restriction of q to the support of y is $g_{(13)}$, and in this form the exceptional vertices are characterized as the unique pair of vertices with $q_{aa'} = 2$ in the component of $\mathcal{G}(13)^{(\omega)}$ with five vertices, cf. Table 5.5).

Thus if $\ell \in \mathbf{supp}(y) - \{a, a'\}$ then

$$0 = q(y|e_\ell) = q(z|e_\ell) - q_{j\ell} + q_{k\ell} = q_{k\ell} - q_{j\ell},$$

and if $\ell \in \{a, a'\}$ then

$$1 = q(y|e_\ell) = q(z|e_\ell) - q_{j\ell} + q_{k\ell} = 1 + q_{k\ell} - q_{j\ell}.$$

In any case, if $\ell \in \mathbf{supp}(y) = (\mathbf{supp}(v) \cup \{k\}) - \{j\}$, we have $q_{j\ell} = q_{k\ell}$, and the same equality holds for $\ell = j$ by hypothesis. This shows (b).

Take now $\ell \notin \mathbf{supp}(v) \cup \{k\}$ and observe by Ovsienko's Theorem 5.25 that $y + e_\ell$ is not a root of q, since otherwise

$$(\sigma_\omega(y + e_\ell))_\omega = y_\omega - q(y + e_\ell|e_\omega) = 6 - q_{\ell\omega} = 7.$$

Hence

$$2 \leq q(y + e_\ell) = 2 + q(y|e_\ell) = 2 + q(v) - q_{j\ell} + q_{k\ell},$$

which shows (c). □

Theorem 5.37. *Let (q, v) be a regular centered pair, and consider vertices $i \in$ $\mathbf{supp}(v)$ and $k \notin \mathbf{supp}(v)$ such that $q_{ik} = 1$ and such that qT_{ik}^+ is a weakly positive unit form. Then there exists a $j \in \mathbf{supp}(v)$ such that $e_j - e_k \in \mathbf{rad}(q)$.*

Proof. Take $q^+ = qT_{ik}^+$ and $v^+ = T_{ik}^-(v) = v + e_k$.

Step 1. *There is a $j \in \mathbf{supp}(v)$ with $q_{jk} = 2$.*

Assume on the contrary that $q_{jk} < 2$ for all $j \in \mathbf{supp}(v)$. It can be shown (Exercise 7 below) that in this case the set

$$\mathcal{N}(k) = \{j \in \mathbf{supp}(v) \mid q_{jk} = 1\}$$

coincides, up to symmetries of vertices, with one of the sets $\{1, 1', 3, 3', 3'', 5\}$. $\{3, 3', 3'', 5, 6, 7\}$ or $\{1, 1', 2, 2', 5, 6\}$ in the following figure (the connected components of the restriction $\mathscr{G}(13)^{(\omega)}$ of bigraph $\mathscr{G}(13)$ in Table 5.5).

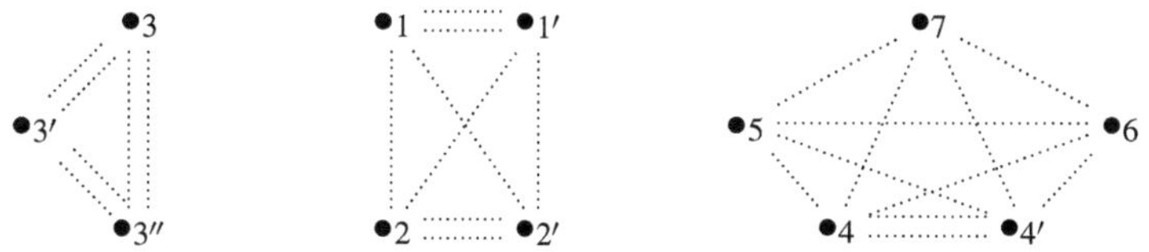

Assume $\mathscr{N}(k) = \{1, 1', 3, 3', 3'', 5\}$ and take $a', a'' \in \mathscr{N}(k)$ such that i, a' and a'' belong to different components $\mathscr{K}$, $\mathscr{K}'$ and $\mathscr{K}''$ of $\mathscr{G}(13)^{(\omega)}$. Then the restriction of q and q^+ to the set of vertices $\{\omega, i, a', a'', k\}$ have the following forms

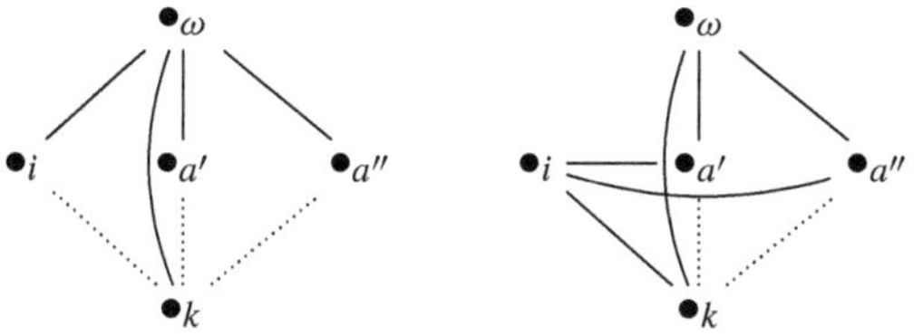

hence the restriction of q^+ to the set $\{\omega, i, a', a''\}$ has the form $\widetilde{\mathbb{A}}_3$, contradicting the weak positivity of q^+.
Assume $\mathscr{N}(k) = \{3, 3', 3'', 5, 6, 7\}$ and $i = 3$. Then the restriction of q^+ to the set $\{1, 2, 3, 3', 5, 6, 7\}$ has the following shape,

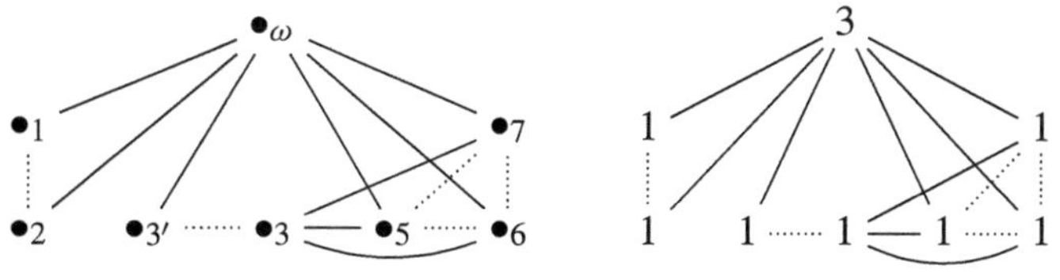

where the positive vector z indicated on the right satisfies $q^+(z) = 0$, a contradiction. Up to symmetry the remaining case is $i = 5$, in which case the restriction of q^+ to the set of vertices $\{\omega, 1, 2, 3, 5, 6, 7\}$ has the shape

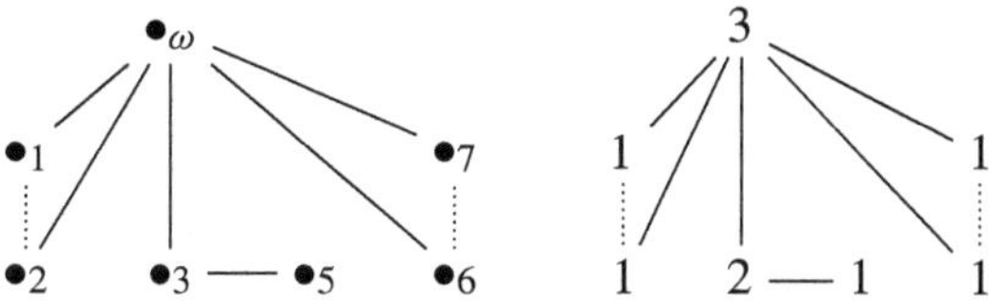

where again the positive vector on the left is radical.
Assume $\mathscr{N}(k) = \{1, 1', 2, 2', 5, 6\}$. Then vertex i is (up to symmetry of vertices) one of vertices $i = 1$ or $i = 5$. In any case the restriction of q^+ to the set

$\{\omega, 1, 1', 2, 2', 5, 6\}$ has the shape

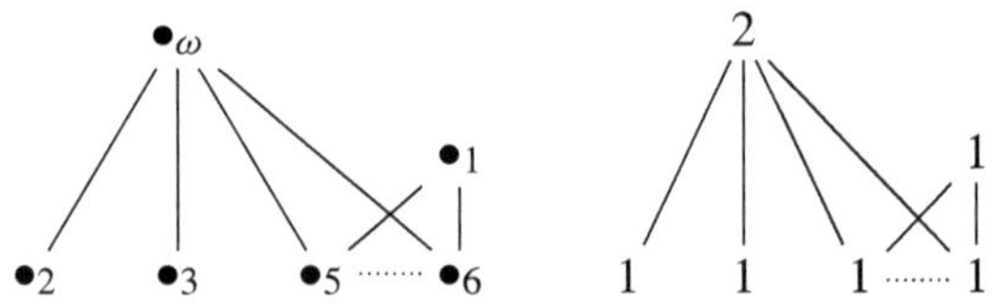

where again on the left we exhibit a positive radical vector. In any case we reach a contradiction, completing the proof of Step 1.

Step 2. *For every ℓ we have $q(e_j - e_k|e_\ell) = 0$.*

Take $z = e_j - e_k$ and let us assume that $q(z|e_\ell) \neq 0$ for some ℓ. By Lemma 5.36(*b*) and (*c*) we have $\ell \notin \mathbf{supp}(v) \cup \{k\}$ and $q(z|e_\ell) = q_{j\ell} - q_{k\ell} < 0$. Consider the following facts:

i) $\mathbf{supp}(v^+) = \mathbf{supp}(v) \cup \{k\}$.
ii) $q^+_{jk} = 2$.
iii) $q^+(v^+|e_\ell) = q(v|e_\ell) = 0$.
iv) $q^+(v^+|e_k) = q(v|e_k) = 0$.

Take now $y = v^+ - e_k + e_j$ and observe that y is a positive root of q^+. Indeed, since $q(v|e_j) = 0$ by Lemma 5.36(*a*), we have

$$\begin{aligned} q^+(y) &= q^+(v^+ - e_k + e_j) = 3 - q^+(v^+|e_k) + q^+(v^+|e_j) - q^+_{kj} \\ &= 1 + q^+(v^+|e_j) = 1 + q(v|e_j) = 1. \end{aligned}$$

Moreover, $q^+(y|e_\ell) = q^+_{j\ell} - q^+_{k\ell} < 0$. Therefore $\sigma_\ell(y)$ is a positive q^+-root with $\ell \in \mathbf{supp}(\sigma_\ell(y))$, and also

$$\begin{aligned} q^+(\sigma_\ell(y)|e_\omega) &= q^+(y - q^+(y|e_\ell)e_\ell|e_\omega) = q^+(y|e_\omega) - q^+(y|e_\ell)q^+_{\ell\omega} \\ &= q^+(v^+|e_\ell) - q^+_{k\ell} + q^+_{j\ell} = q_{j\ell} - q_{k\ell} < 0, \end{aligned}$$

since $q^+(y|e_\omega) = 0$. Hence $\sigma_\omega(\sigma_\ell(y))$ is a positive q^+-root with $\sigma_\omega(\sigma_\ell(y))_\omega = v_\omega - (q_{j\ell} - q_{k\ell}) > 6$, contradicting Ovsienko's Theorem 5.25.

This completes the proof. □

Before we can prove the main result of this section we have to analyze another extreme situation. Let $q : \mathbb{Z}^8 \to \mathbb{Z}$ be a connected positive unit form of Dynkin type $\mathbb{E}_8$ having a maximal positive root v with $v_\omega = 6$. By Theorem 2.20 there exists an iterated inflation T such that $qT = q_{\mathbb{E}_8}$ and $T^{-1}v$ is the maximal root v_8 of $q_{\mathbb{E}_8}$.

Hence $|v| = \sum_{i=1}^{8} \leq \sum_{i=1}^{8} (v_8)_i = |v_8| = 29.$

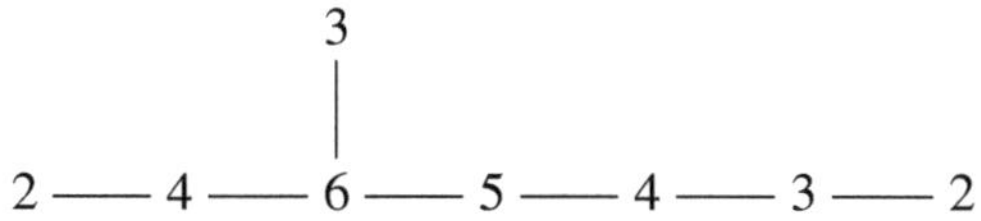

Therefore, if $q^* : \mathbb{Z}^n \to \mathbb{Z}$ and $q^*_{\mathbb{E}_8} : \mathbb{Z}^m \to \mathbb{Z}$ are respectively the full explosion of q and $q_{\mathbb{E}_8}$ with respect of vertex ω, then $n \leq m$ and $m = 8 + 16 = 24$. The bigraph associated to $q^*_{\mathbb{E}_8}$ is shown in Fig. 5.1.

Theorem 5.38. *Let $q : \mathbb{Z}^n \to \mathbb{Z}$ be a weakly positive unit form having a sincere positive root v and a vertex $\omega \in \{1, \dots, n\}$ with $v_\omega = 6$. Then q is a nonnegative unit form with Dynkin type* $\mathbf{Dyn}(q) = \mathbb{E}_8$ *and corank $n - 8$. In particular*

$$8 \leq n \leq 24 \qquad \text{and} \qquad 113 \leq |\mathscr{R}^+(q)| \leq 418\,923\,665 = 5 \cdot 83\,784\,733$$

where the last equality is a prime factorization.

Sketch of Proof. Assume that v is a maximal sincere q-root and take the full explosion $q' : \mathbb{Z}^m \to \mathbb{Z}$ of q with respect to vertex ω (and maximal root v). By Proposition 5.29, the pair (q', v') is sincere and regular, where v' is the root given in Proposition 5.29$(b(ii))$.

We proceed by induction on m. If $m \in \mathbb{N}$ is minimal such that there is a sincere regular pair (q', v'), then (q', v') is a centered pair (for otherwise by Proposition 5.35 there is a deflation T such that the restriction of $(q'T, T^{-1}v')$ to the support of $T^{-1}v$ contradicts the minimality of m). Hence by Lemma 5.32(b) we have $q' \cong g_{(13)}$, which is nonnegative of Dynkin type $\mathbb{E}_8$.

Now, for nonminimal m we have, by Proposition 5.35, an iterated deflation T such that $(q'T, T^{-1}v') = (q'', v'')$ is a centered regular pair. Then T is nontrivial, thus there exist $i \in \mathbf{supp}(v'')$ and $k \notin \mathbf{supp}(v'')$ such that $q''_{ik} = 1$ and qT^+_{ik} is weakly positive. By Theorem 5.37 there is a $j \in \mathbf{supp}(v'')$ with $e_j - e_k \in \mathbf{rad}(q'')$. Consequently q'' is an explosion of the restriction $(q'')^{(k)}$, which by induction is nonnegative of Dynkin type $\mathbb{E}_8$. Then by Proposition 5.29 q'' is nonnegative of Dynkin type $\mathbb{E}_8$, and so are q' (since $q' \cong q''$) and q (cf. Theorem 3.28). In particular, $\mathbf{Dyn}(q) = \mathbb{E}_8$ and $\mathbf{cork}(q) = n - 8$.

For the last claim it is clear that $8 \leq n$. The proof of $n \leq 24$ is briefly sketched: Take n maximal such that a weakly positive unit form $q : \mathbb{Z}^n \to \mathbb{Z}$ has a maximal sincere positive root v with $v_\omega = 6$. By maximality of n the sincere pair (q, v) is regular. By the above, q is a full explosion of a positive unit for $\widetilde{q}$ of Dynkin type $\mathbb{E}_8$ with respect of ω and a maximal $\widetilde{q}$-root $\widetilde{v}$. But a positive unit form with a sincere maximal positive root $\widetilde{v}$ that maximizes the weight $|\widetilde{v}| = \sum_{i=1}^{n} \widetilde{v}_i$ must be precisely $\widetilde{q} = q_{\mathbb{E}_8}$. Therefore $q = q^*_{\mathbb{E}_8}$, the full explosion of $q_{\mathbb{E}_8}$ with respect of the star center

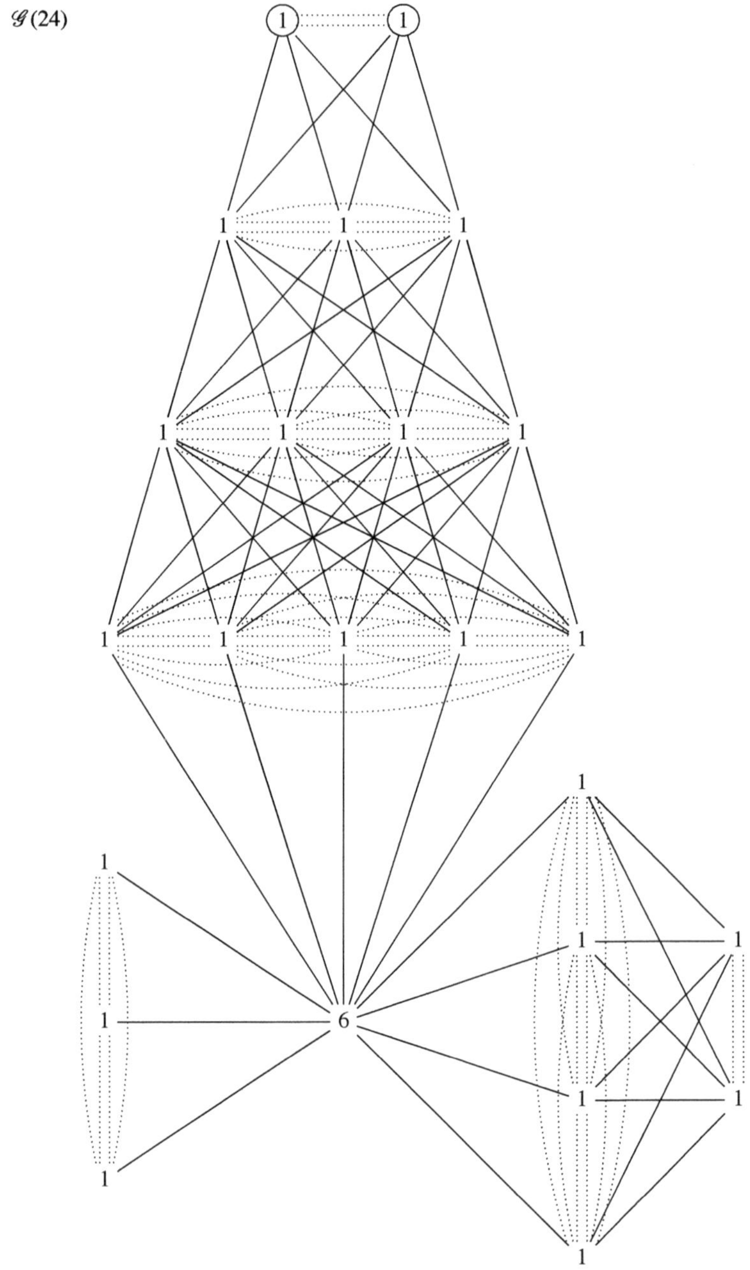

Fig. 5.1 Full explosion $q_{\mathscr{G}(24)} = q^*_{\mathbb{E}_8}$ of $q_{\mathbb{E}_8}$ with respect to the star center. Encircled vertices correspond to exceptional vertices of the indicated (maximal) positive root

(see Fig. 5.1). This shows that $n \leq 24$. The bound for the number of positive roots of q is computed by Ostermann and Pott in [42]. □

Exercises 5.39.

1. Show that if q is a positive centered form with a positive root w, then $q[w]$ is a critical centered form.
2. Determine which of the bigraphs in Table 5.4 correspond to nonnegative forms.
3. How many centered regular pairs (q, v) are there (up to permutation of vertices) with associated bigraph G_q having exactly one double dotted edge?
4. Show that the encircled vertices in the bigraphs of Table 5.5 are in fact exceptional vertices of the corresponding quadratic forms.
5. With the notation of Table 5.5, show that $g_{(13)}$ is a full explosion of $g_{(8)}$ with respect to vertex ω.
6. Prove that if $q : \mathbb{Z}^{24} \to \mathbb{Z}$ is a weakly positive unit form having a sincere root v with $v_\omega = 6$ for some $1 \leq \omega \leq 24$, then $q = q^*_{\mathbb{E}_8}$ as in Fig. 5.1.
7. Let (q, v) be a regular centered pair. Show that if $k \notin \mathbf{supp}(v)$ and $q_{jk} \leq 1$ for all $i \in \mathbf{supp}(v)$, then the set $\{i \in \mathbf{supp}(v) \mid q_{ik} = 1\}$ is (up to symmetry of $\mathbf{supp}(v)$) one of the following subsets of vertices of $g_{(13)}$ (cf. Table 5.5):

   i) $\{1, 1', 3, 3', 3'', 5\}$;
   ii) $\{3, 3', 3'', 5, 6, 7\}$;
   iii) $\{1, 1', 2, 2', 5, 6\}$.

   [Hint: Show that otherwise one of the critical centered forms $\mathscr{C}(2) - \mathscr{C}(6)$ in Table 5.3 is a restriction of q.]

5.7 Thin Forms

In this section we further reduce weakly positive unit forms, following Dräxler, Drozd, Golovachtchuk, Ovsienko and Zeldych [22], to get a so-called *good thin weakly positive unit form.* Since this reduction process is reversible, a classification of such forms determines, in principle, all weakly positive forms. This classification (partially achieved computationally) is presented in [22], cf. Theorem 5.46 and Tables 5.6, 5.7 and 5.8.

A unit form $q : \mathbb{Z}^n \to \mathbb{Z}$ is called *thin* if $q((1, \ldots, 1)) = 1$, that is, if the sincere vector $\tau^{(n)}$ with $\tau_i^{(n)} = 1$ for $i = 1, \ldots, n$ (called the *thin vector* of $\mathbb{Z}^n$) is a q-root. In particular, weakly positive thin forms are sincere. In the following we write τ instead of $\tau^{(n)}$ if no confusion arises.

Proposition 5.40. *For any weakly positive sincere unit form $q : \mathbb{Z}^n \to \mathbb{Z}$ there is an iterated deflation T such that qT is a thin weakly positive unit form having thin vector $\tau^{(n)}$ as unique (thus maximal) sincere root.*

Proof. Let v be a maximal sincere q-root and $\tau = \tau^{(n)}$ the thin vector. We prove the result by induction on $|v| = \sum_{i=1}^{n} v_i$. If $|v| = n$ then $v = \tau$ and we have nothing to do, so assume $v > \tau$. Since v is a root, v cannot be a multiple of τ, and thus using Lemma 5.17 we may find vertices $1 \leq i \neq j \leq n$ with $q_{ij} < 0$ and $v_i < v_j$. By Lemma 5.5(b) we have $q_{ij} = -1$.

Take $q' = qT_{ij}^-$ and $v' = T_{ij}^+ v = v - v_i e_j > 0$. Then q' is a weakly positive unit form (for if $x > 0$ then $T_{ij}^- x > 0$) and has a maximal sincere root v' with $|v'| < |v|$. By the induction hypothesis there is an iterated deflation T' such that $q'T'$ is a weakly positive thin unit form having the thin vector as unique (maximal) sincere root. Take $T = T_{ij}^- T'$ to complete the proof. □

We now restrict our attention to deflations that preserve the thin property. If $q : \mathbb{Z}^n \to \mathbb{Z}$ is a thin weakly positive unit form with $\tau^{(n)}$ a nonmaximal root, then there is a vertex $j \in \{1, \ldots, n\}$ such that $q(\tau^{(n)}|e_j) = -1$ (see Lemma 5.7 and Proposition 5.8). In this case, a deflation T_{ij}^- for q is called a *τ-deflation*. Notice that if $y := \sigma_j(\tau^{(n)}) = \tau^{(n)} + e_j$, then $T_{ij}^+(y) = y - y_i e_j = \tau^{(n)}$. Therefore qT_{ij}^- is again a thin form. An iterated deflation consisting of corresponding τ-deflations will be referred to as an *iterated τ-deflation*. For a τ-deflation T_{ij}^- for q, taking $q^- = qT_{ij}^-$, the inflation T_{ij}^+ for q^- is called a *τ-inflation*, and *iterated τ-inflations* are defined similarly. The following result is evident from the discussion above.

Lemma 5.41. *Let q be a thin weakly positive unit form. Then there is an iterated τ-deflation T such that the thin vector $\tau^{(n)}$ is maximal for the thin weakly positive unit form qT.*

In order to have at hand an effective inductive tool to construct weakly positive unit forms, we define following [22] a new type of extension on weakly positive pairs (q, v). We call a weakly positive pair (q', v') a *reflection-extension* of (q, v) if there exists a vertex i of q (the *extension vertex*) such that $(q')^{(i)} = q$ and $q'(v'|e_i) = v'_i$, and if σ'_i denotes the reflection with respect to the unit form q' and v is identified with its inclusion in $\mathbb{Z}^n$, then $\sigma'_i(v') = v$. If furthermore v' is a maximal q'-root with two exceptional vertices (cf. Lemma 5.9), we say that (q', v') is a *main reflection-extension* of (q, v).

A sincere pair (q, v) is called *bad* if there is a radical vector $\mu \in \mathbf{rad}(q)$ such that both $v + \mu$ and $v - \mu$ are positive q-roots. Otherwise (q, v) is called a *good pair*. Recall that, for a unit form $q : \mathbb{Z}^n \to \mathbb{Z}$ and a q-root z, the one point extension $q[z]$ is defined as the root-induced form $q_{e(z)}$ where $e(z) = (e_1, \ldots, e_n, -z)$ (cf. Sect. 3.5), that is

$$q[z](y_1, \ldots, y_n, y_{n+1}) = q(y_1 e_1 + \ldots + y_n e_n - y_{n+1} z).$$

Proposition 5.42. *Let $q : \mathbb{Z}^n \to \mathbb{Z}$ and $q' : \mathbb{Z}^{n+1} \to \mathbb{Z}$ be weakly positive unit forms, and assume that q is a thin form.*

a) The pair $(q', \tau^{(n+1)})$ is a main reflection-extension of $(q, \tau^{(n)})$ if and only if there is an $i \in \{1, \dots, n\}$ such that $q'(x) = q[\tau^{(n)}](x) + x_i x_{n+1}$.

b) If $(q, \tau^{(n)})$ is a bad pair and $(q', \tau^{(n+1)})$ is a reflection-extension of $(q, \tau^{(n)})$, then $(q', \tau^{(n+1)})$ is a bad pair.

c) If $(q, \tau^{(n)})$ is a good pair, then $(q', \tau^{(n+1)})$ is reflection-extension of $(q, \tau^{(n)})$ and is a bad pair if and only if there is a q-root with $|z_i| \leq 1$ for $i = 1, \dots, n$ such that $q(z|\tau^{(n)}) = -1$ and $q' = q[-z]$.

Proof.

(a) By definition of reflection-extension we have $q'(\tau^{(n+1)}|e_{n+1}) = 1$. By maximality of $\tau^{(n+1)}$ there is exactly one other exceptional vertex for $\tau^{(n+1)}$, say $i \in \{1, \dots, n\}$, that is, $q'(\tau^{(n+1)}|e_j) = \delta_{ij}$ for $j \in \{1, \dots, n\}$. Therefore

$$q'(e_{n+1}|e_j) = q'(\tau^{(n+1)} - \tau^{(n)}|e_j) = q(-\tau^{(n)}|e_j) + \delta_{ij},$$

that is, $q'(x) = q[\tau^{(n)}](x) + x_i x_{n+1}$. Conversely, since $(q')^{(n)} = q$ notice that

$$q'(\tau^{(n+1)}|e_{n+1}) = q'(\tau^{(n+1)}) + q'(e_{n+1}) - q'(\tau^{(n+1)} - e_{n-1}) = 2 - 1 = 1.$$

Now, for $j \in \{1, \dots, n\}$ and $j \neq i$ we have

$$q'(\tau^{(n+1)}|e_j) = q'(\tau^{(n)}|e_j) + q'(e_{n+1}|e_j) = q(\tau^{(n)}|e_j) - q(\tau^{(n)}|e_j) = 0,$$

whereas $q'(\tau^{(n+1)}|e_i) = q(\tau^{(n)}|e_i) - q(\tau^{(n)}|e_i) + 1 = 1$. Hence $\tau^{(n+1)}$ is a maximal q'-root and $(q', \tau^{(n+1)})$ is a reflection-extension of the pair $(q, \tau^{(n)})$.

(b) Take $\mu \in \mathbf{rad}(q)$ with $\mu_i \in \{1, 0, -1\}$ for $i = 1, \dots, n$ and define $\mu' \in \mathbb{Z}^{n+1}$ with $\mu'_i = \mu_i$ for $i = 1, \dots, n$ and $\mu'_{n+1} = 0$. We show that $\mu' \in \mathbf{rad}(q')$.

Since $q'(\mu', e_i) = q(\mu, e_i) = 0$ for $i = 1, \dots, n$, let us assume that $q'(\mu'|e_{n+1}) > 0$ (multiplying μ' by -1 if necessary). Then

$$q'(\mu'|\tau^{(n+1)}) = q'(\mu'|e_{n+1}) \geq 1,$$

and therefore for the positive vector $\tau^{(n+1)} - \mu'$ in $\mathbb{Z}^{n+1}$ we have

$$q'(\tau^{(n+1)} - \mu') = q'(\tau^{(n+1)}) + q'(\mu') - q'(\mu'|\tau^{(n+1)}) = 1 - q'(\mu'|\tau^{(n+1)}) \leq 0,$$

a contradiction.

(c) Assume first that $(q', \tau^{(n+1)})$ is a bad extension of $(q, \tau^{(n)})$, and take $\mu \in \mathbf{rad}(q')$ with $|\mu_i| \leq 1$. If $\mu_{n+1} = 0$ then $(q, \tau^{(n)})$ is itself a bad pair, therefore we may also assume that $\mu_{n+1} = 1$. Then $z := e_{n+1} - \mu$ is a q-root with entries $z_i \in \{1, 0, -1\}$ such that

$$q(z|\tau^{(n)}) = q'(e_{n+1} - \mu|\tau^{(n)}) = q'(e_{n+1}|\tau^{(n+1)}) - q'(e_{n+1}|e_{n+1}) = 1 - 2 = -1.$$

By definition of z we have $q' = q[-z]$, since for $x_1, \ldots, x_n, x_{n+1} \in \mathbb{Z}$, taking $x = \sum_{i=1}^{n} x_i e_i$, we have

$$\begin{aligned} q'(x + x_{n+1}e_{n+1}) &= q'(x) + x_{n+1}^2 + x_{n+1}q'(x|e_{n+1}) \\ &= q'(x) + x_{n+1}^2 + x_{n+1}q'(x|z + \mu) \\ &= q(x) + q(x_{n+1}z) + q(x|z) \\ &= q(x + x_{n+1}z). \end{aligned}$$

Conversely, since $q' = q[-z]$, the restriction of q' to the first n variables is q. Moreover,

$$\begin{aligned} q'(\tau^{(n+1)}|e_{n+1}) &= q'(e_{n+1}|e_{n+1}) + q'(\tau^{(n)}|e_{n+1}) \\ &= 2 + q(\tau^{(n)}|z) = 1. \end{aligned}$$

Hence $(q', \tau^{(n+1)})$ is reflection-extension of $(q, \tau^{(n)})$. □

A small example is in order. Consider the thin unit form $q = q_{\mathbb{D}_4}$, which is positive, hence weakly positive. The thin vector $\tau^{(4)}$ is nonmaximal (we have $q(\tau^{(4)}|e_1) = -1$, see the figure on the left below). The bigraph associated to the one-point extension $q[\tau^{(4)}]$ has the following shape (center):

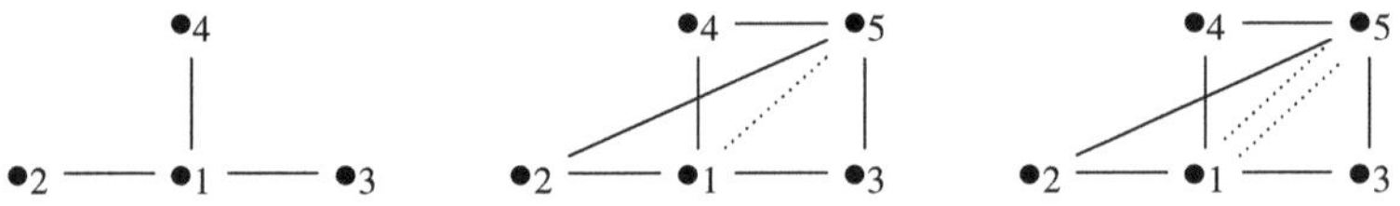

The figure on the right corresponds to a reflection-extension $(q', \tau^{(5)})$ of $(q, \tau^{(4)})$ satisfying both point (a) and (b) of Proposition 5.42. That is, the pair $(q', \tau^{(5)})$ is a bad main reflection-extension of $(q, \tau^{(4)})$, with both q and q' weakly positive unit forms.

Lemma 5.43. *Let $q : \mathbb{Z}^n \to \mathbb{Z}$ be a thin weakly positive unit form. Then there is a sequence of thin weakly positive unit forms $q_i : \mathbb{Z}^i \to \mathbb{Z}$ for $i = 1, \ldots, n$ such that $q_n = q$ and $(q_{i+1}, \tau^{(i+1)})$ is a reflection-extension of $(q_i, \tau^{(i)})$ for $1 \leq i < n$.*

Proof. We proceed by induction on $n \geq 1$. For $n = 1$ there is nothing to show. For $n > 1$ consider the (nonsimple) thin vector $\tau^{(n)} \in \mathbb{Z}^n$ and apply Lemma 5.7(c) to get a vertex $i \in \{1, \ldots, n\}$ with $q(\tau^{(n)}|e_i) = 1$. Then $\sigma_i(\tau^{(n)}) = \tau^{(n)} - e_i$, which is the thin vector τ for the restriction $q^{(i)}$. Then $(q, \tau^{(n)})$ is a reflection-extension of $(q^{(n)}, \tau)$, and the result follows by induction. □

Theorem 5.44. *Let $(q', \tau^{(n+1)})$ be a reflection-extension of $(q, \tau^{(n)})$ with both q and q' weakly positive unit forms. Then there exist an iterated τ-deflation T for q*

and an iterated τ-deflation T' for q' such that $(q'T', \tau^{(n+1)})$ is a main reflection-extension of $(qT, \tau^{(n)})$.

Proof. We proceed by induction on the number $|R^+(q')|$ of positive roots of q'. If $\tau^{(n+1)}$ is a maximal q'-root there is nothing to show (in particular if $|R^+(q')| = 1$). Assume now that $\tau^{(n+1)}$ is a nonmaximal q'-root. By Lemma 5.7 and Proposition 5.8 there is a vertex $j \in \{1, \dots, n\}$ such that $q'(\tau^{(n+1)}|e_j) = -1$. Since q' is weakly positive we get

$$0 < q'(e_j + \tau^{(n)}) = 2 + q'(\tau^{(n+1)} - e_{n+1}|e_j) = 1 - q'_{j,n+1},$$

therefore $q'_{j,n+1} \leq 0$.

If $q'_{j,n+1} = 0$, then by Lemma 5.17 there is a vertex $i \in \{1, \dots n\}$ such that $q'_{ij} = -1$ (hence $i \neq j$). Since $q(\tau^{(n)}|e_j) = q'(\tau^{(n+1)}|e_j) - q'_{j,n+1} = -1$, the deflation T^-_{ij} is a τ-deflation for both q' and q. Observe also that the restriction of $q'T^-_{ij}$ to $\mathbb{Z}^n$ coincides with qT^-_{ij}. Moreover, $(q'T^-_{ij}, \tau^{(n+1)})$ is a reflection-extension of $(qT^-_{ij}, \tau^{(n)})$, since

$$\begin{aligned}(q'T^-_{ij})(\tau^{(n+1)}|e_{n+1}) &= q'(T^-_{ij}(\tau^{(n+1)})|T^-_{ij}(e_{n+1}))\\ &= q'(\tau^{(n+1)} + e_j|e_{n+1})\\ &= q'(\tau^{(n+1)}|e_{n+1}) + q'_{j,n+1} = 1.\end{aligned}$$

If $q'_{j,n+1} < 0$ then $q'_{j,n+1} = -1$ (by Lemma 5.5(*b*)). Then $T^-_{n+1,j}$ is a τ-deflation for q' and the restriction of $q'T$ to $\mathbb{Z}^n$ is q. Again we have

$$\begin{aligned}(q'T^-_{n+1,j})(\tau^{(n+1)}|e_{n+1}) &= q'(T^-_{n+1,j}(\tau^{(n+1)})|T^-_{n+1,j}(e_{n+1}))\\ &= q'(\tau^{(n+1)} + e_j|e_{n+1} + e_j)\\ &= q'(\tau^{(n+1)}|e_{n+1}) + q'(\tau^{(n+1)}|e_j) + q'_{j,n+1} + q'(e_j|e_j)\\ &= 1 - 1 - 1 + 2 = 1,\end{aligned}$$

therefore $(qT^-_{n+1,j}, \tau^{(n+1)})$ is reflection-extension of $(q, \tau^{(n)})$.

To complete the proof we use induction observing that in both cases the number of positive q'-roots decreases (see Lemma 2.19). □

Algorithm 5.45. *Theorem 5.44 is used in [22] to sketch a four step algorithm to produce all good thin weakly positive unit forms in $n + 1$ variables starting from those forms in n variables.*

Step 1. *Apply all possible iterated τ-deflations to the good thin weakly positive unit forms in n-variables.*

Step 2. *Construct all main reflection-extensions (using Proposition 5.42(a)) of the obtained forms.*
Step 3. *Apply all possible iterated τ-inflations to the list obtained in Step 2.*
Step 4. *Filter the final list to sort out any bad thin forms.*

From Theorem 5.44 it is clear that every good thin weakly positive unit form in $n+1$ variables belongs to the list obtained in Step 4 of Algorithm 5.45. For instance, in $n = 1, 2, 3$ variables there is exactly one good thin weakly positive unit form, namely $q_{\mathbb{A}_1}$, $q_{\mathbb{A}_2}$ and $q_{\mathbb{A}_3}$ respectively. For $n = 4$ apply Step 2 to $q_{\mathbb{A}_3}$ to get the two forms on the left below

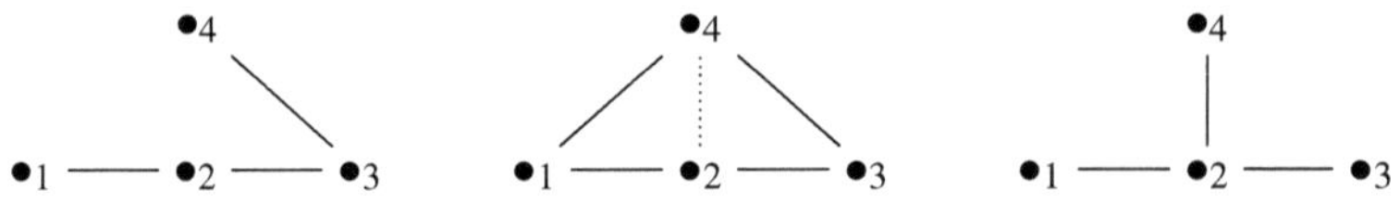

The third form on the right (for which the thin vector $\tau^{(4)}$ is nonmaximal) is obtained after applying Step 3. Case $n = 5$ is sketched in Exercise 4 below.

Before we can state the main classification result of this section we consider yet another construction of unit forms. We say that a point i in a bigraph B is a *linking vertex* if it has exactly two neighbors and is joint to them by simple solid edges. A linking vertex of a unit form q is a linking vertex of its associated bigraph. By a *chain* in a bigraph (or unit form) we mean a sequence of vertices $a_{-1}, a_0 \dots, a_k, a_{k+1}$ where a_i is a linking vertex for $i = 0, \dots, k$ joined precisely to a_{i-1} and a_{i+1}. The number $k+1$ will be referred to as the *length of the chain*.

For $u \geq 1$, the *u-blow up* $q^{(a\star u)}$ of a unit form q with respect to a linking vertex a is the form with bigraph $B^\star$ obtained by replacing vertex a by a chain of length u. To be precise, if a is joined to vertices a_{-1} and a_{u+1} in B_q, we get $B^\star$ from the restriction $B_q^{(a)}$, by adding vertices $a_0, \dots, a_u$ such that a_i is joined by solid simple edges only to a_{i-1} and a_{i+1}, for $i = 1, \dots, u$. Now, if Λ is a set of linking vertices of q and $\overline{u} = (u_\lambda)_{\lambda \in \Lambda}$ is a vector of natural numbers, then the *blow up* of q with respect to $(\Lambda, \overline{u})$ is the unit form $q^{(\Lambda\star\overline{u})}$ defined recursively as

$$q^{(\Lambda\star\overline{u})} = (q^{(\Lambda-\{\lambda\}\star\overline{u}-\{u_\lambda\})})^{(\lambda\star u_\lambda)},$$

for some $\lambda \in \Lambda$. This procedure yields, for a unit form q with a set of linking vertices Λ, a series of unit forms $\{q^{(\Lambda\star\overline{u})}\}$ indexed by $\overline{u} \in \mathbb{N}^\Lambda$. Whether the forms in the series associated to q and Λ are (good and thin) weakly positive, assuming that q is (good and thin) weakly positive, is the subject of investigation in [22, Section 5]. Their outcome leads to the following classification result.

Theorem 5.46. *Every good thin weakly positive unit form in $n \geq 15$ variables is a blow up of one of the* 63 *unit forms in Tables 5.6, 5.7 and 5.8, referred to as* basic *good thin weakly positive unit forms.*

Table 5.6 Basic good thin weakly positive unit forms in n variables for $n = 3, 4, 5, 6$

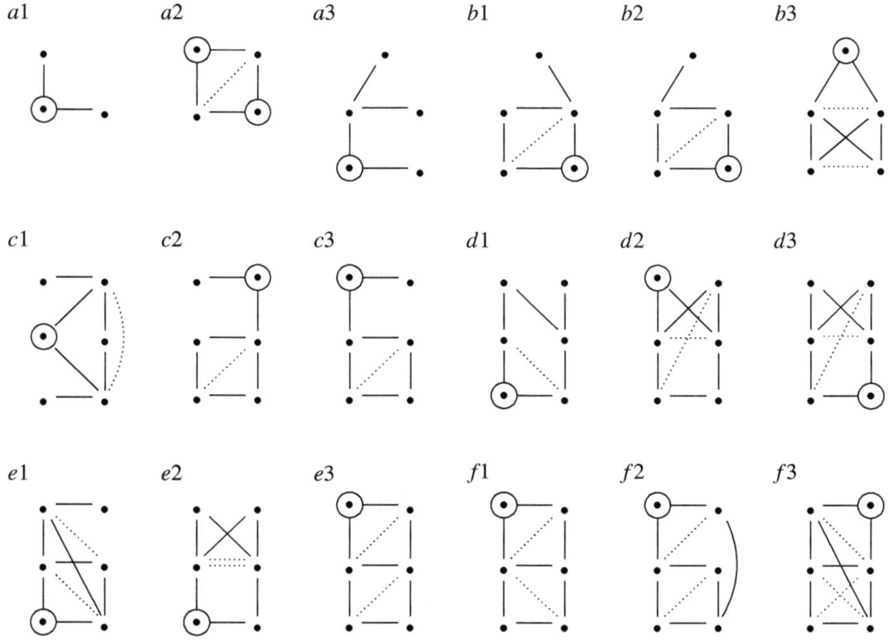

The distinguished set of linking points Λ is denoted by encircled vertices

Table 5.7 Basic good thin weakly positive unit forms in $n = 7$ variables

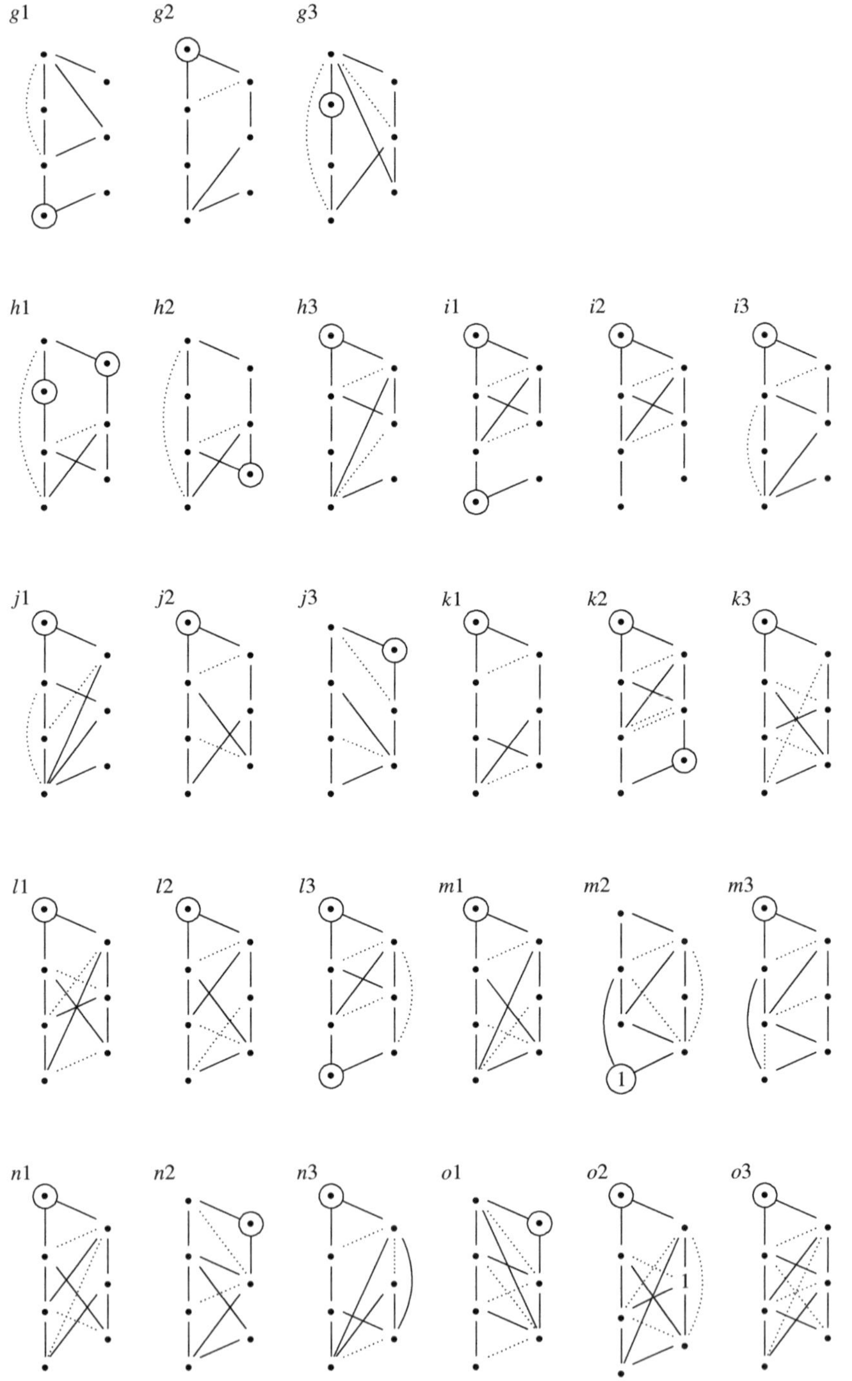

The distinguished set of linking points Λ is denoted by encircled vertices

Table 5.8 Basic good thin weakly positive unit forms in n variables for $n = 8, 9, 10, 11$

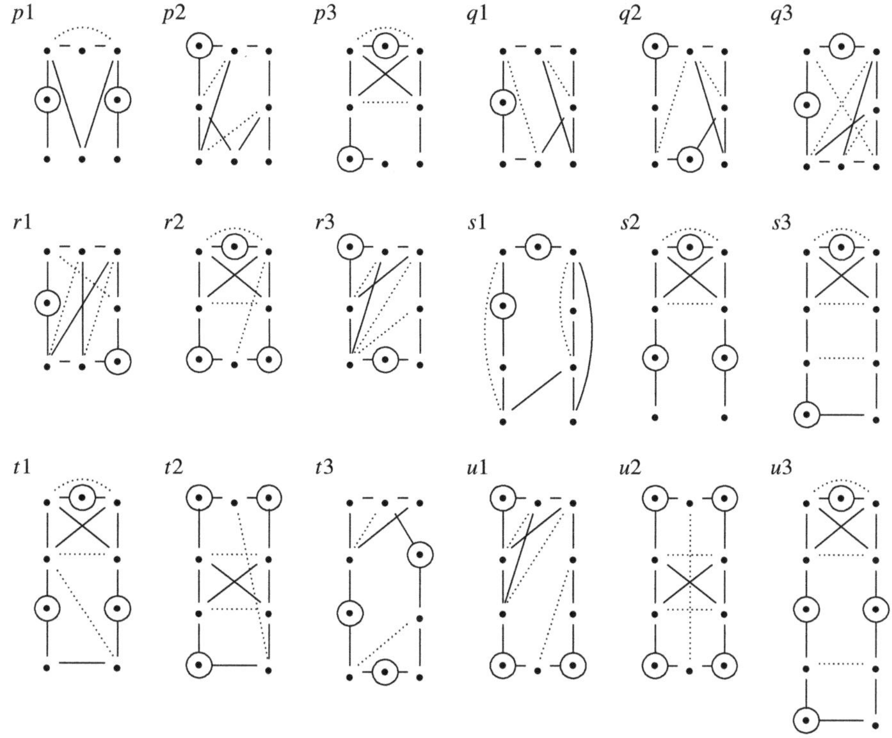

The distinguished set of linking points Λ is denoted by encircled vertices

Exercises 5.47.

1. Show that if (q, v) is a bad sincere pair and (q', v') is a reflection-extension of (q, v), then (q', v') itself is a bad pair.
2. Find a thin weakly positive unit form $q : \mathbb{Z}^{n+1} \to \mathbb{Z}$ with $\tau^{(n+1)}$ a maximal root such that $(q, \tau^{(n+1)})$ is a reflection-extension of $(q^{(n)}, \tau^{(n)})$ and $\tau^{(n)}$ is a nonmaximal $q^{(n)}$-root.
3. Give an example of a good thin weakly positive unit form q and a τ-deflation T_{ij}^{-} for q such that qT_{ij}^{-} is thin weakly positive but not good.
4. Consider the three good thin weakly positive unit forms in 4 variables $q_{\mathbb{A}_4}$, $q_{\mathbb{D}_4}$ and q' with associated bigraph as below.

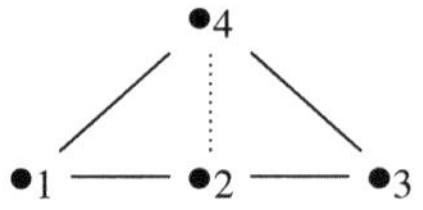

   i) Determine all five main reflection-extensions of the forms $q_{\mathbb{A}_4}$, $q_{\mathbb{D}_4}$ and q'.
   ii) Using τ-inflations determine two remaining good thin weakly positive unit forms in five variables.
   iii) From the seven obtained forms, how many are bad?

5. Use Algorithm 5.45 to produce the complete list of good thin weakly positive unit forms in 6 variables. [Hint: There are exactly 26 such forms.]
6. From the lists obtained in Exercises 4 and 5, how many good thin forms are blow ups of one of the 63 unit forms in Tables 5.6, 5.7 and 5.8 (cf. Theorem 5.46).
7. Show that the quadratic form associated to the following bigraph is a good thin weakly positive unit form which is not blow up of one of the 63 forms in Tables 5.6, 5.7 and 5.8.

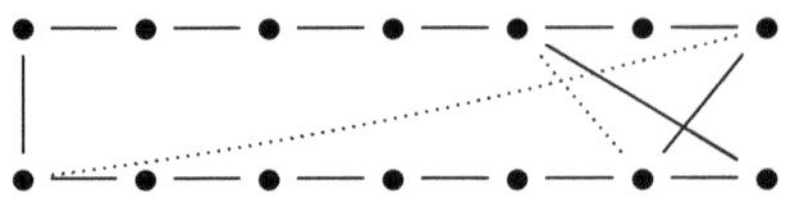

Chapter 6
Weakly Nonnegative Quadratic Forms

In the previous chapters we met notions like positivity, nonnegativity and weak positivity, and applied to them various techniques like deflations, inflations, one-point extensions, reflections and edge reductions. Here we turn our attention to *weakly nonnegative forms*, that is, semi-unit quadratic forms $q : \mathbb{Z}^n \to \mathbb{Z}$ such that $q(v) \geq 0$ for all positive vector v in $\mathbb{Z}^n$. The above-mentioned methods are used to extend earlier results and algorithms to the weak nonnegative context, where now the existence of maximal sincere q-roots plays a key role, and hypercritical forms take the place of critical forms.

6.1 Hypercritical Forms

A quadratic semi-unit form $q : \mathbb{Z}^n \to \mathbb{Z}$ is called *hypercritical* if it is not weakly nonnegative, but every proper restriction q^I is. For instance, the m-Kronecker form $q_m(x_1, x_2) = x_1^2 + x_2^2 - mx_1x_2$ is weakly nonnegative if and only if $m < 3$, and is hypercritical exactly when $m \geq 3$. Theorem 5.2 tells us that if the number of variables is at least three, then a critical (nonweakly positive) form is nonnegative with radical generated by a positive vector z, called a *critical vector*. In Proposition 6.2 below we give an analogous result for hypercritical forms.

Lemma 6.1. *Let $q : \mathbb{Z}^n \to \mathbb{Z}$ be a hypercritical semi-unit form.*

a) If q is also critical, then $n = 2$ and q is the Kronecker form $q_m = x_1^2+x_2^2-mx_1x_2$ for some $m \geq 3$. In particular, q has no critical vector.
b) If q is nonunitary then $n = 2$ and q is (up to order of variables) one of the forms q'_m or q''_m below, with $m > 0$,

$$q'_m(x_1, x_2) = x_1^2 - mx_1x_2 \qquad \textit{and} \qquad q''_m(x_1, x_2) = -mx_1x_2.$$

M. Barot et al., *Quadratic Forms*, Algebra and Applications 25,
https://doi.org/10.1007/978-3-030-05627-8_6

Proof. If q is also critical and $n \geq 3$, by Theorem 5.2 the form q is nonnegative, in particular weakly nonnegative. This is impossible since q is hypercritical.

Then $n = 2$, that is, $q(x_1, x_2) = x_1^2 + x_2^2 - mx_1x_2$ for some $m \in \mathbb{Z}$ (since q is unitary by Lemma 5.5(a)). Observe that if $m \leq 2$ then q is weakly nonnegative, and see Proposition 1.23 for the claim on critical vectors.

To verify (b) observe first that the forms q'_m and q''_m are hypercritical precisely when $m > 0$. Consider a vertex $c \in \{1, \dots, n\}$ such that $q(e_c) = 0$, and take $x^{(c)}$ to be the vector in $\mathbb{Z}^{n-1}$ obtained by deleting the variable x_c. Then

$$q(x) = q^{(c)}(x^{(c)}) + x_c \left(\sum_{i \neq c} q_{ic} x_i \right).$$

Now, if q is hypercritical then there is a positive sincere vector x such that $q(x) < 0$. Moreover, $q^{(c)}(x^{(c)}) \geq 0$ implies that the second summand above is negative. Since x is a positive vector, there must be a $d \neq c$ such that $q_{cd} < 0$. Then the restriction $q' = q^{\{c,d\}}$ is one of the hypercritical forms q'_m or q''_m above. Since q is itself hypercritical, then $q = q'$ and the result follows. □

By Lemma 6.1(b) we may focus only on hypercritical unit forms, which can be characterized as follows.

Proposition 6.2. *For a unit form $q : \mathbb{Z}^n \to \mathbb{Z}$ with $n \geq 3$ the following are equivalent.*

a) *The form q is hypercritical.*
b) *The form q is not weakly nonnegative, and for every critical restriction q^I of q there is an index i with $I = \{1, \dots, n\} - \{i\}$, and a positive critical vector z of q^I such that $q(z|e_i) < 0$.*

Proof. Assume q is hypercritical and consider a positive vector v with $q(v) < 0$. Since any proper restriction q^I is weakly nonnegative, the vector v is sincere. If q^I is critical, since $n \geq 3$ then q^I is a proper restriction of q by Lemma 6.1(a). Moreover, since $q_{ij} \geq -2$ for all $i \neq j$ (for q does not contain any Kronecker form q_m with $m > 2$) by Theorem 5.2 we may take a critical positive vector z for q^I, which we identify with its inclusion in $\mathbb{Z}^n$.

Take positive numbers m and k such that $kv - mz$ is a positive but nonsincere vector, say $(kv - mz)_j = 0$. (Such numbers exist: take an index $j \in \{1, \dots, n\}$ such that $\frac{z_i}{v_i} \leq \frac{z_j}{v_j}$ for all $i \in \{1, \dots, n\}$ and take $k := z_j$ and $m := v_j$.) Therefore

$$0 \leq q^{(j)}(kv - mz) = k^2 q(v) - kmq(z|v) + m^2 q(z) < -km \sum_{i \notin I} v_i q(z|e_i),$$

and since $v_i > 0$ for all $i \in \{1, \dots, n\}$ there must exist an $i \notin I$ with $q(z|e_i) < 0$. Observe now that $2z + e_i$ is a sincere vector for q is hypercritical and

$$q(2z + e_i) = 4q(z) + 2q(z|e_i) + 1 = 2q(z|e_i) + 1 < 0.$$

Hence $I = \{1, \dots, n\} - \{i\}$. For the converse assume that $q^{(i)}$ is not weakly nonnegative for some $i \in \{1, \dots, n\}$ and take $I \subset \{1, \dots, n\} - \{i\}$ such that q^I is a critical restriction of $q^{(i)}$ (thus a critical restriction of q). By hypothesis (b) we have $q^I = q^{(i)}$. Therefore $q^{(i)}$ is hypercritical as well as critical, and by Lemma 6.1, it is the Kronecker form q_m for some $m \geq 3$, which contradicts the existence of a critical vector for $q^{(i)}$. Then $q^{(i)}$ is weakly nonnegative for all $i \in \{1, \dots, n\}$, that is, q is a hypercritical form. □

Lemma 6.3. *Let $q : \mathbb{Z}^n \to \mathbb{Z}$ be a hypercritical unit form with at least three indices i, j, k.*

a) If $q_{ij} = -2$ then $n = 3$. In particular, the bigraph B_q associated to q is one of the following six bigraphs:

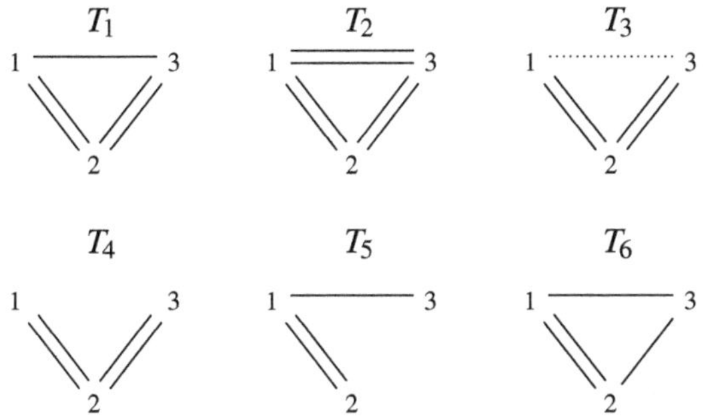

b) If $q_{ij} = q_{ik} = q_{jk} = -1$ and $q^{(k)}$ is critical, then $n = 4$ and the bigraph of q is one of:

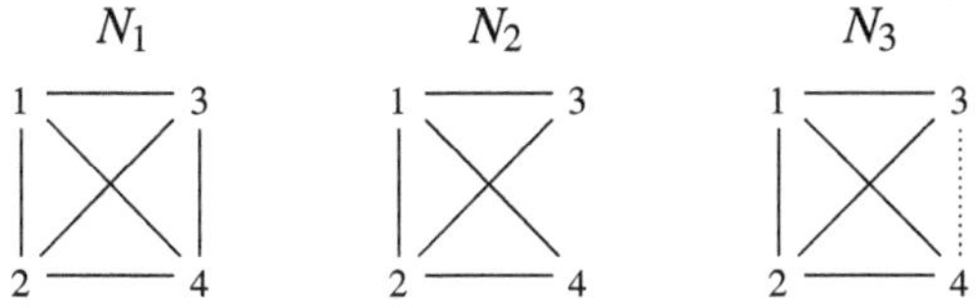

Moreover, the quadratic form q_Δ represents numbers -1 and -3 for

$$\Delta \in \{T_3, T_4, T_5, T_6, N_2, N_3\}.$$

Proof. Since q has at least three vertices, B_q does not contain any Kronecker form q_m as a restriction for $m \geq 3$, that is, $q_{rs} \geq -2$ for all vertices $r < s$.

Assume first that $q_{ij} = -2$ for some vertices $i < j$. Then $q^{\{i,j\}}$ is a critical restriction of q. By Proposition 6.2 we have $n = 3$. Set $(i, j, k) = (1, 2, 3)$ and notice that $z = (1, 1)$ is a critical vector of $q^{(3)}$, thus again by Proposition 6.2 we have

$$0 < q(z|e_3) = q_{13} + q_{23}.$$

This implies that B_q is one of the bigraphs $T_1, \ldots, T_6$. Moreover, the forms q_{T_i} for $i = 1, \ldots, 6$ are all hypercritical (Exercise 1 below).

Assume now that $q_{ij} = q_{ik} = q_{jk} = -1$. Then $q^{\{i,j,k\}}$ is a critical restriction of q, hence $n = 4$ and we may take $(i, j, k, \ell) = (1, 2, 3, 4)$. Since $q^{(3)}$ is critical (with critical vector $e_1 + e_2 + e_4$), we have $q_{14} = q_{24} = -1$ and $0 > q(e_1 + e_2 + e_3 | e_4) = -2 + q_{34}$. Therefore $q_{34} \in \{-1, 0, 1\}$, with corresponding cases N_1, N_2 and N_3.

For the last claim simply verify $q_{T_i}(v_i) = -1 = q_{N_j}(v'_j)$ and $q_{T_i}(w_i) = -3 = q_{N_j}(w'_j)$ for the vectors in the following list (for $i = 3, \ldots, 6$ and $j = 2, 3$)

$$\begin{aligned} v_3 &= (1, 2, 1) \text{ and } w_3 = (2, 5, 2), \\ v_4 &= (1, 1, 1) \text{ and } w_4 = (2, 3, 2), \\ v_5 &= (2, 2, 1) \text{ and } w_5 = (4, 4, 1), \\ v_6 &= (1, 1, 1) \text{ and } w_6 = (2, 2, 1), \\ v'_2 &= (1, 1, 1, 1) \text{ and } w'_2 = (2, 2, 2, 1), \\ v'_3 &= (2, 2, 2, 1) \text{ and } w'_3 = (4, 4, 4, 1). \end{aligned}$$ □

We now show that almost all hypercritical forms represent numbers -1 and -3. The importance of these two numbers will be clear in the proof of Theorem 6.16. In what follows, by a *slender* quadratic form we mean a unit form q with $q_{ij} \geq -1$ for all $i < j$. The bigraph associated to the Kronecker form q_m is denoted by $\mathbb{K}_m$ for $m \neq 0$.

Proposition 6.4. *Let $q : \mathbb{Z}^n \to \mathbb{Z}$ be a hypercritical unit form whose associated bigraph is not $\mathbb{K}_m$ ($m \geq 3$), T_1, T_2 or N_1 (see Lemma 6.3 for notation). Then there are positive (sincere) vectors v and w such that $q(v) = -1$ and $q(w) = -3$.*

Proof. Consider B_q, the bigraph associated to q. Since $\mathbb{K}_m$ is not contained in B_q for $m \geq 3$ we have $q_{ij} \geq -2$ for all $i < j$. If $q_{ij} = -2$ for some $i < j$, by Lemma 6.3(*a*) the bigraph B_q is one of $T_3, \ldots, T_6$, which represent -1 and -3. Therefore we may assume that q is a slender form. We may also assume, using Proposition 6.2, that $q^{(n)}$ is a critical form with critical vector z, and $q(z|e_n) = -s < 0$. Moreover, by Proposition 5.4 we may take $z_1 = 1$.

First we show that $0 < s \leq 3$. By the above assumptions, the vector $x := z - e_1 + e_n$ is positive and not sincere. Thus, since $q(e_1|e_n) = q_{1n} \geq -1$,

$$0 \leq q(x) = q(z) + 2 - q(z|e_1) + q(z|e_n) - q(e_1|e_n) = 2 - q_{1n} - s < 4 - s.$$

Notice now that $s \neq 3$. Indeed, for $s = 3$ and $x = z - e_1 + e_n$ we have $0 \leq q(x) = 2 - q(e_1|e_n) - 3$, that is, $q_{1n} = -1$ and $q(x) = 0$. Then $q^{(1)}$ is not weakly positive, and again by 6.2, the form $q^{(1)}$ is critical. Since $x_n = 1$ the vector x is critical for $q^{(1)}$. We may assume that $q_{2n} = -1$, therefore

$$0 = q(x|e_2) = q(z|e_2) - q_{12} + q_{2n} = -q_{12} - 1,$$

that is, $q_{12} = -1$. Now use Lemma 6.3(*b*) to conclude that $B_q = N_1$, which is impossible.

The proof is completed by showing that in cases $s = 1, 2$ such vectors v and w may be given explicitly:

Case $s = 1$.

$$q(2z + e_n) = 1 + 2q(z|e_n) = -1 \text{ and } q(4z + e_n) = 1 + 4q(z|e_n) = -3.$$

Case $s = 2$.

$$q(z + e_n) = 1 + q(z|e_n) = -1 \text{ and } q(2z + e_n) = 1 + 2q(z|e_n) = -3.$$ □

As an immediate consequence we have:

Corollary 6.5. *Let $q : \mathbb{Z}^n \to \mathbb{Z}$ be a hypercritical unit form with $n \geq 5$. Then q represents numbers -1 and -3.*

For integers $a \leq b$ denote by $[a, b]$ the set of integers ℓ with $a \leq \ell \leq b$.

Lemma 6.6. *For any hypercritical unit form $q : \mathbb{Z}^n \to \mathbb{Z}$ there is a (sincere) vector $v \in [0, 12]^n$ such that $q(v) < 0$.*

Proof. The statement is clear for Kronecker forms $q_m = q_{\mathbb{K}_m}$ with $m \geq 3$, and for the forms with associated bigraphs $T_3, T_4, T_5, T_6, N_2, N_3$ by simple inspection of the proof of the last claim in Lemma 6.3.

If q is not the form associated to graphs T_1, T_2, then by Proposition 6.4 we may assume that $q^{(n)}$ is a critical restriction with positive critical vector z, and -1 is either represented by $2z + e_n$ or by $z + e_n$. Since by Corollary 3.31 we have $z_i \leq 6$ for all i (cf. also Proposition 2.22), then both $2z + e_n$ and $z + e_n$ belong to $[0, 12]^n$.

To deal with cases T_1, T_2 (resp. N_1) evaluate at the vector $v = (1, 1, 1)$ (resp. $v = (1, 1, 1, 1)$) to get $q_{T_1}(v) = -2$, $q_{T_2}(v) = -3$ (resp. $q_{N_1}(v) = -2$). □

Exercises 6.7.

1. Show that all bigraphs in Lemma 6.3 correspond to hypercritical forms.
2. Prove that the form $q_m = q_{\mathbb{K}_m}$ does not represent the number -3 for any $m \geq 2$.
3. Show that the form q_Δ does not represent the number -1 for any
$$\Delta \in \{\mathbb{K}_m, T_1, T_2, N_1\}_{m \geq 4}.$$
4. Which of the forms associated to T_1, T_2 or N_1 represents the number -3?

6.2 Maximal and Locally Maximal Roots

For a weakly nonnegative semi-unit form $q : \mathbb{Z}^n \to \mathbb{Z}$ denote by $\mathbf{rad}^+(q)$ the set of positive vectors x with $q(x|e_i) = 0$ for $i = 1, \dots, n$ (called the *positive radical* of q). Observe that if $x \in \mathbf{rad}^+(q)$ then q has no maximal positive root (with partial order $x \geq y$ if $x - y \in \mathbb{N}_0^n$), since $1 = q(v) = q(v + mx)$ for any positive q-root v and any $m \in \mathbb{N}$. As in the weakly positive case, we say that a weakly nonnegative semi-unit form q is *sincere* if it has a sincere positive root.

Proposition 6.8. *Let $q : \mathbb{Z}^n \to \mathbb{Z}$ be a sincere weakly nonnegative unit form. The following are equivalent.*

a) There are finitely many sincere positive roots of q (and we call q finitely sincere*).*
b) There is a maximal sincere positive root of q.
c) $\mathbf{rad}^+(q) = \emptyset$.

Proof. Clearly we have that (a) implies (b) and that (b) implies (c). To show that (c) implies (a) let us assume that q has infinitely many sincere positive roots. Then we may take a sequence of sincere positive q-roots $y^1, y^2, \ldots$ with $y^m < y^{m+1}$ for $m = 1, 2, \ldots$ (see Lemma 5.12). Notice that $|q(y^m|e_i)| \leq 2$ for all $i = 1, \ldots, n$ and $m \geq 1$. This follows from the sincerity of y^m and the inequality

$$0 \leq q(y^m \pm e_i) = 2 \pm q(y^m|e_i).$$

Hence there are $y^\ell < y^m$ with $q(y^\ell|e_i) = q(y^m|e_i)$ for all $i = 1, \ldots, n$ and $0 \neq y^m - y^\ell \in \mathbf{rad}^+(q)$. □

A positive root v of a semi-unit form q is said to be *locally maximal* if $q(v|e_i) \geq 0$ for all $i = 1, \ldots, n$. For v a maximal positive q-root, since $\sigma_i(v) = v - q(v|e_i)e_i$ is again a q-root where σ_i is the i-th reflection for q (Sect. 1.2), v is a locally maximal root. The converse is false in general, as the following example shows.

Example 6.9. Consider the quadratic form q given by the following bigraph, and selected vectors u and v.

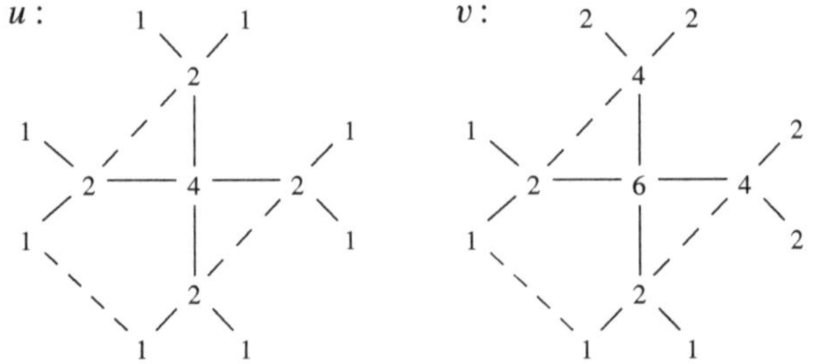

Then q is weakly nonnegative and $u < v$ are positive q-roots with u a locally maximal root.

Proposition 6.10. *Let $q : \mathbb{Z}^n \to \mathbb{Z}$ be a finitely sincere weakly nonnegative unit form. Then a sincere positive root y of q is maximal if and only if y is locally maximal.*

Proof. We only need to show that local maximality implies maximality. Assume y is a locally maximal sincere positive root of q, and that x is a root with $y < x$. Then

$$q(y|x) = \sum_{i=1}^{n} x_i q(y|e_i) \geq \sum_{i=1}^{n} y_i q(y|e_i) = q(y|y) = 2.$$

Thus $0 \leq q(x-y) = 2 - q(y|x) \leq 0$, that is, $q(x-y) = 0$. Notice that $v := x - y$ satisfies $q(v|e_i) \geq 0$ for all $i = 1, \ldots, n$ (for if $q(v|e_i) < 0$ then $q(2v + e_i) = 4q(v) + 1 + 2q(v|e_i) < 0$). We also have $q(v|y) = q(x|y) - 2 = 0$, therefore

$$0 = q(v|y) = \sum_{i=1}^{n} y_i q(v|e_i),$$

which implies that $v \in \mathbf{rad}^+(q)$, in contradiction with Proposition 6.8. □

A vertex i such that $q(y|e_i) > 0$ for a locally maximal positive q-root y of a weakly nonnegative semi-unit form q is called an *exceptional index (or vertex)* for y (cf. Lemma 5.9).

Lemma 6.11. *For a locally maximal positive root y of a weakly nonnegative semi-unit form q, one of the following situations occur:*

a) There are exactly two exceptional indices $i \neq j$ and $q(y|e_i) = y_i = 1 = y_j = q(y|e_j)$.
b) There is only one exceptional index i, and $q(y|e_i) = 1$ and $y_i = 2$.
c) There is only one exceptional index i, and $q(y|e_i) = 2$ and $y_i = 1$.

Furthermore, if y is also maximal then situation (c) never occurs.

Proof. Let y be a sincere locally maximal positive q-root. Then we have

$$2 = q(y|y) = \sum_{i=1}^{n} y_i q(y|e_i),$$

thus clearly one of (a), (b) or (c) occurs. If (c) holds then $q(2y - e_i) = 5 - 2q(y|e_i) = 1$ and $2y - e_i > y$, therefore y is not a maximal root. □

The following lemma will be useful to determine the maximality of positive sincere roots. For instance, this criterion is used below in the proof of Lemma 6.13.

Lemma 6.12. *Let $q : \mathbb{Z}^n \to \mathbb{Z}$ be a semi-unit form with a maximal sincere positive root y. Then for any positive vector v with $q(v) = -1$ we have $q(y|v) = 0$.*

Proof. Since v is positive and y is locally maximal we have $q(y|v) \geq 0$. Then

$$\sigma_v(y) = y - \frac{q(y|v)}{q(v)} v = y + q(y|v)v,$$

which is a positive q-root with $y \leq \sigma_v(y)$. By maximality $q(y|v) = 0$. □

Lemma 6.13. *Let $q : \mathbb{Z}^n \to \mathbb{Z}$ be a unit form such that there are indices $1 \leq i < j \leq n$ with $-5 \leq q_{ij} \leq -3$. Then q has no maximal sincere positive root.*

Proof. Let us assume that y is a sincere maximal positive root of q (hence locally maximal). Consider the triple $r = (e_i, e_j, y)$ and the root induced unit form q_r given by

$$\begin{aligned} q_r(x_1, x_2, x_3) &:= q(x_1 e_i + x_2 e_j + x_3 y) \\ &= x_1^2 + x_2^2 + x_3^2 + x_1 x_3 q(y|e_i) + x_2 x_3 q(y|e_j) - s x_1 x_2, \end{aligned}$$

where $q_{ij} = -s$ for some integer s. Let B be the bigraph associated to q_r. The shape of B depends on the values of $q(y|e_i)$ and $q(y|e_j)$. Since y is a root we have

$$2 = q(y|y) = \sum_{k=1}^{n} y_k q(y|e_k),$$

and since y is sincere, positive and locally maximal then $q(y|e_k) > 0$ for at most two vertices k (and $q(y|e_k) = 0$ for the rest), and in particular $m := q(y|e_i + e_j) \in \{0, 1, 2\}$. Thus we consider four cases: Case 1) $m = 0$; Case 2) $m = 1$; Case 3) $m = 2$ and $q(y|e_i)q(y|e_j) = 0$, and Case 4) $q(y|e_i) = 1 = q(y|e_j)$. These cases correspond to the four possibilities for B as depicted below (from left to right, observe that in all cases we have $(q_r)_{12} = q_{12} = -s$).

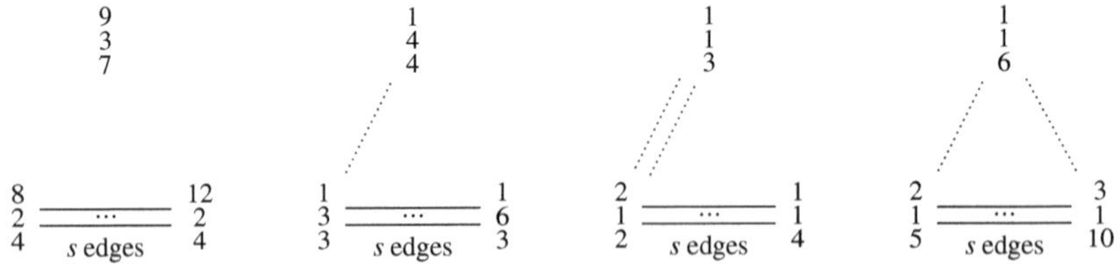

Each vertex of B contains a column with three natural numbers, corresponding to three vectors in $\mathbb{Z}^{B_0}$ which are, from top to bottom, sincere positive roots of q_r for $s = 3, 4, 5$ respectively. Then q_r has a sincere positive root $x = (x_1, x_2, x_3)$ and $y' := x_1 e_i + x_2 e_j + x_3 y$ is a root of q with $y' > y$, which is impossible. □

The following technical lemma imposes restrictions on sincere weakly nonnegative unit forms which fail to be unitary.

Lemma 6.14. *Let $q : \mathbb{Z}^I \to \mathbb{Z}$ be a weakly nonnegative semi-unit form, and $x \in \mathbb{Z}^I$ a positive sincere root. For $r = 0, 1$ take $I^r = \{i \in I \mid q_{ii} = r\}$ and consider the restriction $x^r = x|_{I^r}$ in $\mathbb{Z}^{I^r}$. Then one and only one of the following assertions holds:*

a) $q(x^1) = 1$ and $q_{ij} = 0$ for any $i \in I^0$ and $j \in I$.
b) $q(x^1) = 0$ and there exist $i \neq j$ in I^0 such that $x_i = x_j = q_{ij} = 1$. Moreover, if $s \in I^0$ and $t \in I$ is a different index satisfying $q_{st} \neq 0$ then $\{s, t\} = \{i, j\}$.

c) $q(x^1) = 0$ *and there exist* i *in* I^0 *and* $j \in I^1$ *such that* $x_i = x_j = q_{ij} = 1$. *Moreover, if* $s \in I^0$ *and* $t \in I$ *is a different index satisfying* $q_{st} \neq 0$ *then* $\{s, t\} = \{i, j\}$.

Furthermore, if x *is maximal and* $I^0 \neq \emptyset \neq I^1$*, then* (*c*) *holds and* I^0 *contains exactly one element.*

Proof. Let us suppose that $I = \{1, \dots, n\}$ and $I^0 = \{1, \dots, m\}$ for $m \leq n$. Then

$$1 = q(x) = q(x^1) + q(x^0) + q(x^1|x^0), \quad \text{where} \quad q(x^1|x^0) = \sum_{\substack{i \in I^1 \\ j \in I^0}} q_{ij} x_i x_j.$$

Now, by Lemma 6.1 we have $q_{ij} \geq 0$ for $i \in I^0$ and $j \in I$ (for q'_m and q''_m are hypercritical if $m > 0$). Hence the three summands on the right of the equation are nonnegative, therefore exactly one of them is nonzero. This leads to the three assertions above, since x is sincere.

For the last claim we give a root $y > x$ for both cases (*a*) and (*b*). For (*a*) take $y = 2x^0 + x^1$, whereas for (*b*) take $y = x^0 + 2x^1$. Notice that $y > x$ since $I^r \neq \emptyset$ for $r = 0, 1$. Thus if x is a maximal root then (*c*) holds, that is, $1 = q(x) = q(x^1|x^0)$. Further, if $k \in I^0$ with $k \neq i$ then

$$q(x + e_k) = q(x^1|x^0) + q(x^1|e_k) + q(x^0|e_k) = q(x^1|x^0) = 1,$$

that is, $x + e_k$ is a root of q larger than x. □

Exercises 6.15.

1. Do hypercritical unit forms have to be connected?
2. Show that the quadratic form q in Example 6.9 is weakly nonnegative.
3. Show that if q contains a bigraph with shape T_1, T_2 or N_1 (as in Lemma 6.3), then q does not have a maximal sincere positive root.
4. In Example 6.9, verify that u and v are roots of q.
5. For a weakly nonnegative semi-unit form q, a positive q-root x and a positive isotropic vector z of q, show that the following assertions hold:

   i) $q(x|e_i) \geq -2$ and $q(z|e_i) \geq -1$ for $i = 1, \dots, n$.
   ii) If $x_i > 0$ then $q(x|e_i) \leq 2$, and if $z_i > 0$ then $q(z|e_i) \leq 1$.
   iii) $q_{ij} \leq 3$ if $x_i \neq 0 \neq x_j$, and $q_{ij} \leq 2$ if $z_i \neq 0 \neq z_j$.

6. Let q be a nonzero connected weakly nonnegative semi-unit form. Must q be unitary?
7. Consider the unit form in three variables $q(x_1, x_2, x_3) = x_1^2 + x_2^2 + x_3^2 - s x_1 x_2$. Show that if $s \geq 3$ and $s - 2$ is not the square of an integer, then q has a sincere positive root.

6.3 Criteria for Weak Nonnegativity

Here we prove a **Weak Nonnegativity Criterion** due to Happel and de la Peña in [31]. Ovsienko showed in [44] that this result also holds without the condition $q_{ij} \geq -5$ for all $i < j$.

Theorem 6.16 (Happel–de la Peña). *Let $q : \mathbb{Z}^n \to \mathbb{Z}$ be a unit form with $q_{ij} \geq -5$ for all $1 \leq i < j \leq n$. If q has a maximal sincere positive root, then q is weakly nonnegative.*

Proof. Let y be a maximal sincere positive root of q. By Lemma 6.13 and Exercise 6.15.3, the form q does not contain bigraphs of type N_1, T_1, T_2, $\mathbb{K}_3$, $\mathbb{K}_4$, $\mathbb{K}_5$, nor, by assumption, bigraphs $\mathbb{K}_m$ for $m \geq 6$. If q is not weakly nonnegative, there is a hypercritical restriction q^I of q, and by Proposition 6.4 there exist positive vectors v and w with support I and with $q(v) = -1$ and $q(w) = -3$. It follows from Lemma 6.12 that $q(y|v) = 0$. Since v and w are positive vectors with same support and y is locally maximal, then $q(y|w) = 0$. Therefore

$$q(2y + w) = 4q(y) + q(w) = 1,$$

in contradiction with the maximality of y. □

Lemma 6.17. *Let q be a hypercritical unit form and i an index such that $q^{(i)}$ is not critical. Then $q^{(i)}$ is a positive form.*

Proof. Observe first that $q^{(i)}$ is weakly positive, since otherwise it would contain a critical restriction q^I, contradicting Proposition 6.2. Again by Proposition 6.2 there must exist a vertex c such that $q^{(c)}$ is a critical restriction of q (hence $c \neq i$), with critical positive vector z such that $q(z|e_c) < 0$. Then $q^{(c)(i)}$ is a positive unit form by Corollary 5.3.

If $q^{(i)}$ is not positive, there is a nonzero vector v such that $q^{(i)}(v) \leq 0$. In particular $v_c \neq 0$ since $q^{(c)(i)}$ is positive, so we may assume that $v = v' + v_c e_c$ with $v'_c = 0$ and $v_c > 0$. Notice that for $\alpha, \beta > 0$ we have

$$q(\alpha v + \beta z) = \alpha^2 q^{(i)}(v) + \alpha\beta q(z|v' + v_c e_c) \leq \alpha\beta v_c q(z|e_c) < 0.$$

Since $q^{(i)}$ is weakly positive the vector v' has a negative entry. But z is a critical positive vector of $q^{(c)}$, therefore we may find $\alpha, \beta > 0$ such that $\alpha v + \beta z$ is a positive nonsincere vector (take for instance $\alpha = z_a$ and $\beta = -v_a$ where a is an index such that $\frac{v_a}{z_a}$ is minimal among all fractions $\frac{v_j}{z_j}$ for $j \in \mathbf{supp}(z) = \{1, \dots, n\} - \{c\}$). This is impossible since $q(\alpha v + \beta z) < 0$ and q is hypercritical, hence $q^{(i)}$ is a positive unit form. □

The following immediate consequence may be considered as a partial analogue of Theorem 5.2 (see also Corollary 5.3).

Corollary 6.18. *Any proper restriction of a hypercritical unit form is nonnegative.*

Proof. The result is clear for Kronecker forms q_m with $m \geq 3$. Therefore we may assume the hypercritical form q has at least three vertices (in particular $q_{ij} \geq -2$ for all $i < j$). By Lemma 6.17, if $q^{(i)}$ is not positive then $q^{(i)}$ is critical, thus nonnegative by Theorem 5.2. □

We now prove a generalization of the (Jacobi-like) Zeldych Criterion 5.26 given in [55]. Again we do not assume the quadratic form to be unitary. Let $\mathbf{ad}(B)$ denote the *adjugate* of a square matrix B.

Proposition 6.19. *Let $q : \mathbb{Z}^n \to \mathbb{Z}$ be an integral quadratic form with associated symmetric matrix A (that is, $q(x) = x^t Ax$ for any $x \in \mathbb{Z}^n$). The following assertions are equivalent:*

a) The form q is weakly nonnegative.
b) For every principal submatrix B of A we have either $\det(B) \geq 0$, *or* $\mathbf{ad}(B)$ *has a negative entry.*

Proof. Let B be a principal submatrix of A and assume that $\mathbf{ad}(B)$ is nonnegative (that is, it has no negative entry). By Perron–Frobenius Theorem 1.36 there exists a positive eigenvector $v \in \mathbb{R}^n$ of $\mathbf{ad}(B)$ with eigenvalue $\rho > 0$. Assuming that q is weakly nonnegative and considering q as a real function $q_{\mathbb{R}} : \mathbb{R}^n \to \mathbb{R}$ we have by continuity

$$0 \leq q_{\mathbb{R}}(v) = v^t Bv = \frac{1}{\rho} v^t B(\mathbf{ad}(B)v) = \frac{1}{\rho} \det(B)\|v\|^2,$$

and therefore $\det(B) \geq 0$.

Suppose now that q satisfies (b) but is not weakly nonnegative. Since property (b) is preserved by principal minors, by induction on n we may assume that q is hypercritical. By Corollary 6.18, every proper restriction of q is nonnegative, therefore by Proposition 1.33 we have $\det(B) \geq 0$ for each proper principal submatrix B of A.

Thus $\det(A) < 0$ since otherwise q would be nonnegative. Take $\mathbf{ad}(A) = (v_{ij})$. By hypothesis there must exist i, j with $v_{ij} < 0$. Let v be the j-th column of $\mathbf{ad}(A)$, so that $Av = \det(A)e_j$ and $q(v) = \det(A)v_{jj}$. Further, let $w > 0$ be a sincere positive vector with $q(w) < 0$. For $\lambda = -\frac{v_{ij}}{w_i} > 0$ we have $(v + \lambda w)_i = 0$ and (since the restriction $q^{(i)}$ is nonnegative)

$$\begin{aligned} 0 &\leq q(v + \lambda w) \\ &= q(v) + 2\lambda w^t Av + \lambda^2 q(w) \\ &< \det(A)[v_{jj} + 2\lambda w_j] \\ &= \frac{\det(A)}{w_i}[v_{jj}w_i - 2v_{ij}w_j]. \end{aligned}$$

As in the proof of Proposition 5.26, if $v_{jj} < 0$ we take $i = j$, thus

$$0 \leq q(v + \lambda w) < \det(A)(-v_{jj}) \leq 0,$$

and if $v_{jj} \geq 0$ then $v_{jj}w_i - 2v_{ij}w_j \geq 0$ and we have

$$0 \leq q(v + \lambda w) < \frac{\det(A)}{w_i}[v_{jj}w_i - 2v_{ij}w_j] \leq 0.$$

Both cases yield a contradiction. □

The following practical criterion is useful for the computational verification of weak nonnegativity.

Theorem 6.20. *A semi-unit form $q : \mathbb{Z}^n \to \mathbb{Z}$ is weakly nonnegative if and only if $q(z) \geq 0$ for every $z \in [0, 12]^n$.*

Proof. If q is weakly nonnegative then $q(z) \geq 0$ for all $z \in [0, 12]^n$. If q is not weakly nonnegative, then there is a hypercritical restriction q' of q, and by Lemmas 6.1 and 6.6 there is a vector $z \in [0, 12]^n$ with $q(v) < 0$. □

We say that a weakly nonnegative semi-unit form q is 0-*sincere* if there exists a sincere vector $y \in \mathbf{rad}^+(q)$. We point out that in this case any isotropic vector $y \in \mathbb{N}_0^n$ belongs to the positive radical $\mathbf{rad}^+(q)$ of q. In fact, we have the following more general result.

Lemma 6.21. *Let $q : \mathbb{Z}^I \to \mathbb{Z}$ be a weakly nonnegative semi-unit form and take $\mu \in q^{-1}(0)$.*

a) If $x \in \mathbf{rad}^+(q)$ and $\mathbf{supp}(\mu) \subset \mathbf{supp}(x)$, then $\mu \in \mathbf{rad}(q)$.
b) If μ is positive and $z \in \mathbb{Z}^I$ is such that $q(z|\mu) = 0$ and $z + n\mu$ is a positive sincere vector for some $n \geq 0$, then $\mu \in \mathbf{rad}^+(q)$.

Proof. Assume there is an index $i \in I$ such that $q(\mu|e_i) \neq 0$ and take $\epsilon = \pm 1$ such that $\epsilon q(\mu|e_i) > 0$. Taking $y = e_i - 2\epsilon\mu$, we observe that

$$q(y) = q(e_i) - 2\epsilon q(\mu|e_i) \leq -1.$$

By the requirement on the supports in (a), notice that there exists a $k \geq 0$ such that $y + kx$ is a positive vector, thus we arrive at the contradiction

$$0 \leq q(y + kx) = q(y) \leq -1.$$

This shows (a). For (b) assume that $\mu \notin \mathbf{rad}(q)$, thus there exists $i \in I$ with $q(\mu|e_i) > 0$ (for μ is a positive vector). In particular, there is $k \geq 0$ such that

$$q(z + k\mu|e_i) = q(z|e_i) + kq(\mu|e_i) \geq q(z) + 2.$$

Take $m := \max(k, n)$ and $y := z + m\mu$. Then $q(y|e_i) \geq q(z) + 2$ and $q(y) = q(z)$ since $q(z|\mu) = 0$ and $q(\mu) = 0$. Therefore

$$q(y - e_i) = q(y) + q(e_i) - q(y|e_i) = q(z) + 1 - q(y|e_i) \leq -1,$$

which is impossible since $y \geq z + n\mu$ is positive and sincere. □

For a semi-unit form q with a sincere positive radical vector z, we trivially observe that any vector x may be taken into a positive vector $x + kz$ with $k \in \mathbb{N}$, so that $q(x) = q(x + kz)$. This proves the following lemma.

Lemma 6.22. *Any* 0*-sincere form is nonnegative.*

6.4 Iterated Edge Reductions

Recall from Sect. 5.3 that for a unit form $q : \mathbb{Z}^n \to \mathbb{Z}$ and indices $i \neq j$ with $q_{ij} < 0$ we construct a quadratic form $q'(x) = q(\rho(x)) + x_i x_j$, with $\rho : \mathbb{Z}^{n+1} \to \mathbb{Z}^n$ given by

$$\rho(e_k) = \begin{cases} e_k, & \text{if } 1 \leq k \leq n; \\ e_i + e_j, & \text{if } k = n + 1, \end{cases}$$

called the *edge reduction of q with respect to i and j*. The same construction can be performed when q is a semi-unit form (or even a pre-unit form, that is, an integral quadratic form q with $q(e_i) \leq 1$ for all indices i) satisfying $q(e_i) = 1 = q(e_j)$ and $q_{ij} < 0$.

The quadratic form q can be recovered from q' using the nonlinear map $\pi : \mathbb{Z}^n \to \mathbb{Z}^{n+1}$ defined as follows,

$$\pi(x)_k = x_k, \quad \text{for } k \notin \{i, j, n+1\} \text{ and}$$

$$(\pi(x)_i, \pi(x)_j, \pi(x)_{n+1}) = \begin{cases} (0, x_j - x_i, x_i), & \text{if } x_i \leq x_j, \\ (x_i - x_j, 0, x_j), & \text{if } x_i > x_j. \end{cases}$$

Since $\rho \circ \pi = \mathbf{Id}$ we have $q(x) = q'(\pi(x))$ for any vector $x \in \mathbb{Z}^n$.

Example 6.23. Consider the unit form q with associated bigrah B_q as shown below (left). Its edge reduction with respect to vertices a, b is the form q' with bigrah $B_{q'}$ (right).

$B_q =$ 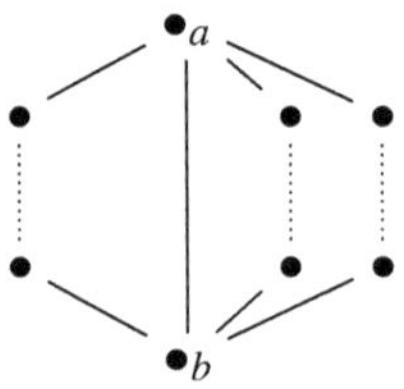

$B_{q'} =$ 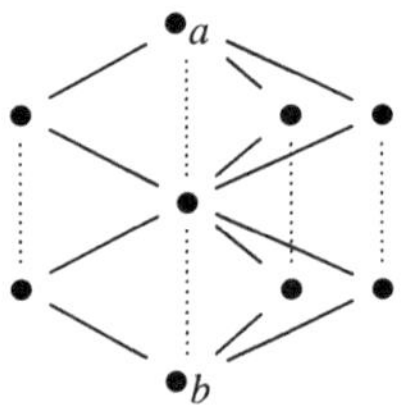

For a quadratic form q denote by $\Sigma^+(q)$ the set of isotropic vectors of q with nonnegative entries.

Proposition 6.24. *Let $q : \mathbb{Z}^n \to \mathbb{Z}$ be a semi-unit form and $q' : \mathbb{Z}^{n+1} \to \mathbb{Z}$ be obtained from q by edge reduction with respect to indices i and j. Then q is weakly nonnegative if and only if q' is weakly nonnegative. In this case the maps ρ and π are bijections (inverse to each other) between the sets $\Sigma^+(q)$ and $\Sigma^+(q')$.*

Proof. Take a positive vector y in $\mathbb{Z}^{n+1}$. If q is weakly nonnegative, since $\rho(y) > 0$ we have

$$q'(y) = q(\rho(y)) + y_i y_j \geq 0.$$

Conversely, if $0 < x \in \mathbb{Z}^n$ and q' is weakly nonnegative, then $\pi(x) > 0$ and

$$q(x) = q'(\pi(x)) \geq 0.$$

Assume that q and q' are weakly nonnegative. By the identity $q(x) = q'(\pi(x))$ the mapping π restricts to a function $\pi : \Sigma^+(q) \to \Sigma^+(q')$. If $y \in \Sigma^+(q')$ then

$$0 = q'(y) = q(\rho(y)) + y_i y_j.$$

Since both summands on the right are nonnegative, it follows that $y_i y_j = 0$ (thus $y \in \mathbf{Im}(\pi)$) and that $\rho(y) \in \Sigma^+(q)$. In particular, $\pi : \Sigma^+(q) \to \Sigma^+(q')$ is a surjective mapping, and the result follows since $\rho \circ \pi = \mathbf{Id}$. □

Even though there is a bijection between $\Sigma^+(q)$ and $\Sigma^+(q')$ when q is a weakly nonnegative semi-unit form and q' is an edge reduction of q, it is not always true that q and q' have the same number of critical vectors (a vector z is critical for q if the restriction of q to the support $\mathbf{supp}(z)$ of z is critical having z has positive generator of its radical). For instance, if q and q' are the forms shown in Exercise 6.23, then the following vectors v_1 and v_2 are critical vectors for q,

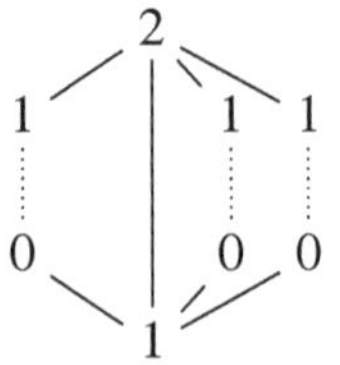

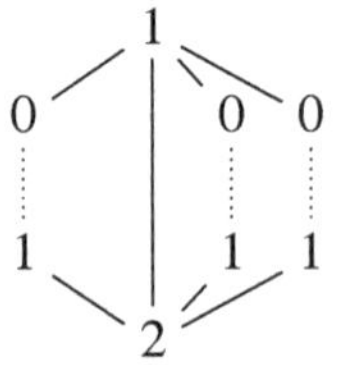

while q' has the following different critical vectors, $\pi(v_1)$, $\pi(v_2)$ and a third vector w with an entry 3.

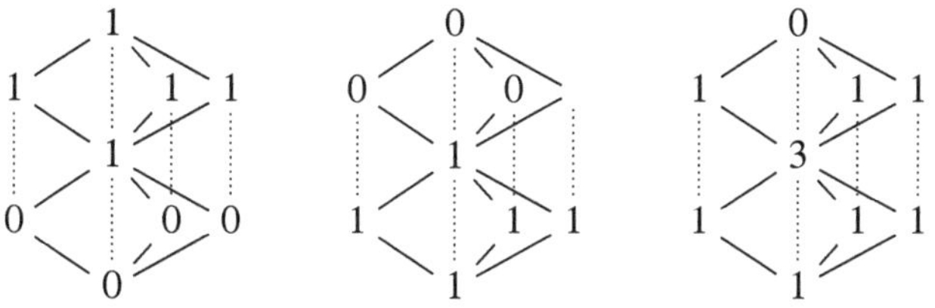

This behavior, along with notions like *positive corank* and *conformality* for edge reductions, are further explored in [54].

For a unit form q which is not weakly positive there might be an arbitrarily long iterated edge restriction for q, which is evident from the following example,

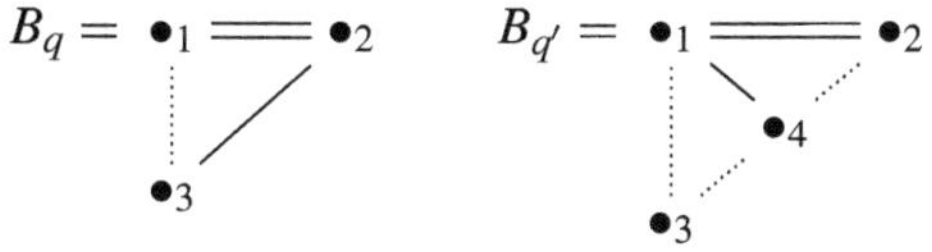

where B_q is a subbigraph of the bigraph $B_{q'}$ associated to the edge reduction q' of q with respect to the vertices 2 and 3. Notice that this example is actually weakly nonnegative.

An *iterated edge reduction* for a semi-unit form $q' : \mathbb{Z}^n \to \mathbb{Z}$ is a quadratic form $q : \mathbb{Z}^m \to \mathbb{Z}$ with $m \geq n$ that is obtained iteratively from q by a sequence of edge reductions. For instance, for the example in three variables q above, consider the iterated edge reductions q'' by edges $\{1, 2\}$, $\{1, 2\}$ and $\{2, 3\}$, and the reduction q''' by edges $\{2, 3\}$, $\{1, 2\}$ and $\{1, 2\}$ respectively, as shown below.

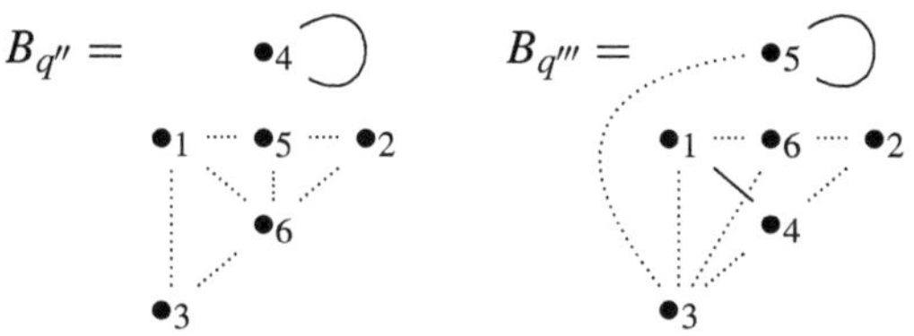

The following is a suitable generalization of Theorem 5.24 to the weakly nonnegative setting.

Theorem 6.25. *A semi-unit form $q : \mathbb{Z}^n \to \mathbb{Z}$ is weakly nonnegative if and only if any iterated edge reduction q' of q is semi-unitary.*

Proof. The necessity follows from Proposition 6.24.

For the converse assume that q is a semi-unit form which is not weakly nonnegative. If there are vertices $a \neq b$ with $q_{ab} < -2$, then the edge reduction of q with respect to a and b is not semi-unitary. Therefore we may assume that $q_{ab} \geq -2$ for all vertices $a \neq b$. By Proposition 6.2 there is a critical vector z and $i \notin \mathbf{supp}(z)$ such that $q(z|e_i) < 0$. In particular,

$$q(2z + e_i) = q_{ii} + 2q(z|e_i) < 0.$$

Take vertices a and b with $q_{ab} < 0$ and consider the reduction q' of q with respect to a, b. First we notice that there exists a $j \notin \mathbf{supp}(z)$ such that

$$q'(2\pi(z) + e_j) < 0.$$

If $a = i$ and $b \in \mathbf{supp}(z)$ then take $j = n + 1$, so that

$$q'(2\pi(z) + e_{n+1}) = q(\rho[2\pi(z) + e_{n+1}]) = q(2z + e_i + e_b) \leq q(2z + e_i) < 0.$$

If $i \notin \{a, b\}$ or $\{a, b\} \cap \mathbf{supp}(z) = \emptyset$ then take $j = i$ and observe that

$$q'(2\pi(z) + e_i) = q(2z + e_i) < 0.$$

Now, if the weight $|z| = \sum_i |z_i|$ of z is greater than one, taking $a, b \in \mathbf{supp}(z)$ we have $|\pi(z)| < |z|$. By the above argument, replacing q for some iterated reduction of q, we may assume that $|z| = 1$, that is, $z = e_k$ for some $k \in \{1, \ldots, n\}$. Hence

$$0 > q(2e_k + e_i) = 4q_{kk} + q_{ii} + 2q_{ki}.$$

Since $q_{ii}, q_{kk} \in \{0, 1\}$ we have $q_{kk} = 0 > q_{ki}$. Then the bigraph B associated to the restriction $q^{\{k,i\}}$ has one of the following forms,

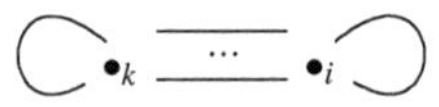

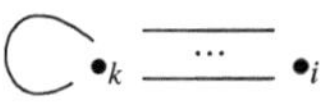

corresponding to cases $q_{ii} = 0$ (left) and $q_{ii} = 1$ (these restrictions are the hypercritical semi-unit forms q'_m and q''_m from Lemma 6.1). For the reduction q' of q with respect to k and i we have

$$q'_{n+1,n+1} = q_{kk} + q_{ii} + q_{ki} = q_{ii} + q_{ki},$$

thus q' is not a semi-unit form unless B has the form . In this case the restriction $(q')^{\{k,i,n+1\}}$ has the following associated bigraph,

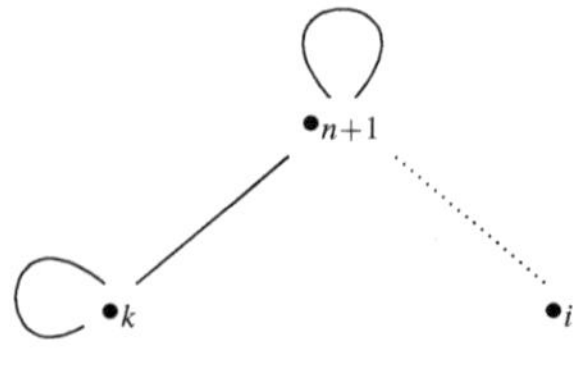

hence the reduction of q' with respect of k and $n+1$ is not a semi-unit form, which completes the proof. □

Following [54], by an *exhaustive reduction* for a semi-unit form $q : \mathbb{Z}^n \to \mathbb{Z}$ we mean an iterated edge reduction q' of q satisfying the following conditions:

i) Every edge reduction involved in the construction of q' is with respect to vertices i and j satisfying $1 \leq i < j \leq n$.
ii) For any $1 \leq i < j \leq n$ we have $q'_{ij} \geq 0$.

Notice that all exhaustive reductions involve the same number K of edge reductions, namely

$$K = (-1) \sum_{i<j \text{ and } q_{ij}<0} q_{ij}.$$

The forms q'' and q''' right before Theorem 6.25 are examples of exhaustive reductions of the quadratic form

$$q(x_1, x_2, x_3) = x_1^2 + x_2^2 + x_3^2 - 2x_1x_2 + x_1x_3 - x_2x_3.$$

Furthermore, we may consider a sequence $q^0, q^1, q^2, \ldots$ of semi-unit forms such that $q^0 = q$ and for $k > 0$ the form q^k is obtained from q^{k-1} by an exhaustive reduction. Then we say that q^k is obtained from q by an *iterated exhaustive reduction* (of length k). Notice that there is a sequence of integers

$$n = n_0 < n_1 < n_2 < \ldots < n_k$$

such that q^i is a semi-unit form in n_i variables for $i = 0, \ldots, k$. It is not known whether a semi-unit form q is weakly nonnegative if and only if any iterated exhaustive reduction of q stops, after finitely many steps, in a quadratic form having only nonnegative coefficients. However, the following criterion (which is an alternative version of Theorem 6.25) was proved in [54].

Remark 6.26. Let $q : \mathbb{Z}^n \to \mathbb{Z}$ be a semi-unit form, and $q^k : \mathbb{Z}^{n_k} \to \mathbb{Z}$ be a sequence of iterated exhaustive reductions of q for $k = 0, 1, 2, \ldots$ Then q is weakly nonnegative if and only if q^k is semi-unitary for all $k \leq 31$.

6.5 Semi-Graphical Forms

The following result, known as the *reduction theorem by deflations* of weakly nonnegative forms, gives the main procedure to obtain graphical forms from weakly nonnegative semi-unit forms, which is one of the main tools in next section. We present a useful generalization.

Theorem 6.27. *Let $q : \mathbb{Z}^I \to \mathbb{Z}$ be a weakly nonnegative semi-unit form with a maximal sincere positive root x. If $I = J \cup K$ is a nontrivial partition of the index set I, then there is an iterated deflation T for q concentrated in J such that the form $q' = qT$ satisfies the following.*

a) The form q' is a weakly nonnegative semi-unit form.
b) The form q' has a maximal positive root x' with $x = T(x')$.
c) We have $q'_{ij} \geq 0$ for all $i, j \in J \cap \mathbf{supp}(x')$.
d) There are inclusions,

$$R^+(q') \xrightarrow{\quad T \quad} R^+(q) \quad \textit{and} \quad \Sigma^+(q') \xrightarrow{\quad T \quad} \Sigma^+(q).$$

Proof. Take a deflation T_{ij}^- for q and the form $q^- = qT_{ij}^-$. Consider a positive vector $y \in \mathbb{Z}^n$ and take $y^- = T_{ij}^-(y) = y + y_i e_j$. Then y^- is a positive vector and

$$q^-(y) = q(T_{ij}^- y) = q(y^-) \geq 0,$$

which shows (a). For (b) we take i and j with $x_j \geq x_i$ so that $x^- := (T_{ij}^-)^{-1} x$ is a positive q^--root. If y^- is a positive q^--root with $y^- \geq x^-$, then $y := T_{ij}^-(y^-)$ is a positive q-root with $y \geq x$. Hence $y = x$, that is, the vector x is a maximal root. The claim (d) follows as in Lemma 2.19. Therefore points (a), (b) and (d) hold for iterated deflations.

For (c), as long as there are vertices i and j such that $q_{ij} < 0$ we may take a deflation T_{ij}^- or T_{ji}^- and continue with the reduction. The process must stop since in each step the weight $|x^-| = \sum_i x_i^-$ of x^- is smaller than the weight $|x|$ of x. □

Following Dräxler, Golovachtchuk, Ovsienko and de la Peña [22], we say that a semi-unit form $q : \mathbb{Z}^I \to \mathbb{Z}$ is *semi-graphical* if there exists a vertex $\omega \in I$ such that $q_{\omega i} < 0$ for all $i \neq \omega$, and $q_{ij} \geq 0$ for all $i, j \neq \omega$. As defined by Ringel [46], a *graphical form* is a semi-graphical unit form q such that $|q_{ij}| \leq 1$ for all $i \neq j$. According to Sect. 5.5, a centered form q is a semi-graphical unit form with $q_{\omega i} = -1$ for all $i \neq \omega$. Therefore graphical forms are centered.

Lemma 6.28. *Let q be a finitely sincere weakly nonnegative semi-unit form. Then B_q is a connected bigraph. Moreover, $q_{ii} = 0$ for a vertex i if and only if $q_{ij} \geq 0$ for all $j \neq i$.*

Proof. If B_q has a nontrivial partition supported by the sets of vertices I^1 and I^2, and x is a sincere positive q-root, then $x = x^1 + x^2$ with $\mathbf{supp}(x^i) = I^i$ for $i = 1, 2$. Since $1 = q(x^1) + q(x^2)$ we may assume that $q(x^1) = 1$ and $q(x^2) = 0$, and thus conclude that all vectors $x^1 + mx^2$ are sincere positive q-roots for $m > 0$, in contradiction with q being finitely sincere.

For the second assertion notice that if $q_{ii} = 0$ and $q_{ij} < 0$, then $q(2e_i + e_j) = q_{jj} + 2q_{ij} < 0$. Conversely, assume that $q_{ij} \geq 0$ for all $j \neq j$ and that $q_{ii} = 1$. Since B_q is connected, there exists $j \neq i$ such that $q_{ij} > 0$. Then for any sincere positive root x we have

$$q(x|e_i) = 2x_i + \sum_{k \neq i} x_k q_{ki} \geq 3,$$

which is impossible since $0 \leq q(x - e_i) \leq 2 - q(x|e_i)$. □

The Kronecker form q_m for $m \geq 2$ is a semi-graphical form which is critical and hypercritical for $m \geq 3$. All other critical semi-graphical forms are actually graphical.

Lemma 6.29. *Any critical semi-graphical form q in $n \geq 3$ variables is a graphical form.*

Proof. Since $n \geq 3$ we have $q_{ij} \geq -1$ for all $i, j \neq \omega$. We show that $q_{ij} \leq 1$ for $i, j \neq \omega$. Since the vector $e_i - e_j$ is not sincere, and proper restrictions of critical forms are positive (cf. Corollary 5.3), we have

$$0 < q(e_i - e_j) = 2 - q_{ij},$$

thus the result. □

The list of critical semi-graphical forms with $n \geq 3$ is precisely that of Table 5.3. It will be useful to have a classification of centered hypercritical forms (equivalently, hypercritical semi-graphical forms with $n > 3$ variables). In Table 6.1 we exhibit such forms.

Recall that by a 0-sincere form we mean a weakly nonnegative semi-unit form q having a sincere positive radical vector. We say that a 0-sincere (weakly nonnegative) unit form is *reduced* provided $q_{ij} \leq 1$ for all vertices i, j (compare to slender forms). The following lemma justifies this definition. Recall from Sect. 5.5 that a unit form q is obtained from q' by doubling a vertex k if q is the one-point extension $q = q'[-e_k]$ (cf. also Exercise 3.32.4).

Lemma 6.30. *Let q be a 0-sincere (weakly nonnegative) unit form. Then q is not reduced if and only if there is a vertex i such that q can be recovered from the restriction $q^{(i)}$ by doubling a vertex.*

Proof. Assume $q : \mathbb{Z}^n \to \mathbb{Z}$ and take for simplicity $i = n$ and $q' = q^{(n)}$. Then clearly $q'[-e_k]_{kn} = 2$, thus $q = q'[-e_k]$ is not reduced.

For the converse assume that $q_{ij} > 1$ for some vertices $i \neq j$, and take z to be a sincere positive radical vector of q. Then we have

$$0 \leq q(z + e_i - e_j) = q(e_i - e_j) = 2 - q_{ij},$$

that is, $q_{ij} = 2$. In particular $q(e_i - e_j) = 0$, and since q is a nonnegative unit form (Lemma 6.22), by Lemma 3.2(a) the vector $e_i - e_j$ is radical, that is, q is obtained from q' by doubling vertex j. □

Table 6.1 Hypercritical graphical forms

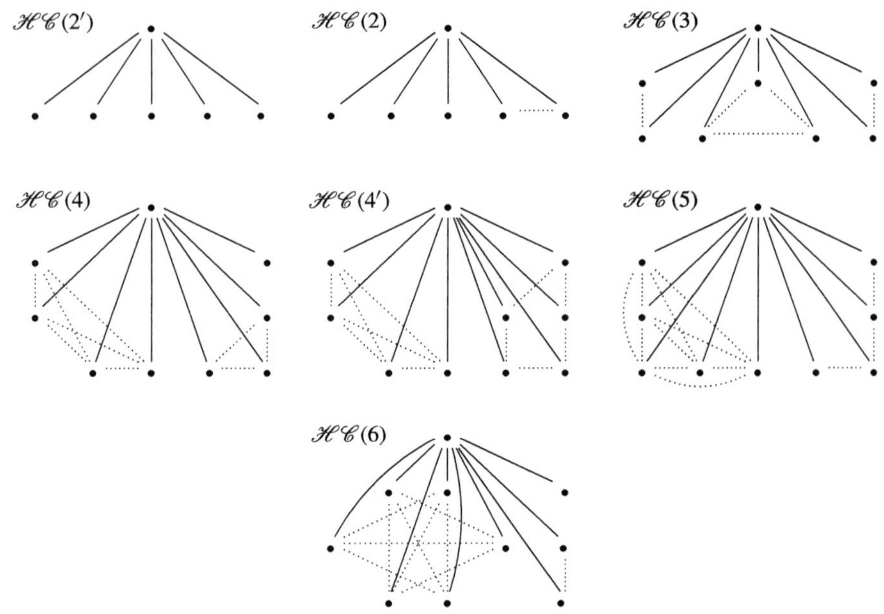

In the last part of this section we begin with technical preparations to end our discussion on integral quadratic forms with a generalization of Ovsienko's Theorem 5.25 to the weakly nonnegative context. In Table 6.2 we show some 0-sincere forms of small corank. The reason why we exclude those forms associated to bigraphs $\mathscr{C}(1)$, $\mathscr{C}(2)$, $\mathscr{C}(3)$ and $\mathscr{C}(4')$ is the content of the following result (cf. Table 5.3 and the figure below).

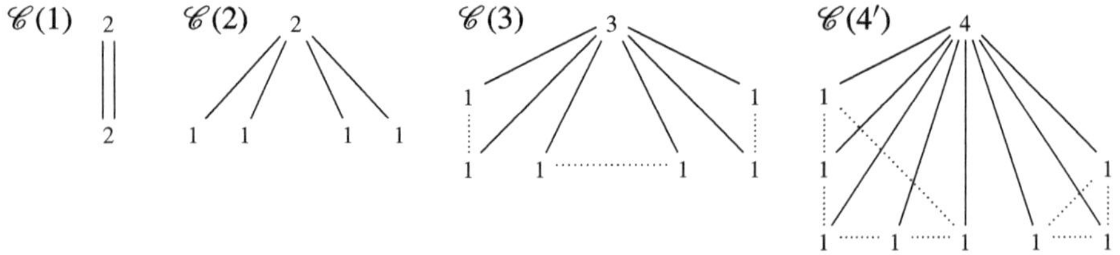

The following classification result of graphical weakly nonnegative unit forms of small corank, due partially to Ringel [46] (cf. [23] for comments and proofs), will be used in the last steps in the proof of our last result Theorem 6.37.

We say that a 0-sincere graphical form q is *triangular* if there are precisely three critical restrictions q^{I_1}, q^{I_2} and q^{I_3} of q such that for any $i \neq j$ in $\{1, 2, 3\}$ the restriction $q^{I_i \cup I_j}$ is a 0-sincere form of corank 2.

Table 6.2 Reduced 0-sincere semi-graphical forms of corank one or two, without the forms associated to $\mathscr{C}(1)$, $\mathscr{C}(2)$, $\mathscr{C}(3)$ and $\mathscr{C}(4')$ appearing as critical restrictions

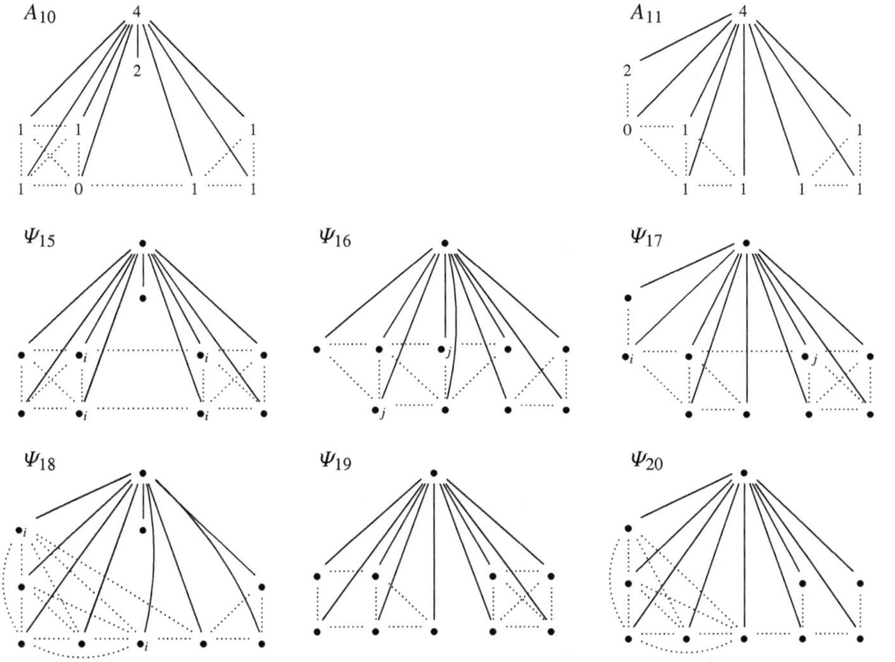

In cases A_{10} and A_{11} the vector shown as integers at the vertices is the positive generator of the radical. A vertex marked as $\bullet_i$ or $\bullet_j$ represents a critical restriction $q^{(i)}$ or $q^{(j)}$ of shape A_{10} or A_{11}, respectively

Theorem 6.31. *Let $q : \mathbb{Z}^I \to \mathbb{Z}$ be a 0-sincere graphical form without critical restrictions having associated bigraph of the shape $\mathscr{C}(1)$, $\mathscr{C}(2)$, $\mathscr{C}(3)$ or $\mathscr{C}(4')$.*

a) If $\mathbf{cork}(q) = 3$ *then q is either triangular or one of the forms associated to Θ_1 or Θ_2 (see Table 6.3).*
b) $\mathbf{cork}(q) = 2$ *if and only if $q = q_{\Psi_\ell}$ for $\ell = 15, \ldots, 20$ (see Table 6.2).*

Remark 6.32. Let q be one of the forms q_{Φ_ℓ} for $\ell = 15, \ldots, 20$. If $\mu^{(1)}$ and $\mu^{(2)}$ are critical vectors of q one can show by inspection that there are vertices i and j such that $\mu_i^{(1)} = 1$ and $\mu_j^{(1)} = 0$, and $\mu_i^{(2)} = 0$ and $\mu_j^{(2)} = 1$. In particular, for any positive radical vector μ of q, there are positive numbers m_1 and m_2 such that $\mu = m_1\mu^{(1)} + m_2\mu^{(2)}$.

Similarly, it can be shown that if q is a triangular 0-sincere form, then there are vertices $\{i, j, k\}$ such that the restriction of the critical vectors $\mu^{(1)}$, $\mu^{(2)}$ and $\mu^{(3)}$ are the canonical vectors with three entries. Therefore, for any sincere positive radical vector μ there are positive numbers m_1, m_2 and m_3 such that $\mu = m_1\mu^{(1)} + m_2\mu^{(2)} + m_3\mu^{(3)}$.

Exercises 6.33.

1. Show that the solid star $\mathbb{T}_{r_1,\dots,r_s}$ is equivalent to the centered form q where $q^{(\omega)}$ has associated bigraph $B = \sqcup I(r_i - 1)$ and where $I(m)$ is the complete dotted bigraph on m vertices.
2. Let $q : \mathbb{Z}^I \to \mathbb{Z}$ be a weakly nonnegative centered unit form with center ω, and for $i \in I$ consider the set $S_i = \{j \in I \mid q_{ij} > 0\}$. Show that if x is a positive sincere vector, $S \subset S_i$ with $i \neq \omega$ and
$$x_i - q(x|e_i) \geq x_\omega - \sum_{j \in S} x_j,$$
then $S = S_i$ and $q_{ij} = 1$ for all $j \in S_i$.
3. Let $q : \mathbb{Z}^I \to \mathbb{Z}$ be a weakly nonnegative semi-graphical form with center ω. Suppose that x is a maximal sincere positive root with $x_\omega \geq 7$ and only one exceptional vertex.

   a) Show that q is a centered form, and that $q_{ij} \leq 1$ for all $j \neq \omega$.
   b) Set $S_i' = \{j \neq i \mid q_{ij} > 0\}$ and show that the restriction of B_q to S_i' is a complete graph with dotted edges. Moreover, $x_j = 1$ for all $j \in S_i'$ and if $j \in S_i$ and $k \in I$ satisfy $q_{jk} > 0$, then $k \in S_i$.
   c) Prove that S_i has exactly $x_\omega - 2$ elements.
   d) Notice that q is not weakly positive (why?) and show that if $J \subset I$ and the restriction q^J is critical, then $S_i \subset J$.
   e) Conclude that $x_\omega = 7$. [Hint: use (c) and (d) to verify that the restriction q^I may be identified with the critical form $q_{\mathscr{C}(6)}$, see Table 5.3].

4. Which of the hypercritical centered forms in Table 6.1 have as restriction the following bigraphs?

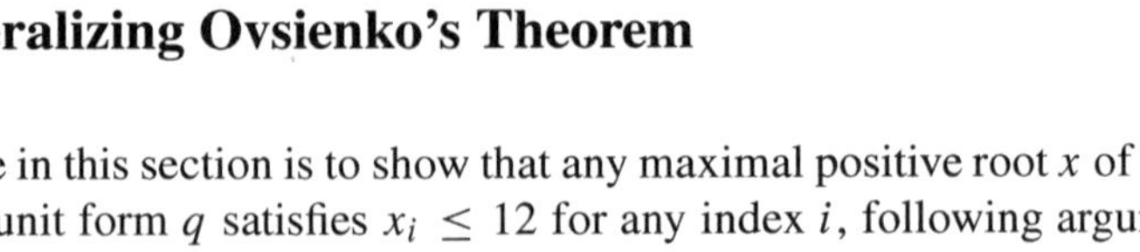

6.6 Generalizing Ovsienko's Theorem

Our objective in this section is to show that any maximal positive root x of a weakly nonnegative unit form q satisfies $x_i \leq 12$ for any index i, following arguments by Dräxler, Golovachtchuk, Ovsienko and de la Peña in [23]. We say that $x \in \mathbb{Z}^n$ is a *2-layer root* of an integral quadratic form $q : \mathbb{Z}^n \to \mathbb{Z}$ if x is a positive q-root and there exist positive isotropic vectors μ and μ' such that $x = \mu + \mu'$ (in particular $1 = q(x) = q(\mu|\mu')$).

Theorem 6.34. *Let $q : \mathbb{Z}^n \to \mathbb{Z}$ be a weakly nonnegative semi-unit form with a maximal positive root x.*

a) If there is a positive isotropic vector μ with $\mu < x$ then x is a 2-layer root.
b) If x is a 2-layer root then $x_i \leq 12$ for all $i = 1, \dots, n$.

Proof. Without loss of generality we may assume that x is a sincere vector. To show (a), by maximality of x we have $\mu \notin \mathbf{rad}(q)$, and therefore $q(x|\mu) \neq 0$ by Lemma 6.21(b). That $q(x|\mu) = 1$ follows from the equations

$$0 \leq q(x - \mu) = q(x) - q(x|\mu) = 1 - q(x|\mu),$$
$$0 \leq q(x + m\mu) = q(x) + mq(x|\mu) = 1 + mq(x|\mu), \qquad \text{for all } m \geq 0.$$

Hence $q(x - \mu) = q(x) - q(x|\mu) + q(\mu) = 0$, that is, $\mu' := x - \mu$ is an isotropic vector.

We now turn to the proof of (b), which we illustrate with an example. Take $x = \mu + \mu'$ with μ and μ' positive isotropic vectors of q.

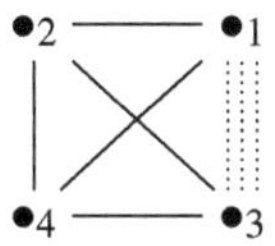

$$x = {\begin{smallmatrix}1&1\\1&0\end{smallmatrix}} + {\begin{smallmatrix}1&0\\1&1\end{smallmatrix}}.$$

Step 1. First we *double all vertices* $I = \{1, \dots, n\}$ of the form $q : \mathbb{Z}^n \to \mathbb{Z}$ (cf. Exercises 3.32.4 and 5) to get a weakly nonnegative form $\overline{q} : \mathbb{Z}^{I \cup J} \to \mathbb{Z}$, where $J = \{n+1, \dots, 2n\}$. Consider μ as a vector in $\mathbb{Z}^{I \cup J}$ and define $\overline{\mu} = \sum_{i=1}^n \mu'_i e_{i+n}$. Then the projection $\pi : \mathbb{Z}^{I \cup J} \to \mathbb{Z}^I$ given by $\pi(e_{i+n}) = e_i = \pi(e_i)$ for $i \in I$ satisfies $\pi(\overline{x}) = x$ where $\overline{x} = \mu + \overline{\mu}$ is a maximal positive root of $\overline{q}$ (see Exercise 3.32.4(d)).
Take $I' = \mathbf{supp}(\mu)$ and $J' = \mathbf{supp}(\overline{\mu})$, and replace $\overline{q}$ by its restriction to $I' \cup J'$ (figure below for our example).

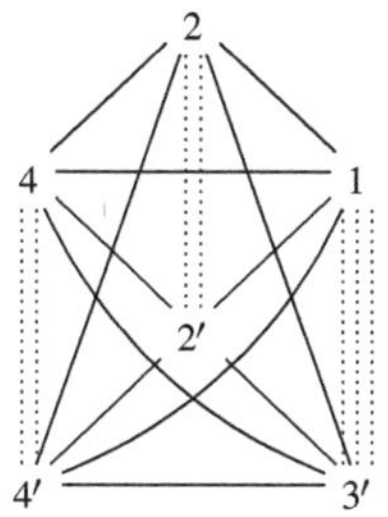

Step 2. Apply now the Reduction Theorem 6.27 to $\overline{q}$ with respect to I' to get an iterated deflation T (concentrated in I') and a weakly nonnegative quadratic form $\overline{q}' = \overline{q}T$ with a positive maximal root η such that $T(\eta) = \overline{x}$, and $\overline{q}'_{ij} \geq 0$ for all $i, j \in I' \cap \mathbf{supp}(\eta)$.

By Lemma 6.14, there is a vertex $\omega \in I'$ such that the support of η is $J' \cap \{\omega\}$, $\overline{q}'_{\omega\omega} = 0$ and the restriction q' of $\overline{q}'$ to J' is a unit form (in the example below the iterated flation is $T = T_{12}^{-}T_{14}^{-}$). Moreover, there exists a $j \in J'$ such that $\eta_\omega = \eta_j = \overline{q}'_{\omega j} = 1$ (in particular $\eta = \overline{\mu} + e_\omega$), and j is the unique element in J' satisfying $\overline{q}'_{\omega j} \neq 0$.

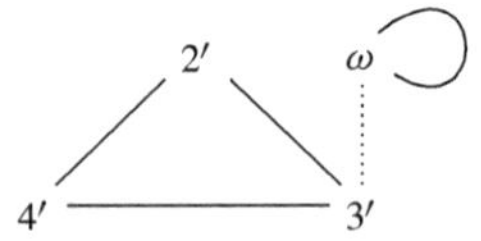

Step 3. By Lemma 6.21(b) we have $\overline{\mu} \in \mathbf{rad}^+(q')$, thus $\overline{\mu}$ belongs to the set

$$U = \{y \in \mathbf{rad}^+(q') \mid y_j = 1\}.$$

If U has infinitely many elements, there exist $y' < y \in U$, therefore $y - y' \in \mathbf{rad}^+(\overline{q}')$. This contradicts the maximality of η.

We conclude by pointing out that U is a finite set, thus by Lemma 6.35 below we have $\overline{\mu}_i \leq 6$ for $i \in I$. Since by symmetry we also have $\mu_i \leq 6$, then $x_i \leq 12$ for all $i \in I$. □

Lemma 6.35. *Suppose $q : \mathbb{Z}^I \to \mathbb{Z}$ is a (weakly nonnegative) 0-sincere semi-unit form such that there is an index $i \in I$ with $q^{(i)}$ unitary. If the set U of positive radical vectors y of q with $y_i = 1$ is finite, then $y_i \leq 6$ for any $y \in U$ and $i \in I$.*

Proof. We claim that the restriction $q^{(i)}$ is a weakly positive unit form. Otherwise there exists a positive isotropic vector μ with $i \notin \mathbf{supp}(\mu)$. By Lemma 6.21$(a)$ the vector μ is radical, contradicting the finiteness of U.

If $y \in U$ then $q(y - e_i) = q(e_i) = 1$, thus $y - e_i$ is a positive root of the weakly positive form $q^{(i)}$. The result follows from Ovsienko's Theorem 5.25. □

The following example shows that the bound 12 in Theorem 6.34 is optimal. The example is constructed by identifying all but the exceptional vertices of two copies of $q_{\widetilde{\mathbb{E}}_8}$, where the vector shown (a maximal positive root) is the sum of the corresponding positive generators of the radicals of $q_{\widetilde{\mathbb{E}}_8}$ (one for each copy).

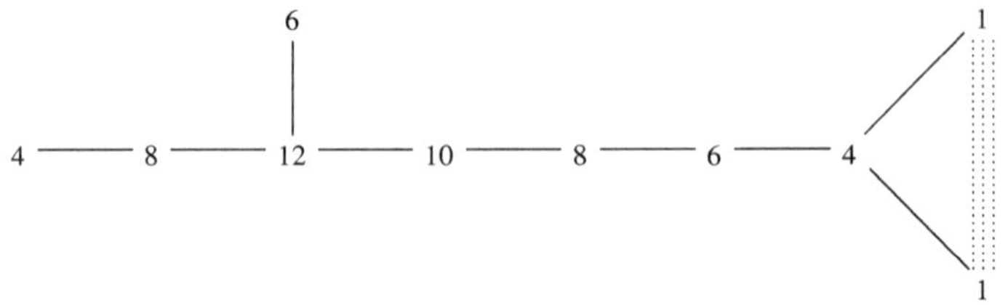

The example above is not a 0-sincere form, which is a direct consequence of the following lemma.

Lemma 6.36. *Suppose $q : \mathbb{Z}^n \to \mathbb{Z}$ is a (weakly nonnegative) 0-sincere unit form. Then $q_{ij} > 1$ if and only if q is obtained from $q^{(i)}$ by doubling vertex j.*

Proof. Assume that $q_{ij} > 1$. By Exercise 6.15.5 we have $q_{ij} = 2$. Since $e_i - e_j$ is an isotropic vector ($0 = 2 - q_{ij} = q(e_i - e_j)$), by Lemma 6.21($a$) the vector $e_i - e_j$ is radical. Therefore by Exercise 3.32.6 the form q is equal to $q^{(i)}[j]$ (up to a reordering of vertices if necessary). The converse is evident. □

The following generalization of Ovsienko's Theorem is the main result in [23].

Theorem 6.37. *Let $q : \mathbb{Z}^n \to \mathbb{Z}$ be a weakly nonnegative semi-unit form with a maximal positive root x. Then $x_i \leq 12$ for all $i = 1, \ldots, n$.*

Sketch of Proof. Suppose on the contrary that x is a maximal positive root of q with $x_\omega > 12$ for some $\omega \in \{1, \ldots, n\}$.

Step 1. *We may assume that $q : \mathbb{Z}^I \to \mathbb{Z}$ is a weakly nonnegative centered form without critical restriction of shape $\mathscr{C}(1), \mathscr{C}(2), \mathscr{C}(3)$ or $\mathscr{C}(4')$. In this case, the maximal root x has two exceptional vertices. We may further assume that the cardinality $|I|$ is minimal among all such forms.*
Apply the Reduction Theorem 6.27 with respect to the set $I' = I - \{\omega\}$ and the maximal root x to get an iterated deflation T concentrated in I' and a maximal positive root x' of $q' = qT$ such that $x = T(x')$. Deleting some vertices if necessary, we may assume that x' is sincere, thus $q'_{ij} \geq 0$ for all $i, j \neq \omega$.
If there exists an $i \neq \omega$ such that $q'_{i\omega} \geq 0$ then by Lemma 6.28 we have $q'_{ii} = 0$. In particular e_i is a positive isotropic vector of q' with $e_i < x'$, therefore by Theorem 6.34, x' is a 2-layer root and $x_\omega = x'_\omega \leq 12$, a contradiction.
Moreover, if $q'_{i\omega} < -1$ then $q_{i\omega} = -2$ and the vector $e_\omega + e_i$ is isotropic for q' with $e_\omega + e_i < x'$, which is again impossible. Hence q' is a centered form.
Observe from Table 5.3 (see also the graphs after Lemma 6.30) that if q^J is a critical restriction of q' with associated bigraph $\mathscr{C}(2), \mathscr{C}(3)$ or $\mathscr{C}(4')$, then there is a positive isotropic vector $\mu < x$, which once more by Theorem 6.34 yields a contradiction.
Finally, the statement about the exceptional vertices of x' is worked out in Exercise 6.33.3. Write q for q' and x for x'.

Step 2. *Let i and j be the exceptional vertices of x and consider the quadratic form $\overline{q}(y) = q(y) - y_i y_j$. Then $\overline{q}$ is a 0-sincere centered form with sincere positive radical vector x.*
By Lemma 6.11 we have $x_i = 1 = x_j$, therefore $i, j \neq \omega$.
First notice that the restriction $q^{(i)(j)}$ is weakly positive (otherwise there is a critical restriction with a critical positive vector μ, and $q(\mu + x) = q(x) = 1$ since $i, j \notin \mathbf{supp}(\mu)$, contradicting the maximality of x). This implies that $2 \leq q_{ij} \leq 3$. Indeed, by Exercise 6.15.5 the inequality $0 \leq q_{ij} \leq 3$ holds. If $q_{ij} \leq 1$ then $q(e_i + e_j + e_\omega) \leq 2$ and the claim below yields a contradiction with $z = e_i + e_j + e_\omega$.

Claim. If z is a positive vector with $q(z) \leq 2$ satisfying $z_k \leq 1$ for all $k \neq \omega$ and $z_i = 1 = z_j$, then $z_\omega > 6$.

Proof. If $z_\omega \leq 6$ then $x - z$ is a positive vector and since $q^{(i)(j)}$ is weakly positive we have

$$0 < q^{(i)(j)}(x - z) = q(x - z) = q(x) + q(z) - q(x|z)$$
$$= 1 + q(z) - (z_i q(x|e_i) + z_j q(x|e_j)) = q(z) - 1 \leq 1.$$

Then $x - z$ is a positive root of $q^{(i)(j)}$, and by Theorem 5.25 we have $x_\omega - z_\omega \leq 6$, in contradiction with $x_\omega > 12$. □

Observe that the bilinear form associated to $\overline{q}$ has the following shape,

$$\overline{q}(v|w) = q(v|w) - v_i w_j - v_j w_i,$$

hence $\overline{q}(x|e_k) = 0$ for all k since $q(x|e_k) = x_k = 1$ for $k = i, j$. Then x is a sincere positive radical vector for $\overline{q}$, and we only need to show that $\overline{q}$ is weakly nonnegative. Observe that $\overline{q}^{(i)} = q^{(i)}$ and $\overline{q}^{(j)} = q^{(j)}$. If $\overline{q}$ is not weakly nonnegative, then there is a hypercritical restriction $\overline{q}^J$ where $J \subset I$ contains both i and j. From Table 6.1 we see that $q_{ij} \neq 3$. Furthermore, if $q_{ij} = 2$ then $\overline{q}^J$ has a restriction including i and j with one of the following bigraphs (see Exercise 6.33.4)

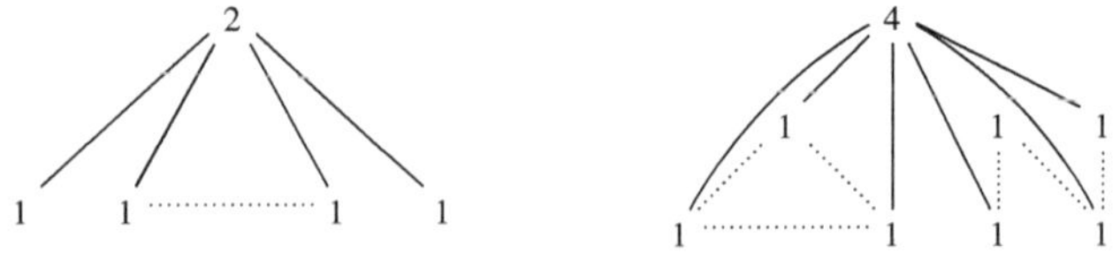

Using the claim above with the vector z as indicated by the vertices in the figure, which satisfies $q(z) \leq 2$, we get a contradiction. Then $\overline{q}$ is a 0-sincere form with sincere positive radical vector x.

Step 3. *If for some vertices $s, t \in I$ we have $q_{st} > 1$, then $\{s, t\} = \{i, j\}$.*
Assume on the contrary that i does not belong to the set $\{s, t\}$ and consider the restriction $q' = q^{(i)}$, which has the vector $y = x - e_i$ as positive root. If q' is weakly positive, then $y_\omega = x_\omega \leq 6$, contradicting Ovsienko's Theorem. Then there is a critical restriction $(q')^J$ of q' with critical positive vector μ.
Since $\overline{q}_{st} = q_{st} > 1$, by Lemma 6.36 the 0-sincere form $\overline{q}$ is obtained from its restriction $\overline{q}^{(t)}$ by doubling vertex s. Consequently the vector $w = \mu - \mu_t e_t + \mu_t e_s$ is a positive isotropic vector for q' (thus also for q). Since $i, j \notin \mathbf{supp}(w)$ implies that $q(w|x) = 0$ we get the equation

$$q(x + w) = q(x) = 1,$$

which contradicts the maximality of x.

For a weakly nonnegative unit form $q : \mathbb{Z}^I \to \mathbb{Z}$ consider the union I^+ of the supports of all positive radical vectors of q. By Lemma 6.21, the restriction $q^+ := q^{I^+}$ is a 0-sincere form, called the 0-*sincere kernel* of q.

Step 4. *Let $\xi^+ : \mathbb{Z}^K \to \mathbb{Z}$ be the 0-sincere kernel of the restriction $q^{(i)}$. Then ξ^+ is nontrivial and satisfies* $\mathbf{cork}(\xi^+) \leq 2$.

Notice that $y = \sigma_i(x) = x - e_i$ is a sincere positive root of $q' := q^{(i)}$. Since $y_\omega > 12$, the form q' is not weakly positive, thus the 0-sincere kernel ξ^+ is nontrivial.

Now, since $q' = \overline{q}^{(i)}$, by Step 3 the form q' is graphical. Assume that $\mathbf{cork}(\xi^+) \geq 3$. Then we may take a 0-sincere restriction ξ of ξ^+ such that $\mathbf{cork}(\xi) = 3$ (cf. Lemma 6.22 and Remark 3.21).

Apply Theorem 6.31 to the form ξ, and notice first of all that Θ_1 is not the bigraph associated to ξ (by Theorem 6.34, since the vector z with $z_\omega = 5$ and $z_k = 1$ for all other vertices is isotropic with $z < x$). Thus if ξ is triangular or B_ξ is Θ_2, it can be seen that there exist critical vectors μ_1, μ_2 and μ_3 of ξ such that

$$|(\mu_s - \mu_t)_\omega| \leq 2, \quad \text{and} \quad |(\mu_s - \mu_t)_k| \leq 1 \quad \text{for } k \neq \omega,$$

for any $s \neq t$ in $\{1, 2, 3\}$ (see Exercise 2 below). Hence $x - (\mu_s - \mu_t) > 0$. Suppose that there are $s \neq t$ such that $q(x|\mu_s - \mu_t) \geq 2$. Then

$$q(x|\mu_s - \mu_t) = q(y + e_i|\mu_s - \mu_t) = q(e_i|\mu_s - \mu_t) \geq 2,$$

and since $x - (\mu_s - \mu_t) > 0$, we get the contradiction

$$0 \leq q(x - (\mu_s - \mu_t)) = q(x) + q(\mu_s - \mu_t) - q(x|\mu_s - \mu_t) < 0.$$

In particular, in the set $\{q(e_i|\mu_k)\}_{k=1,2,3}$ there are at least two equal elements, say $q(e_i|\mu_1) = q(e_i|\mu_2)$. We may also assume that $(\mu_1 - \mu_2)_\omega \geq 0$. Then $\mu_s - \mu_t$ is a radical vector of q, and taking $d = \min(x_k \mid (\mu_s - \mu_t)_k = -1)$ we get a nonsincere positive q-root $z = x + d(\mu_s - \mu_t)$ satisfying $z_\omega > 12$. Using Exercise 1 below, the vector z is a sincere maximal positive root of the restriction of q to the (proper) support of z, obtaining in this way a contradiction to the minimal choice of $|I|$ established in Step 1.

Step 5. *The form $\overline{q}$ admits no critical restriction with associated bigraph of shape $\mathscr{C}(1), \mathscr{C}(2), \mathscr{C}(3)$ or $\mathscr{C}(4')$.*

Since the form $\overline{q}$ is centered (Step 2), its bigraph does not contain the bigraph $\mathscr{C}(1)$ as a restriction. In all other cases notice that the support of the critical vector μ must contain both i and j (otherwise it would be a critical vector for q). Thus μ would be a positive root of q, and using the claim in Step 2 we get $\mu_\omega > 6$, a contradiction.

Step 6. *Final analysis of the case* $\mathbf{cork}(\xi^+) = 2$.

Consider that $\xi^+ : \mathbb{Z}^K \to \mathbb{Z}$ is a 0-sincere graphical form with $\mathbf{cork}(\xi^+) = 2$, which is by construction a restriction of the quadratic form $q^{(i)} = \overline{q}^{(i)}$ where

$q : \mathbb{Z}^I \to \mathbb{Z}$ is our original form. First we notice that $K = I - \{i\}$, that is, that $\xi^+ = q^{(i)}$. Indeed, if there is a $k \neq i$ in $I - K$ then the restriction of $q^{(i)}$ to the set $K \cup \{k\}$ has corank 3 by Exercise 5 below. This is impossible since ξ^+ has corank 2. Hence $K = I - \{i\}$. We will reach a contradiction by considering two cases.
Case $\overline{q}_{ij} = 2$. By Lemma 6.30 the form $\overline{q}$ is obtained from $\xi^+ = q^{(i)}$ by doubling vertex j. Define the vector $u := x - e_i + e_j$, which can be shown to be a sincere isotropic vector for ξ^+ with $u_j = 2$. Indeed, we have

$$\xi^+(u) = \overline{q}(x - e_i + e_j) = \overline{q}(x) = q(x) - x_i x_j = 0.$$

Taking $\mu^{(1)}$ and $\mu^{(2)}$ to be critical vectors of the two critical restrictions of ξ^+, there are positive integers m_1 and m_2 such that $u = m_1\mu^{(1)} + m_2\mu^{(2)}$ (see Remark 6.32). Since $u_j = 2$, up to exchanging the roles of $\mu^{(1)}$ and $\mu^{(2)}$ we may suppose that $\mu^{(1)}_j = 0$ or $\mu^{(1)}_j = 1$. But notice that in both cases we have $\mu^{(1)} < x$, therefore x is a 2-layer root by Theorem 6.34(a). This contradicts $x_\omega > 12$ by part (b) of that theorem.
Case $\overline{q}_{ij} = 1$. Again by Exercise 5 and Theorem 6.31, either $\overline{q}$ is a triangular form, or the form associated to one of the bigraphs Θ_1 or Θ_2. If $\overline{q}$ is triangular, then by Remark 6.32 there are positive integers m_1, m_2, m_3 such that

$$x = m_1\mu^{(1)} + m_2\mu^{(2)} + m_3\mu^{(3)},$$

where $m_1\mu^{(1)}$, $m_1\mu^{(2)}$ and $m_1\mu^{(3)}$ are critical vectors of $\overline{q}$. Since $x_i = 1$ we may assume that $\mu^{(1)}_i = 0$, therefore $\mu^{(1)} < x$. This is again impossible by Theorem 6.34. A similar argument can be formulated for case Θ_2 (see Exercise 3 below). Finally, if $\overline{q} = q_{\Theta_1}$, then the vector z given by $z_\omega = 5$ and $z_i = 1$ for $i \neq \omega$ is a positive q-root, contradicting the claim in Step 2 (see Table 6.3 and Exercise 4).

Step 7. *Final analysis of the case* **cork**$(\xi^+) = 1$.
We assume now that ξ^+ is itself a critical form, and let μ be its critical vector. Suppose first that ξ^+ is the form associated to one of the graphs $\mathscr{C}(5)$ or $\mathscr{C}(6)$. It can be shown then (see Exercise 6(b) and (c) below) that ξ^+ is the (one-point) restriction of a form of corank 2. Therefore we have again $\xi^+ = q^{(i)} = \overline{q}^{(i)}$. As before we consider separately the cases $\overline{q}_{ij} = 2$ and $\overline{q}_{ij} = 1$.
Case $\overline{q}_{ij} = 2$. By Lemma 6.30 the form $\overline{q}$ is obtained from ξ^+ by doubling vertex j. Hence $u := x - e_i + e_j$ is a sincere positive radical vector of ξ^+. Because $u_j = 2$ we have $u = m\mu$ for some $m \in \{1, 2\}$. However, recall from Proposition 5.4 that $\mu_\omega \leq 6$, therefore $x_\omega \leq 2\mu_\omega \leq 12$, a contradiction.
Case $\overline{q}_{ij} = 1$. A direct inspection of the bigraphs $\Psi_{17}, \ldots, \Psi_{20}$ given in Exercise 6 shows that, since $x_i = 1$, we may find a critical restriction of $\overline{q}$ avoiding vertex i, and such that its critical vector μ satisfies $\mu < x$. The contradiction is again derived from Theorem 6.34.

By Step 5 and the discussion above we may finally suppose that ξ^+ is the form associated to the graph $\mathscr{C}(4)$. Let us first assume that $K = I - \{i\}$ (that is, that $\xi^+ = q^{(i)}$). Then if $\overline{q}_{ij} = 2$ we can argue as above, while if $\overline{q}_{ij} = 1$ then by Exercise 6(a) the form $\overline{q}$ fails to be 0-sincere, in both cases a contradiction.
Therefore we may fix a vertex $k \neq i$ in the set $I - K$. Let us now assume that $I - K = \{i, k\}$. If $\overline{q}_{ij} = 2$ then one can check that $\overline{q}$ is not 0-sincere, and if $\overline{q}_{ij} = 2$ then by Exercise 6(d) the form $\overline{q}$ is associated to one of the bigraphs $\Psi_{15}, \dots, \Psi_{18}$, and one can proceed as above to find a critical vector μ with $\mu < x$, obtaining a contradiction using Theorem 6.34.
Assume now that we can find a second vertex $\ell \neq i$, different from k, in the set $I - K$. Consider the restriction $\widetilde{q} = q^{K \cup \{k,\ell\}}$, and take $q' = \widetilde{q}^{(\ell)}$. Hence $(q')^{(k)} = \xi^+$, which is the form associated to the graph $\mathscr{C}(4)$. By Exercise 6(a), the graph associated to q' has shape A_{10} or A_{11}. By Exercise 6(d), the form $\widetilde{q}$ is either not 0-sincere, or is associated to one of the bigraphs Ψ_{15}, Ψ_{16}, Ψ_{17} or Ψ_{18}. It is shown in [23, Sect. 9.9] that all these cases imply that $\overline{q}$ itself is not 0-sincere, a contradiction.

This completes the proof. □

Exercises 6.38.

1. Let $q : \mathbb{Z}^I \to \mathbb{Z}$ be a weakly nonnegative semi-unit form with a maximal sincere positive root x. If $\mu \in \mathbf{rad}(q)$ and $x + \mu$ is a positive vector, show that $x + \mu$ is a maximal sincere positive root of the restriction of q to the support of $x + \mu$.
2. Let q be a 0-sincere graphical form of corank 3 without having as restriction a form associated to the bigraphs $\mathscr{C}(1)$, $\mathscr{C}(2)$, $\mathscr{C}(3)$ or $\mathscr{C}(4')$.

   a) If q is a triangular form, let μ_1, μ_2 and μ_3 be the positive critical vectors of q. Show that for $s \neq t$ in $\{1, 2, 3\}$ we have

   $$|(\mu_s - \mu_t)_\omega| \leq 2, \quad \text{and} \quad |(\mu_s - \mu_t)_k| \leq 1, \quad \text{for } k \neq \omega.$$

   [Hint: Use Theorem 6.31.]

   b) If $B_q = \Theta_2$ consider the vectors

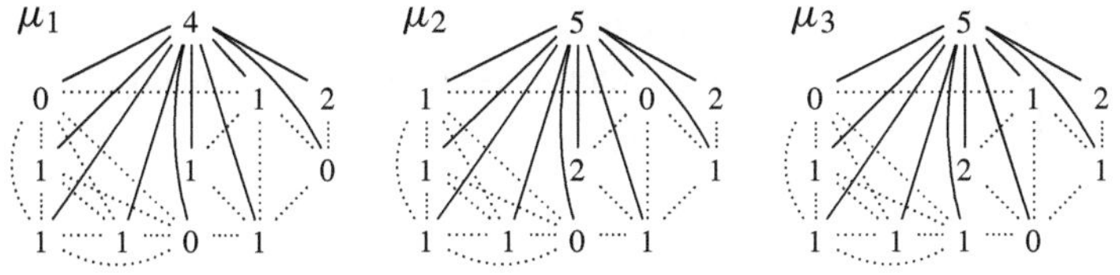

   Show that μ_1, μ_2 and μ_3 are critical vectors of q, and that for $s \neq t$ in $\{1, 2, 3\}$ we have

   $$|(\mu_s - \mu_t)_k| \leq 1, \quad \text{for all } k.$$

   Why is q not a triangular form?

Table 6.3 Some 0-sincere semi-graphical forms of corank 3

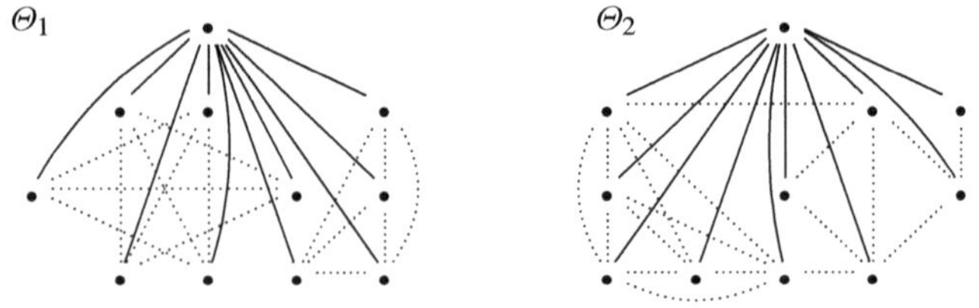

3. Show that if q is the quadratic form associated to Θ_2 (Table 6.3), and $\mu^{(1)}$, $\mu^{(2)}$, $\mu^{(3)}$ and $\mu^{(4)}$ are its critical vectors, then there are nonnegative integers $m_1, \ldots, m_4$ such that any sincere positive radical vector μ can be written as

$$\mu = m_1\mu^{(1)} + m_2\mu^{(2)} + m_3\mu^{(3)} + m_4\mu^{(4)}.$$

   Show also that we may assume, up to a reordering of variables, that m_1 and m_2 are positive integers.
4. Consider the quadratic form q with bigraph Θ_1 with center ω (Table 6.3), and let z be the vector with $z_\omega = 5$ and $z_i = 1$ for all other vertices. Show that z is an isotropic vector for q. Is it a radical vector?
5. Let $q : \mathbb{Z}^K \to \mathbb{Z}$ be a 0-sincere graphical form without critical restriction of shape $\mathscr{C}(1)$, $\mathscr{C}(2)$, $\mathscr{C}(3)$ or $\mathscr{C}(4')$, and take $k \in K$.

   a) Show that if the restriction $q^{(k)}$ is a 0-sincere form of corank 2, then q is 0-sincere of corank 3.
   b) Show that in the situation of point (a), either q is a triangular form, or q is one of the forms Θ_1 or Θ_2 shown in Table 6.3.

6. Let $q : \mathbb{Z}^J \to \mathbb{Z}$ be a weakly nonnegative graphical form having no critical restriction of shape $\mathscr{C}(1)$, $\mathscr{C}(2)$, $\mathscr{C}(3)$ or $\mathscr{C}(4')$. Consider a vertex $j \in J$.

   a) Show that if $q^{(j)} = q_{\mathscr{C}(4)}$, then q is the form associated to one of the bigraphs A_{10} or A_{11} below, and $\mathbf{cork}(q) = 1$.
   b) Show that if $q^{(j)} = q_{\mathscr{C}(5)}$, then q is the form associated to Ψ_{17}, Ψ_{19} or Ψ_{20}, and $\mathbf{cork}(q) = 2$ (see Table 6.2).
   c) Show that if $q^{(j)} = q_{\mathscr{C}(6)}$, then q is the form associated to Ψ_{18} or Ψ_{20}, and $\mathbf{cork}(q) = 2$ (see Table 6.2).

d) Show that if $q^{(j)} = q_{A_{10}}$ or $q^{(j)} = q_{A_{11}}$, then either q is the form associated to Ψ_{15}, Ψ_{16}, Ψ_{17} or Ψ_{18}, or q is not 0-sincere.

7. Let $q : \mathbb{Z}^J \to \mathbb{Z}$ be a graphical 0-sincere form having no critical restriction of shape $\mathscr{C}(1)$, $\mathscr{C}(2)$, $\mathscr{C}(3)$ or $\mathscr{C}(4')$. Show that if there is a vertex $j \in J$ such that $q^{(j)}$ has associated bigraph A_{10} or A_{11}, then B_q is Ψ_{15}, Ψ_{16}, Ψ_{17} or Ψ_{18}. In particular q has corank 2.

References

1. A'Campo, N.: Sur les valeurs propres de la transformation de Coxeter. Inventiones Math. **33**, 61–67 (1976)
2. Assem, I., Simson, D., Skowroński, A.: Elements of the Representation Theory of Associative Algebras: Techniques of Representation Theory. London Mathematical Society Student Texts. Cambridge University Press, Cambridge (2006)
3. Auslander, M., Reiten, I., Smalø, S.O.: Representation Theory of Artin Algebras. Cambridge Studies in Advanced Mathematics, vol. 36. Cambridge University Press, Cambridge (1997)
4. Barbeau, E.J.: Pell's Equation. Problem Books in Mathematics. Springer, New York (2003)
5. Barot, M.: A characterization of positive unit forms. Bol. Soc. Mat. Mexicana (3) **5**, 87–93 (1999)
6. Barot, M.: A characterization of positive unit forms, Part II. Bol. Soc. Mat. Mexicana (3) **7**, 13–22 (2001)
7. Barot, M., de la Peña, J.A.: The Dynkin type of a non-negative unit form. Expo. Math. **17**, 339–348 (1999)
8. Barot, M., de la Peña, J.A.: Algebras whose Euler form is non-negative. Colloq. Math. **79**, 119–131 (1999)
9. Barot, M., de la Peña, J.A.: Root-induced integral quadratic forms. Linear Algebra Appl. **412**, 291–302 (2006)
10. Bernstein, I.N., Gelfand, I.M., Ponomarev, V. A.: Coxeter functors and Gabriel's theorem, Uspiehi Mat. Nauk, **28** (1973) 19–33 (in Russian). English translation in Russ. Math. Surv. **28** (1973) 17–32
11. Bhargava, M.: On the Conway–Schneeberger fifteen theorem. In: Quadratic Forms and Their Applications (Dublin, 1999). Contemporary Mathematics, vol. 272. American Mathematical Society, Providence, RI (2000)
12. Birkhoff, G.: Linear Transformations with Invariants Cones. Am. Math. Mon. **74**, 274–276 (1967)
13. Boldt, A.: Methods to determine Coxeter polynomials. Linear Algebra Appl. **230**, 151–164 (1995)
14. Brechenmacher, F.: Lagrange and the secular equation. Lett. Mat. Int. **2**, 79 (2014). https://doi.org/10.1007/s40329-014-0051-3
15. Brechenmacher, F.: Algebraic generality vs arithmetic generality in the 1874 controversy between C. Jordan and L. Kronecker. In: Chemla, K., Chorlay, R., Rabouin, D. (eds.) The Oxford Handbook of Generality in Mathematics and the Sciences. Oxford University Press, Oxford (2016)

M. Barot et al., *Quadratic Forms*, Algebra and Applications 25,
https://doi.org/10.1007/978-3-030-05627-8

16. Brouwer, A.E., Haemers, W.H.: Spectra of Graphs. Springer, Universitext. Springer, New York, NY (2012)
17. Buell, D.A.: Binary Quadratic Forms, Classical Theory and Modern Computations. Springer, New York (1989)
18. Chartrand, G., Lesniak, L., Zhang, P.: Graphs and Digraphs, 5th edn. Chapman and Hall/CRC Press, Boca Raton, FL (2015)
19. Conway, J.H., Sloane, N.J.A.: Sphere packings, lattices and groups. (English summary) With additional contributions by E. Bannai, R.E. Borcherds, J. Leech, S.P. Norton, A.M. Odlyzko, R.A. Parker, L. Queen and B.B. Venkov. Grundlehren der Mathematischen Wissenschaften, vol. 290. Springer, New York (1999)
20. Cvetković, D., Rowlinson, P., Simić, S.: An Introduction to the Theory of Graph Spectra. London Mathematical Society Student Texts, vol. 75. Cambridge University Press, Cambridge (2010)
21. de la Peña, J.A.: Periodic Coxeter matrices. Linear Algebra Appl. **365**, pp. 135–142 (2003). Special issue on linear algebra methods in representation theory
22. Dräxler, P., Drozd, A., Golovachtchuk, N.S., Ovsienko, S.A., Zeldych, M.V.: Towards the classification of sincere weakly positive unit forms. Eur. J. Combin. **16**, 1–16 (1995)
23. Dräxler, P., Golovachtchuk, N.S., Ovsienko, S.A., de la Peña, J.A.: Coordinates of maximal roots of weakly non-negative unit forms. Colloq. Math. **78**(2), 163–193 (1998)
24. Erdös, P.: Some remarks on Euler's ϕ function. Acta Arith. **4**, 10–19 (1954)
25. Ford, K.: The number of solutions of $\psi(x) = m$. Ann. Math. **150** (1), 283–311 (1999)
26. Gabriel, P., Roiter, A.V.: Representations of Finite Dimensional Algebras, Algebra VIII, Encyclopedia of Mathematical Science, vol. 73. Springer, Berlin, Heidelberg, New York (1992)
27. Gantmacher, F.R.: The Theory of Matrices. vols. 1 and 2 (Translated by K.A. Hirsch). Chelsea Publishing Company, New York, NY (1960)
28. Gauss, C.F.: Disquisitiones Arithmeticae. Fleischer, Leipzig (1801); English translation, Yale University Press, New Haven, CT (1966)
29. Gustafson, W.H.: The history of algebras and their representations. In: Representations of Algebras, Proceedings (Workshop), Puebla, Mexico (1980). Lecture Notes in Mathematics, vol. 944, pp. 1–28. Springer, Berlin, Heidelberg, New York (1982)
30. Happel, D.: The converse of Drozd's theorem on quadratic forms. Commun. Algebra **23**(2), 737–738 (1995)
31. Happel, D., de la Peña, J.A.: Quadratic forms with a maximal sincere root. Can. Math. Soc. **18**, 307–315 (1996)
32. Happel, H., Preisel, U., Ringel, C.M.: Vinberg's characterization of Dynkin diagrams using subadditive functions with applications to DTr-periodic modules. In: Proceedings of the ICRA II, Ottawa 1979. Lecture Notes in Math, vol. 832, pp. 280–294. Springer, Berlin, Heidelberg, New York (1980)
33. Hoffman, K., Kunze, R.: Linear Algebra, 2nd edn. Prentice-Hall, Englewood Cliffs, NJ (1971)
34. Howlett, B.B.: Coxeter groups and M-matrices. Bull. Lond Math. Soc. **14** (2), 127–141 (1982)
35. Jiménez González, J.A.: Incidence graphs and non-negative integral quadratic forms. J. Algebra **513**, 208–245 (2018)
36. Katz, V.J., Parshall, K.H.: Taming the Unknown, a History of Algebra from Antiquity to the Early Twentieth Century. Princeton University Press, Princeton (2014)
37. Kirillov Jr., A.: Quiver Representations and Quiver Varieties. Graduate Studies in Mathematics, vol. 174. American Mathematical Society, Providence, RI (2016)
38. Ladkani, S.: On the periodicity of Coxeter transformations and the non-negativity of their Euler forms. Linear Algebra Appl. **428**, 742–753 (2008)
39. Lang, S.: Algebra. Graduate Texts in Mathematics, vol. 211 (Revised 3rd edn.). Springer, New York, NY (2002)
40. Mróz, A.: Coxeter energy of graphs. Linear Algebra Appl. **506**, 279–307 (2016)
41. Newman, M.: Integral Matrices. Academic, New York (1972)
42. Ostermann, A., Pott, A.: Schwach positive ganze quadratische Formen, die eine aufrichtife, positive Wurzel mit einem Koeffizienten 6 besitzen. J. Algebra **126**, 80–118 (1989)

43. Ovsienko, A.: Integral weakly positive forms (Russian). In: Schurian Matrix Problems and Quadratic Forms, pp. 3–17. Mathematics Institute of the Academy of Sciences of the Ukrainian SSR, Kiev (1978)
44. Ovsienko, A.: Maximal roots of sincere weakly nonnegative forms. Lecture at the workshop on Quadratic Forms in the Representation Theory of Finite-Dimensional Algebras, Bielefeld, November 9–12 (1995)
45. Prasolov, V.V.: Polynomials. Algorithms and Computation in Mathematics. Springer, Berlin, Heidelberg, NewYork (2010)
46. Ringel, C.M.: Tame Algebras and Integral Quadratic Forms. Lecture Notes in Mathematics, vol. 1099. Springer, New York (1984)
47. Sato, M.: Periodic Coxeter matrices and their associated quadratic forms. Linear Algebra Appl. **406**, 99–108 (2005)
48. Schlafly, A., Wagon, S.: Carmichael's conjecture on the Euler function is valid below $10^{10,000,000}$. Math. Comput. **207**(63), 415–419 (1999)
49. Sierpiński, W.: In: Schinzel, A. (ed.) Elementary Theory of Numbers. North-Holland, Amsterdam (1988)
50. Vandergraft, J.S.: Spectral properties of matrices which have invariant cones. SIAM J. Appl. Math. **16**, 1208–1222 (1968)
51. Vinberg, E.B.: Discrete linear groups generated by reflections. Izv. Akad. Nauk SSSR **35** (1971). Transl.: Math. USSR Izvestija **5**, 1083–1119 (1971)
52. von Höhne, H.-J.: On weakly positive unit forms. Comment. Math. Helvetici **63**, 312–336 (1988)
53. von Höhne, H.-J.: Edge reduction for unit forms. Arch. Math. **65**, 300–302 (1995)
54. von Höhne, H.-J., de la Peña, J.A.: Isotropic vectors of non-negative integral quadratic forms. Eur. J. Combin. **19**, 621–638 (1998)
55. Zeldych, M.V.: A criterion for weakly positive quadratic forms (Russian). In: Linear Algebra and the Theory of Representations. SSR, Kiev (1983)

Index

M. Barot et al., *Quadratic Forms*, Algebra and Applications 25,
https://doi.org/10.1007/978-3-030-05627-8

MIX
Papier aus verantwortungsvollen Quellen
Paper from responsible sources
FSC® C105338

If you have any concerns about our products,
you can contact us on
ProductSafety@springernature.com

In case Publisher is established outside the EU,
the EU authorized representative is:
Springer Nature Customer Service Center GmbH
Europaplatz 3, 69115 Heidelberg, Germany

Printed by Libri Plureos GmbH
in Hamburg, Germany